Proceedings of the Circular Building Sector Conference 2025 (CBSC2025)

Hosted by LTH Profile Area Circular Building Sector
1-3 June 2025, Lund, Sweden

Editors:
Riikka Kyrö
Erik Serrano

LUND UNIVERSITY | LTH FACULTY OF ENGINEERING

ISBN: 979-8-3313-2166-6
eISBN: 979-8-3313-2165-9
DOI: https://doi.org/10.52202/080684

Printed from e-media with permission by:

Curran Associates, Inc.
57 Morehouse Lane
Red Hook, NY 12571

Some format issues inherent in the e-media version may also appear in this print version.

Lund University
LTH, Faculty of Engineering
Box 118
SE-221 00 LUND, Sweden

Phone: +46 46 222 72 00

info@lth.se

Additional copies of this publication are available from:

Curran Associates, Inc.
57 Morehouse Lane
Red Hook, NY 12571 USA
Phone: 845-758-0400
Fax: 845-758-2633
Email: curran@proceedings.com
Web: www.proceedings.com

EDITORIAL

You are reading the Proceedings of the first ever Circular Building Sector Conference, held in Lund, Sweden in the early days of June 2025. The conference, and consequently these proceedings, encompass a wide array of circularity. The papers presented at the conference cover circularity on all scales, from materials and components to whole buildings to the neighborhood and city scale. Circular strategies covered comprise a spectrum of the R strategies from recycling of building materials to reducing the need for new construction. Conference participants hail from the host country Sweden, Norway, Finland, Denmark, Italy, the Netherlands, Singapore, and Taiwan. Their fields of research range from building physics to business and economics.

The proceedings comprise 23 papers, which are organized thematically. The first three papers discuss circular design. The following five papers consider the reuse of elements, components, and materials. Six papers focus on lifecycle assessments, and assessments of reusability and adaptability potential. Seven papers focus on different circular use strategies, and business models aligned with circularity principles. Finally, two papers focus on circularity in higher education. We wish to thank all authors for their contribution to advancing knowledge in the field of circular building.

An additional eight brilliant presentations at the conference were based on abstracts, which are not included in these proceedings but deserve recognition as a valuable contribution to the conference. Most importantly, the proceedings would not have been possible without the altruistic work of academics who joined the scientific committee and contributed with reviews. Our scientific committee comprises 24 scholars from Sweden, Norway, Finland, Denmark, the Netherlands, and the UK. Our warmest thanks for your invaluable contribution to the scientific quality of the conference.

The proceedings present the state-of-the-art research in circularity in the building sector. The research emphasizes the urgency to act, highlights challenges, proposes solutions, and gives hope for the future. We wish you insightful and enjoyable moments reading!

Lund, June 2025

Riikka Kyrö
Proceedings Editor
Co-chair of Scientific Committee

Erik Serrano
Chair of Scientific Committee
Proceedings Co-editor

SCIENTIFIC COMMITTEE

ORGANIZING COMMITTEE

Conference Chair, Profile area coordinator

Ulla Janson, Faculty of Engineering at Lund University (LTH)

Conference Co-Chair, Profile area deputy coordinator

Dennis Johansson, Faculty of Engineering at Lund University (LTH)

Profile area administrator

Karin Lundholm, Faculty of Engineering at Lund University (LTH)

Profile area leadership team

Riikka Kyrö, Faculty of Engineering at Lund University (LTH)

Jonas Niklewski, Faculty of Engineering at Lund University (LTH)

Erik Serrano, Faculty of Engineering at Lund University (LTH)

TABLE OF CONTENTS

CIRCULAR MATERIALS AND COMPONENTS

CIRCULARITY ASSESSMENT

CIRCULAR USE AND BUSINESS MODELS

CIRCULAR DESIGN

CIRCULARITY IN HIGHER EDUCATION

Proceedings of the Circular Building Sector Conference 2025, Lund, Sweden, 1-3 June 2025

EARLY CASES OF PRECAST CONCRETE REUSE IN SWEDISH CONSTRUCTION (1984-2002): REPURPOSING THE MILLION HOMES PROGRAMME

Helena Westerlind [1*] and Erik Stenberg[1]

[1]KTH Royal Institute of Technology, School of Architecture

Corresponding Author: helena.westerlind@arch.kth.se

ABSTRACT

Background and aim: Precast concrete elements are structural components with significant potential to support circular construction practices, and several initiatives are currently underway to address the technical, regulatory, and economic challenges associated with scaling up element reuse from a niche practice to mainstream application. Although the reuse of concrete elements is gaining renewed interest within the broader framework of circularity, the earliest initiatives to reuse precast concrete elements in Sweden can be traced back to the 1980s. During this period, widespread vacancies in newly constructed mass housing developments under the Million Homes Programme prompted several municipal housing companies to explore deconstruction as a strategic alternative to conventional demolition. This paper examines the lessons learned from these early pioneering projects and investigates how insights from past reuse efforts can inform and advance current and future circular practices in the building sector.

Methods and data: The paper presents a comparative study of early cases involving the reuse of prefabricated concrete elements in Sweden over a twenty-year period. The analysis draws on a combination of literary sources, architectural drawings, and interviews with key individuals involved in the original projects. Through this multi-source approach, the study conducts a structured examination of the deconstruction and reuse processes associated with each identified case. Particular emphasis is placed on the architectural transformations between the donor buildings and their corresponding recipient buildings, providing deeper insights into the potential of precast concrete systems to be repurposed in new construction projects.

Findings: Between 1984 and 2002, seven building projects in Sweden were completed incorporating salvaged precast concrete elements from deconstructed Million Programme developments. Although the original structures were widely criticized for their systematization and repetition, these very characteristics made the precast concrete systems particularly well-suited for deconstruction and reuse. The recipient projects demonstrate that, through relatively simple design interventions, precast systems can be effectively adapted to meet diverse site contexts, building types, and spatial requirements.

Theoretical/practical/societal implications: Gaining a deeper understanding of the early cases of concrete element reuse in Sweden and the reasons why this seemingly successful approach failed to lead to broader implementation can support current reuse initiatives in fostering a more systematic and lasting transformation of the construction sector, extending beyond the scope of isolated pilot projects

KEYWORDS: Reuse, precast concrete elements, architecture, reuse potential, the Million Homes Programme.

1 INTRODUCTION

To mitigate the environmental impact of construction, circular solutions are increasingly promoted as alternatives to the traditional linear economic model in building production. The Circular Economy (CE) approach seeks to extend the service life of products and materials by preserving their highest possible value across multiple life cycles. When the reuse of entire buildings is not feasible, repurposing building components in new construction offers the most effective strategy for prolonging the lifespan of materials. When successfully implemented, this approach contributes to reducing demolition waste, decreases reliance on virgin natural resources, and offers considerable potential for lowering the carbon emissions associated with new construction. Given that the structural frame accounts for approximately 60% of a building's embodied carbon

https://doi.org/10.52202/080684-0001

during the construction phase (Malmqvist et al., 2023), the reuse of structural components represents one of the most effective strategies for reducing carbon emissions in building production. Precast concrete elements (PCEs) are structural components with significant potential for circularity (Huuhka et al., 2023). Unlike cast-in-place concrete structures, precast concrete buildings are designed as modular systems, with concrete elements typically prefabricated off-site and subsequently assembled on location. The types and configurations of elements in a precast system are determined by the spatial requirements of the building's intended function, ranging from room-sized slabs and walls in residential wall-frame systems to long-spanning beams and columns in portal frame systems (Hernández Vargas & Stenberg, 2024). The building technology was widely adopted in the post-war period, and as a result, a significant portion of Europe's building stock consists of precast concrete elements (Alonso & Palmarola, 2019). Although these buildings were not originally intended to be disassembled, several projects have successfully demonstrated the deconstruction and reuse of prefabricated concrete elements in new construction. A study by Küpfer et al. (2023) identifies approximately 50 completed projects incorporating reclaimed precast concrete elements between 1967 and 2022. The majority of these projects were carried out in Germany, with a smaller number of initiatives implemented in the Netherlands, Sweden, Belgium, France, the United States, and Finland.

One of the earliest documented cases of large-scale reuse of precast concrete elements was carried out in the Swedish city of Gothenburg during the mid-1980s. Faced with a severe surplus of vacant apartments, the municipal housing company Göteborgsbostäder made a groundbreaking decision: instead of resorting to conventional demolition, they opted for partial deconstruction, salvaging precast concrete elements for reuse in four new housing developments in the Gothenburg region. Approximately two decades later, another reuse project was initiated in the Swedish city of Linköping, involving precast concrete elements salvaged from a mass housing area in the neighbouring city of Norrköping. In addition to these internationally recognized Swedish cases commonly cited in the discourse on concrete element reuse, at least two further initiatives involving the repurposing of precast concrete elements were undertaken during the same period. In total, the early history of precast concrete reuse in Sweden, which forms the focus of this study, comprises at least four donor buildings and seven receiver buildings, spanning a period of approximately twenty years, from 1984 to 2002. All four donor buildings were located in large-scale housing areas developed as part of so-called Million Homes Programme—a national initiative implemented between 1965 and 1974 that led to the construction of over one million housing units, representing approximately twenty percent of Sweden's current housing stock (SCB, 2025). Approximately one-third of these units were high-rise multifamily buildings with four or more storeys, and around fifteen percent were constructed using precast concrete systems (Vidén & Lundahl, 1992). Although they constitute a relatively modest share of the total housing stock, these large-scale developments incorporating precast concrete have become the most emblematic representations of the Million Programme and are often associated with its perceived shortcomings (Johansson, 2012).

Although the early pilot projects examined in this study demonstrated the technical feasibility of precast concrete reuse, they did not lead to its widespread adoption within the construction sector. Today, reuse rates for concrete remain negligible, highlighting a significant gap between the theoretical potential of precast concrete reuse and its practical implementation in contemporary construction practices. Currently, several research initiatives are actively working to scale up the reuse of precast concrete for broader implementation. Notable examples include the EU project ReCreate (Huuhka et al., 2023) and the Swedish research initiative Återhus (Återhus, 2023). As part of these efforts, two pilot projects were completed in Sweden in 2022: a temporary exhibition pavilion in Helsingborg for the City Fair H22 with repurposed concrete elements from three different donor buildings (Westerlind et al., 2025), and a temporary building known as Hållbarhetshuset in Stockholm, which incorporated repurposed hollow-core slabs sourced from a nearby office (Återhus, 2023). Alongside these research-driven efforts, initiatives to promote the reuse of concrete have also been undertaken by the building sector itself. In 2023, approximately 300 square meters of hollow-core slabs were salvaged from an office building in Lund and reused in a new office development in Karlskrona (Ikanobostad, 2023). The following year, 3,000 square meters of hollow-core slabs were salvaged from a decommissioned IKEA department store to be reused in a new housing development in Gothenburg (Framtiden Byggutveckling, 2024). This growing momentum reflects an emerging shift in both research and industry practices toward the integration of circular strategies in concrete construction. However, current efforts in concrete reuse appear to be developing in relative isolation, with limited recognition or reference to the early reuse initiatives implemented several decades earlier. This historical disconnect is further underscored by the limited availability of sources documenting these initial projects. Although several of the deconstruction and reuse projects received substantial media coverage at the time, first-hand accounts remain scarce. Within the discourse on the Swedish Million Programme housing stock, the reuse of precast concrete has received only limited attention, with a few notable exceptions (Botta & Vidén, 2006; Huuhka et al., 2019). It is primarily within the expanding international research field on precast concrete reuse that the memory of these early projects is preserved, with several of the early Swedish cases regularly cited (Mettke, 1995; Asam, 2005; Addis, 2006; Huuhka, 2010; Fischer et al., 2011; Küpfer et al., 2023). Yet, to date, no comprehensive study has been undertaken to systematically examine this initial period of concrete element reuse in Sweden. As a result, while the reuse of concrete elements is currently gaining renewed attention in the Swedish building sector within the framework of the Circular Economy, many of the experiences and lessons learned from these early projects risk being overlooked and lost.

This paper seeks to address this knowledge gap by presenting a comparative study of early cases of precast concrete reuse in Sweden, with a particular focus on the architectural transformations that occurred between donor and recipient buildings. The objective is to compare the context, reuse process, and outcomes of these pioneering projects, in order to better understand the factors that contributed to their reuse potential. Drawing on written sources, architectural drawings, and interviews with individuals involved in the original projects, the study aims to compile valuable experiences and insights from these early reuse efforts, making them accessible to a new generation of practitioners and researchers. Gaining a deeper understanding of these projects—and the factors that inhibited their wider implementation—can support current reuse initiatives in achieving a more systematic and enduring transformation of the construction sector, beyond the scope of isolated pilot projects.

2 METHODOLOGY

The study employs a three-step methodological approach. First, a literature search was conducted to identify early cases of precast concrete reuse in Sweden. Given the limited documentation available in academic sources, the search scope was expanded to include news articles, trade magazines, and other forms of grey literature. In the case of Norra Bergsjön, the study benefited from a collection of news clippings generously provided by retired KTH researcher Sonja Vidén. This expanded approach led to the identification of two additional early reuse projects not included in the international case compilation by Küper et al.

In the second step, the collected material was reviewed to extract relevant data on the deconstruction and reuse processes of each identified case, with particular emphasis on the architectural transformations that occurred between the donor buildings and the resulting recipient buildings.

Given the often fragmented and limited nature of the available literature, architectural drawings for both donor and receiver buildings have been obtained from local planning offices to support the analysis. Efforts were also made to conduct interviews with individuals directly involved in these projects. In several cases, these first-hand accounts serve as the only remaining sources of meaningful insight into the reuse efforts. The data collection for each case was guided by the following research questions: What were the primary incentives and driving factors behind the reuse projects? What were the specific characteristics of the original precast systems? How did the deconstruction and reuse processes unfold? What were the main design features of the receiving buildings, and what adaptations were necessary to integrate the reclaimed elements for their new intended purpose? The results are summarized in Section 3, beginning with a description of the donor buildings, followed by an account of the reuse outcomes in terms of the corresponding receiver projects for each case.

In the final step, the similarities and differences among the projects are analysed and discussed in Section 4, with an emphasis on their relevance for future reuse efforts. This methodological approach enables a structured examination of the early reuse projects, facilitating a deeper understanding of their successes, limitations, and potential applicability within contemporary circular construction practices.

3 REUSE CASES

3.1 Norra Bergsjön

The municipal housing company Göteborgsbostäder AB was one of four municipal housing companies active in Gothenburg during the Million Programme Era. Under the leadership of its director, Inge Hjertén, a strong proponent of industrialized construction methods, the company began using prefabricated concrete elements as early as the late 1950s (Hjertén, 1969). Over the subsequent 20-year period, the company was involved in the development of several new mass housing areas on the outskirts of Gothenburg (Hjertén, 1969). However, by the 1980s, the demand for apartments in Gothenburg had declined drastically, leaving approximately 3000 of the company's 40,000 apartments vacant (Axelsson, 1983).

The Norra Bergsjön area, located northeast of Gothenburg, was one of the Million Programme neighbourhoods particularly affected by high vacancy rates. Originally completed 1969, the development comprised ten four-storey housing blocks, containing approximately 650 apartments. By the early 1980s, 90% of these units were unoccupied, contributing to substantial social challenges in the area and placing considerable financial strain on the housing company (Huuhka et. al., 2019). The discontinuation of a government subsidy for vacant apartments in 1982 ultimately prompted the company to take action (Huuhka et al., 2019). Rather than opting for demolition, an idea emerged to partially dismantle the buildings and reuse the concrete elements in areas with higher housing demand. The concept was initially proposed to Göteborgsbostäder by architect Lars Broberg and Lars Jonsson, head of development at the contractor ABV, and was met with great interest by the company (Kubu, 1983; Huuhka et al., 2019). Before fully committing to the concept, the company decided to test the approach through a pilot project, initiating the deconstruction of the first housing block in 1984.

A collaboration was established between Göteborgsbostäder, the contractor ABV, and the architectural group CFL Arkitekter, which included

https://doi.org/10.52202/080684-0001

architect Bengt Forser, who had also been responsible for designing the original buildings eighteen years earlier (Forser & Sundbom, 1986). In addition, engineer Helmut Junker, who had contributed to the development of the precast concrete system, also became involved in the project, bringing valuable continuity and technical insight (Huuhka et al., 2019).

In the project's first phase, 107 apartments were dismantled by removing the top three floors of a four-storey, U-shaped residential block. Many of the salvaged elements were reused in the construction of a new residential building in central Gothenburg. The remaining ground floor was converted into 32 row houses, incorporating design modifications such as bay windows, a new wooden facade, and a pitched roof (Fig. 1). Following the successful completion of this initial pilot project, the transformation of Norra Bergsjön continued with the deconstruction of additional buildings in subsequent phases. Salvaged elements from these deconstructions were reused in three new housing developments in and around Gothenburg.

Figure 1: View of the Norra Bergsjön residential area showing the partial deconstruction and conversion into row houses during Phase 2. (Photo: Lars Mongs, 1985. Used with permission).

3.1.1 The Elementhus 65 System

The original housing blocks in Norra Bergsjön were constructed using the *Elementhus 65* system, one of seven precast concrete systems employed by Göteborgsbostäder between 1956 and 1974 (Hjertén, 1969) The system consisted of room-length floor slabs supported by load-bearing internal walls positioned transversely relative to the building block, forming what is referred to as a cross-wall system or 'bookshelf' structure. The primary element types of the system included hollow-core slabs in two lengths (3.0, and 4.8 meters) and 2-meter-wide load-bearing inner wall elements (Hjertén, 1969). The facade elements of the building were non-load bearing, except for the gable walls. In the case of Norra Bergsjön, the facade elements were produced in white concrete with marble aggregate and featured a distinct geometric relief pattern (Hjertén, 1969). Like Göteborgsbostäder's previous systems, this system was based on standardized apartment layouts comprising one to four bedrooms, which could be varied and configured around a central stairwell (Hjertén, 1969).

3.1.2 Deconstruction of Donor Building

The partial deconstruction of the housing blocks in Norra Bergsjön involved the removal of the upper floors of each building. The process began with cutting the roof into sections and lifting it off, together with its insulation. A specialized forklift was then employed to carefully remove the top floor slabs. To maintain structural stability during disassembly, both exterior and interior walls were braced prior to the detachment of the wall elements. These elements were subsequently lifted down using the original iron lifting loops that had been utilized during the initial assembly eighteen years earlier (Tibblin, 1986). The system's connectors proved particularly suitable for reuse, as the elements were neither welded nor screwed together and could simply be 'unhooked' from their positions (Huuhka et al., 2019). Once disassembled, the elements were transported to Göteborgsbostäder's vacant element factory in Ingebäck, where they underwent a thorough cleaning process (Tibblin, 1986). High-pressure washing was used to remove wallpaper, dirt, and loose mortar, restoring the components to near-original condition. After cleaning, the elements were temporarily stored at the factory, awaiting reuse in future construction projects It was estimated that only 5% of the elements were damaged during the disassembly process (Forser & Sundbom, 1986).

3.1.3 Receiver Project 1: Olivedal (1986)

Approximately two-thirds of the disassembled elements from the first pilot project at Norra Bergsjön were reused in the construction of a new residential building in central Gothenburg (Nyström, 1984) (Fig. 2). The urban context of the city block site differed significantly from that of the original location, necessitating adaptations to both the building's overall form and its internal spatial configuration. CFL Arkitekter served as the project's architects, while ABV acted as the main contractor.

In contrast to the original four-storey structure, the new building comprises both six- and seven-storey sections, accommodating a total of 108 apartments arranged around an internal courtyard on a horseshoe-shaped building plot. Integration of the salvaged elements into this irregular urban site was made possible through the addition of newly fabricated elements, which were used to form the chamfered corners of the building (Fig. 6) (Nyström, 1984). Salvaged facade elements were used for the courtyard-facing elevations, while the external facades were clad in brick to harmonize with the surrounding architectural context (Nyström, 1984).

Figure 2: The new residential building in Olivedal. (Photo: Lars Mongs, 2025. Used with permission).

Similar to the original building, the new structure comprises a mix of one-, two-, three-, and four-room apartments. However, the salvaged elements were reassembled in new configurations, resulting in apartment layouts that, in some cases, differed significantly from the original (Fig. 3). The new floor plan also accommodated the integration of an elevator into each service shaft, necessitated by the increased number of storeys. To reconfigure the elements according to the new layout, slight modifications to their original dimensions were sometimes necessary to ensure a proper fit, as evidenced by the original construction drawings for the project (Fig. 4).

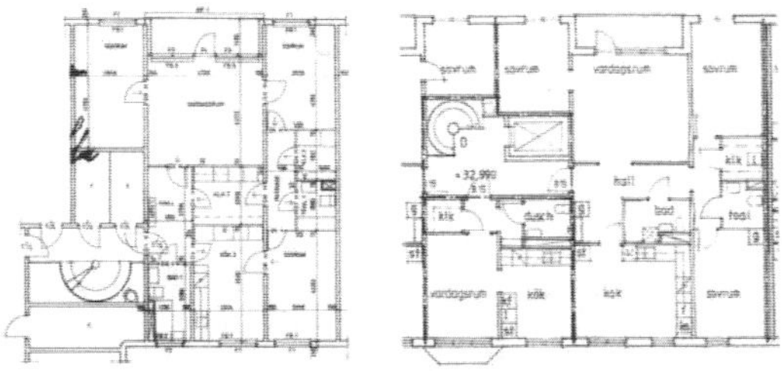

Figure 3: Floor plans showing the layout of a three-bedroom apartment in Norra Bergsjön (left) and Olivedal (right).

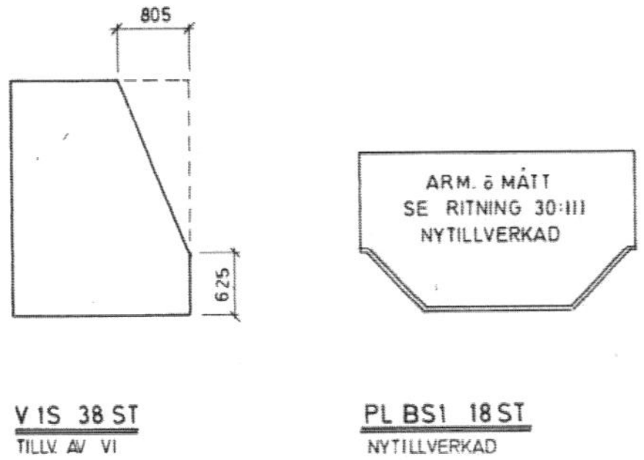

Figure 4: Details from the construction drawings of the new building in Olivedal, showing modifications to reclaimed element V 1S (left) and specifications for a new element (right).

3.1.4 Receiver project 2: Lerum (1986)

A smaller selection of salvaged elements from Norra Bergsjön was repurposed in a new housing development in the neighbouring municipality of Lerum, where ABV was also involved as the contractor (Fig. 5). The project comprised four two-storey buildings, providing a total of 29 new apartments, and was situated approximately 12 kilometres from the original donor building. The housing development represented a significantly smaller scale compared to the housing blocks in Norra Bergsjön and the salvaged elements were reconfigured to accommodate substantially altered floor plans (Fig. 6).

Figure 5: The new two-storey residential buildings in Lerum. (Photo: Lars Mongs, 2025. Used with permission).

The new buildings are narrower in depth compared to the original mass housing blocks and feature hipped roofs. In certain sections of the exterior envelope, wood stud infill walls were incorporated to complete the facade. The development comprises a mix of one- to three-bedroom apartments, which are accessed through private entrances at the front of each building. The apartments on the upper floor are reached via wooden access balconies.

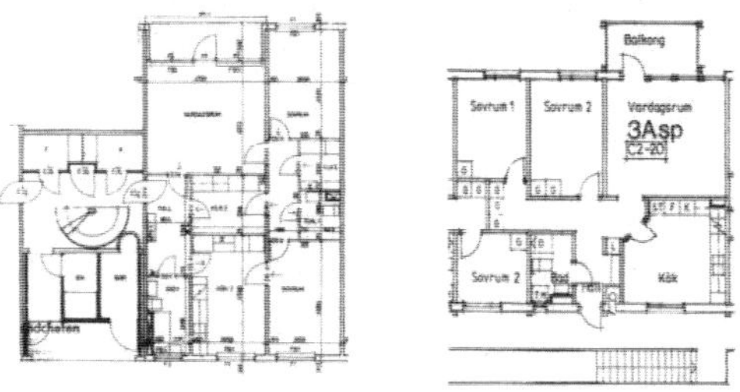

Figure 6: Floor plans showing the layout of a two-bedroom apartment in Norra Bergsjön (left) and Lerum (right).

3.1.5 Receiver project 3: Ytterby (1985)

Concrete elements from Norra Bergsjön were also repurposed in a residential development in Ytterby, located in Kungälv Municipality, approximately 15 kilometres from the original site in Norra Bergsjön. In this project, ABV again served as the main contractor. The development comprises a mix of row houses and low-rise multifamily buildings, each containing up to four apartments. Although the buildings were originally

https://doi.org/10.52202/080684-0001

Figure 7: Installation of concrete slabs from Norra Bergsjön at the construction site in Ytterby. (Photo: Lars Mongs, 1985. Used with permission).

designed with wooden structural frames, precast concrete slabs from Norra Bergsjön were used to construct the foundation slabs, as shown in Fig. 7.

3.1.6 Receiver project 4: Backatorp (1989)

Salvaged elements from Norra Bergsjön were also utilized in the construction of a new low-rise residential area in Backatorp, located approximately 6 kilometres from the original donor building. The development comprised 41 two-storey buildings accommodating in total 150 new apartments (Isemo, 1989).

Figure 8: One of the new receiver buildings in Backatorp. (Photo: Lars Mongs, 2025. Used with permission).

The three-bedroom apartments in the new buildings retain a general resemblance to the original floor plan of the donor building, though several key modifications were introduced (Fig. 9).

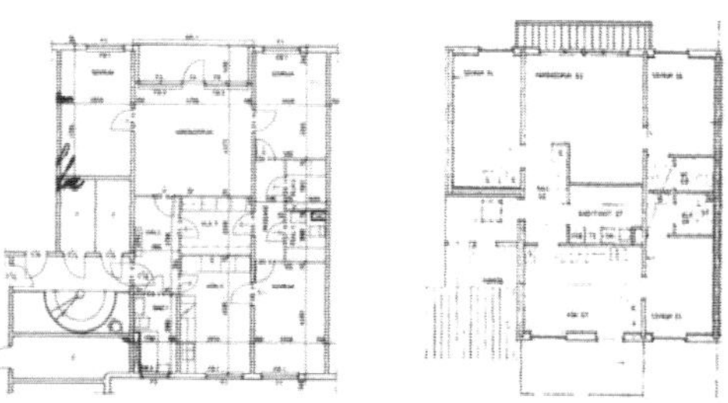

Figure 9: Floor plans showing the layout of a three-bedroom apartment in Norra Bergsjön (left) and Backatorp (right).

The new buildings are slightly narrower, incorporating only 12 floor slabs across their depth, compared to 13 in the original structure. The recessed balconies featured in the initial design were replaced with a larger living room area, marking a notable departure from the original layout. Another significant alteration was the integration of the kitchen with the former bathroom area, creating a more spacious cooking and dining zone that benefits from two windows. The main bathroom was relocated to the center of the building, replacing the space that previously served as a walk-in closet. The apartment entrances remain in their original positions and are thus accessed from the two gable facades of the new buildings, with apartments on the second floor accessed via an external staircase. In total, approximately 6,400 elements from Norra Bergsjön were reported to have been reused in the construction of the new residential area at Backatorp (Isemo, 1989).

3.2 LÖVGÄRDET

Approximately a decade after the partial deconstruction of Norra Bergsjön, the same municipal housing company, by then renamed Bostads AB Poseidon, once again turned to deconstruction as a strategy to address the problem of vacant apartments, this time in the Lövgärdet residential area. Located approximately 13 kilometres northeast of central Gothenburg, Lövgärdet was built between 1972 and 1974 during the final phase of the Million Programme. The area was never fully completed according to its original plans and had faced persistent vacancy challenges since its inception. In 1997, a project was initiated to remove six of the twelve existing nine-storey tower blocks in the area, with the aim of creating a more open and appealing living environment (Kraenzmer, 1999; Vidén & Botta, 2006). There were advanced plans to donate the dismantled elements to Poland as part of a Swedish aid package initiated by Prime Minister Göran Persson following a severe flooding (Ekenstam, 1997). When these plans fell through, Bostads AB Poseidon shifted its primary focus from the reuse of elements to concrete recycling (Ekenstam, 1997). Approximately 25,000 tons of dismantled concrete elements were crushed on-site and repurposed for various applications, including use as infill material for the construction of a new football pitch in the area (Kraenzmer, 1999; Vidén & Botta, 2006). However, a smaller number of concrete elements were salvaged and reused in a nearby municipal building project at Lärjeåns Trädgårdar (Vidén & Botta, 2006).

3.2.1 The Byggtema System

The tower blocks in Lövgärdet were originally constructed using the *Byggtema* building system, the last of seven precast concrete systems developed by Göteborgsbostäder (Hjertén, 1969). At the time of construction, the elements were produced at the company's newly established factory in Ingebäck, which had a production capacity of 2,000 apartment units per year (Wallinder, 1969). Similar to the Elementhus 65 system, the Byggtema system employed a cross-wall structural approach, utilizing a limited set of precast elements that could be configured into various standardized apartment layouts. The system also featured a hook fastening solution for connecting the elements, which significantly facilitated the disassembly process (Kraenzmer, 1999). The main differences compared to the earlier system were the design of the slabs, which were now produced as solid elements up to 3 meters wide, and the increased length of the load-bearing wall elements, which reached up to 4.8 meters (Wallinder, 1969). These modifications resulted in fewer joints and enabled a more rapid assembly process.

3.2.2 Deconstruction of Donor Building

As with the Bergsjön project, deconstruction at Lövgärdet began with stripping the buildings down to their structural frames. The roofs were cut into sections and removed using a crane. To expose the hook joints of each element, a small robot was employed, and approximately 1,300 holes had to be drilled per floor of each tower block (Kraenzmer, 1999). The five upper storeys of each tower were then disassembled with the aid of a crane. Aside from the limited number of elements salvaged for reuse, the disassembled elements were subsequently broken down into square-meter sections using an excavator equipped with a demolition shear, and then transported to a concrete crushing facility (Kraenzmer, 1999).

3.2.3 Receiver Project 5: Lärjeån (1998)

Approximately 3 kilometres from Lövgärdet, a municipal initiative to develop a commercial garden and cultivation centre on the outskirts of Angered Centre had been underway since the early 1990s (Engelbrektson, 1997). As part of the development, a new building was planned to accommodate a café and meeting rooms, serving as the central hub for the garden. Reflecting the project's strong ecological orientation, a significant portion of the construction materials was sourced from demolition sites within the Gothenburg region. In this context, a small number of salvaged concrete elements from Lövgärdet were repurposed for the construction of the warehouse's basement (Kvist, 1998; D. Björklund Jonsson, personal communication, February 14, 2025) (Fig. 10). The superstructure was subsequently built using reclaimed timber components from a dismantled wooden warehouse formerly located in Gothenburg's harbour (Heyman, 2000)

Figure 10: Construction of basement at Lärejåns Trädgårdar. (Photo by courtesy of D. Björklund Jonsson).

3.3 HAMMARKULLEN

In 1996, Göteborgs Stads Bostads AB, another municipal housing company in Gothenburg, initiated the most ambitious deconstruction effort to date in Hammarkullen, a Million Programme neighbourhood located approximately 12 kilometres north of the city centre (Fig. 11). Originally constructed between 1968 and 1972, the area had been planned to accommodate 2,700 new apartments. However, as part of a broader neighbourhood redevelopment strategy implemented during the 1990s, a 225-metre-long housing block comprising 176 apartments and located near the central square was designated for demolition (Lövkvist, 1996). This intervention, along with other planned measures, sought to enhance the area's attractiveness by reducing housing density and fostering a more open and varied living environment. Sixteen semi-detached houses were later constructed on the same site, incorporating salvaged elements from the original building (Sahlberg, 1997).

Figure 11: Facade element disassembled at Hammarkullen. (Bengtson, 1997).

3.3.1 The Göteborgs Stads Bostäder System

The original nine-storey housing block in Hammarkullen was constructed using Göteborgs Stads Bostäders' own precast concrete system. This system was developed based on a Danish system by Larsen & Nielsen (Wallinder, 1969). Similar to the two previously mentioned systems, it employed a cross-wall structural design with load-bearing gable and internal walls. However, this system was distinguished by significantly larger room-sized wall elements, measuring up to 7 meters in length. The floor slabs were produced as solid precast elements, ranging from 3.5 to 5.1 meters in length (Bengtson, 1997; Wallinder, 1969). A distinctive feature of this system was the use of volumetric kitchen and bathroom modules, which were prefabricated with cabinets and installations already in place at the factory. The floor plans were based on standardised two-, three- and four-room apartments, which could be configured in various arrangements around a central stairwell (Wallinder, 1969). The non-load-bearing sandwich facade elements were typically room-sized and featured an exposed aggregate surface finish (Wallinder, 1969).

https://doi.org/10.52202/080684-0001

3.3.2 Deconstruction of Donor Building

The proposal to deconstruct the housing block, rather than proceed with conventional demolition, was initiated by Stabilator, the demolition contractor commissioned to carry out the removal (Blomquist, 1997). Before implementing the plan at full scale, Stabilator conducted a test dismantling and prepared a demonstration apartment, in which materials and structural components were carefully separated to gain a clearer understanding of their composition and assembly. The tests revealed that the concrete was of high quality, exhibiting a compressive strength that significantly exceeded the original specification (Bengtson, 1997). It took approximately one month to determine the optimal method and necessary tools for the deconstruction and to facilitate the reuse process. Part of this work involved devising a method to separate the walls from the slabs, which had originally been cast together. The selected technique employed high-powered demolition hammers to achieve the necessary separation (Lövkvist, 1996). This method also allowed the original metal loops, used during the initial assembly 25 years earlier, to be exposed and reused during disassembly (Lövkvist, 1996).

The initial phase of the deconstruction process consisted of stripping the building to its structural frame and systematically labelling all elements prior to dismantling (Bengtson, 1997). Disassembly then commenced at the gable ends and proceeded in a stepwise manner to maintain the stability of the remaining structure. Notably, the facade elements were dismantled with the windows and balcony doors still attached. The kitchen and bathroom interiors were also preserved, as they were deemed to be in good condition (Bengtson, 1997). Once removed, the prefabricated elements were transported to an intermediate storage site in Surte, where they occupied an area equivalent to three football fields (Bengtson, 1997; Kraenzmer 1999). In total, approximately 3,800 concrete elements were dismantled at Hammarkullen, corresponding to 16,000 square meters of living space and 18,000 tons of concrete (Bengtson, 1997).

3.3.3 Receiver project 6: Hammarkullen (1998)

Despite ambitious plans to export the salvaged elements for use abroad, the construction of 16 semi-detached houses on the same site in Hammarkullen remains the only documented recipient project resulting from the large-scale deconstruction effort. The project was completed with financial support from the government's newly established *'Kretsloppsfond'*, which aimed to support refurbishment and reconstruction initiatives with a clear environmental focus (Sahlberg, 1997).

The new two-storey buildings, featuring pitched roofs and private gardens, bear little visual resemblance to the original housing block. Although facade elements from the original structure were reused, their exposed aggregate finish was concealed beneath a new layer of plaster to modernize the appearance (Blomquist, 1997). The architect responsible for the new design was Gunnar Werner of White Arkitekter, who had also contributed to the design of the original housing block three decades earlier (Lövkvist, 1996).

The shift in building typology—from a multi-family residential block to semi-detached houses—also necessitated substantial changes to the original floor plan (Fig. 12). Most notably, the depth of the building was reduced from approximately 13 meters to 9 meters. Reclaimed wall elements were installed on new slab foundations, alongside the salvaged kitchen and bathroom modules, and subsequently covered with a layer of floor slabs. As the new dwellings were constructed as two-storey structures, internal staircases were incorporated into each unit to accommodate the vertical layout. The main entrance to each house is positioned at the gable facades, consistent with the original design in which entrances were arranged around a central stairwell.

Figure 12: Floor plans showing a building section from the original donor building (left) and the new ground-floor plan of the semi-detached houses completed on the same site (right).

3.4 NAVESTAD

In 1999, the municipal housing company Hyresbostäder initiated an extensive refurbishment project in Navestad, a Million Programme residential area located on the outskirts of Norrköping (Fig. 13). The objective was to reduce the number of apartments from 1,600 to 950 by removing the upper floors of several buildings and converting around 200 of the remaining units into offices and educational facilities (Vidén & Botta, 2006). At the same time as Norrköping, a former industrial town, faced a surplus of housing, the neighbouring university city of Linköping was experiencing a severe shortage of student accommodation (Eklund et al., 2003). A few years earlier,

Figure 13: Assembly of precast concrete elements during the construction of the Navestad estate.

the municipal housing company AB Stångåstaden had successfully completed a student housing project in Linköping, utilizing reclaimed concrete elements from a cast-in-place residential building in Finspång (Rapport Profilen–Ryd, 1999). Gunnar Sundbaum, the initiator and project manager of the earlier reuse project, saw an opportunity to improve the reuse process in a new initiative by utilizing salvaged precast concrete elements from the ongoing work at Navestad (G. Sundbaum, personal communication, February 21, 2024). For Hyresbostäder, the idea to reuse the disassembled elements aligned with the high environmental ambitions of the Navestad refurbishment project, for which the company had received a substantial government grant (Vidén, & Botta, 2006; Eklund et al., 2003). Consequently, an agreement was reached between the two municipal housing companies to facilitate the transfer and reuse of concrete elements for a new project.

3.4.1 The Norrköpings Byggelement System

The Navestad estate was originally designed by architect Eric Ahlin and conceived as multiple buildings arranged in concentric rings around two circular courtyards. The buildings were designed with varying heights, reaching up to eight storeys, and featured a high concentration of two-bedroom apartments, complemented by a smaller number of one-bedroom and studio units. Completed in 1972, the estate was constructed using the *Norrköpings Byggelement* system, a precast concrete system employed by the company in new housing developments between 1965 and 1972 (Nilsson & Eliaeson, 1998). This highly rationalized system was developed by Eric Ahlin in collaboration with structural engineer Arne Johnson, specifically for Hyresbostäder. Like the previous three donor buildings, the system followed a cross-wall structural logic, with load-bearing walls positioned transversely relative to the building block. The wall elements were room-sized to minimize assembly time and restrict joints to the corners of rooms (Johnson, 1965). The floor slabs were solid concrete slabs, produced in widths of up to 3 meters and spanning approximately 3.5 meters (Wallinder, 1969).

3.4.2 Deconstruction of Donor Building

The deconstruction of elements at Navestad was completed prior to the commencement of construction at the Linköping site. Following disassembly, the elements were transported to a nearby intermediate storage site. Each component was specially marked with a designation and identity number to facilitate tracking and reuse components (AB Stångåstaden, 2002). Loading and unloading operations were carried out using a mobile crane (AB Stångåstaden, 2002). To verify that the reclaimed concrete elements met the structural requirements of the new buildings, pressure tests were conducted on selected components (Eklund et al., 2003).

3.4.3 Receiver project 7: Ryd (2001)

Salvaged wall elements, floor slabs, and staircases from Navestad were repurposed in the construction of new student housing in the Ryd area of Linköping, located approximately 57 kilometres from the original site. The architectural design was led by P.O. Kelpe of AG Architects, the same design office responsible for the refurbishment project at Navestad (P.O. Kelpe, personal communication, January 21, 2025). The development consisted of one four-storey and one two-storey building, together providing 54 new student apartments. In total, the buildings were completed using reclaimed elements amounting to approximately 1,400 tonnes of salvaged concrete (Eklund et al., 2003). In addition to the reuse of the concrete structural frame, the project incorporated other salvaged components, including windows, windowsills, and stair railings also sourced from the Navestad refurbishment (Eklund et al., 2003).

Figure 14: Receiver buildings accommodating student housing in Ryd. (Photo: Mikael Damkier. Courtesy of AG Arkitekter).

The concrete elements were reassembled in the same curved layout as the original structure, but the buildings were updated with a new metal mono-pitched roof, new entrances, and canopy roofs. The facade elements received new insulation to meet current standards and were plastered in white and red, lending the buildings a modern aesthetic (Eklund et al., 2003). Internally, the apartments range in size from 29 to 32 square meters and feature either an open-plan layout with a small kitchen or a separate bedroom with a kitchenette (Fig. 15). This adaptation of the original two-bedroom apartment configuration into single-room flats required the incorporation of new wrought-iron details to reinforce the connections between walls and slabs (AB Stångåstaden,

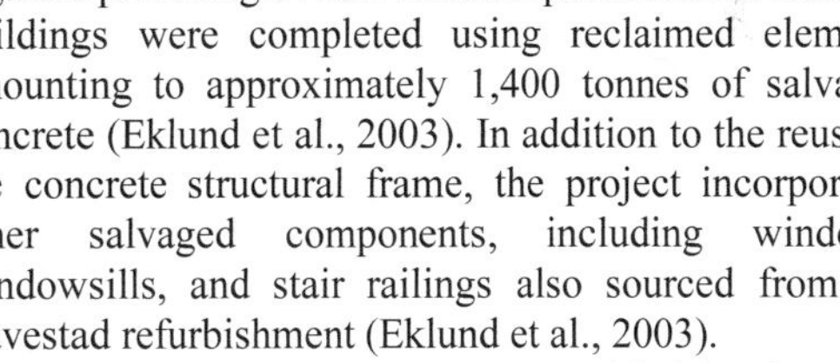

Figure 15: Floor plans showing a building section from the original donor building with two tow-storey apartments (left), and the new student housing block with five apartments (right).

https://doi.org/10.52202/080684-0001

2002). Since the time of the original construction, building regulations concerning acoustic performance had evolved, and the existing elements no longer complied with current standards. Consequently, partition walls between apartments were upgraded with improved sound insulation and the addition of new gypsum boards (Eklund et al., 2003).

4 DISCUSSION

4.1 THE PUBLIC HOUSING SECTOR: PIONEER IN PRECAST CONCRETE REUSE

4.1.1 Reuse as a Strategy for Addressing Uneven Housing Availability and Demand

All reuse cases examined in this study originate from the deconstruction of mass housing blocks constructed during the so-called Million Homes Programme (1965–1974). During this period, demand for new housing began to decline even before the programme was fully completed, contributing to widespread vacancies and a perception of socioeconomic decline in many of the newly developed areas (Hall, 1999). In both Gothenburg and Norrköping, this housing surplus was largely driven by the decline of local industries, which significantly reduced the anticipated need for housing. The reuse initiatives that characterize the initial phase of precast concrete reuse in Sweden were thus closely tied to the public housing sector and can be understood as pragmatic responses to an uneven distribution of available housing in relation to housing demand. A similar context shaped the adoption of precast concrete reuse in Germany, where most known reuse initiatives have taken place. Following the reunification of the country, many mass housing areas in former East Germany, commonly referred to as Plattenbau, faced high vacancy rates. This situation prompted efforts to dismantle and repurpose precast concrete elements in smaller-scale residential projects, such as single-family homes, for which there was a larger demand (Asam, 2005; 2007).

4.1.2 Economic and Ecological Incentives for Reuse

While the deconstruction of the donor buildings examined in this study share clear common economic and social incentives for reducing vacancy rates in Million Programme areas, a shift in the motivation to pursue reuse rather than conventional demolition can be observed over time. While resource efficiency is commonly promoted as a key benefit of the reuse process, the earlier projects primarily framed efficiency in economic terms, whereas later initiatives reflect a more pronounced ecological orientation. By the late 1990s, the concept of circularity, referred to in Swedish as *'kretsloppsprincipen'*, had begun to gain traction in Swedish society (Johansson, 1995). This shift is clearly exemplified in the Navestad reuse case. The government grant that supported the refurbishment of the Navestad donor building was closely linked to environmental objectives. Furthermore, collaborations with researchers at Linköping University enabled the first attempt to assess the environmental benefits of reuse compared to conventional construction methods. These studies indicated that the reuse approach reduced greenhouse gas emissions by two-thirds and decreased construction waste by 82% relative to a standard concrete building (AB Stångåstaden, 2002). Upon completion, the Swedish National Board of Housing, Building, and Planning recommended the project for the highest available subsidy level for ecological construction. As a result, the project received a green building grant of 2,000 SEK per square meter, bringing the total cost in line with that of a conventional project (Eklund et al., 2003)

4.2 REUSE POTENTIAL OF PRECAST CONCRETE SYSTEMS

The reuse potential of precast concrete elements largely depends on two key factors: the feasibility of disassembling the elements in a safe and economically viable manner, and the adaptability of the elements to meet the design requirements of a new building. A key aspect of the deconstruction and reuse process therefore involves understanding the structural and spatial logic of a specific system and exploiting its potential for new applications. By 1968, at least sixteen different precast concrete systems for multi-family residential buildings were in use across Sweden (Andersson, 1968b). Each system had distinct characteristics and followed different structural logics, depending on the spanning direction of the slabs and the placement of the load-bearing walls. The four precast concrete systems used to construct the donor buildings in this study were all cross-wall structural systems, featuring room-length slabs supported on interior load-bearing walls. However, the spans of the floor slabs varied between systems, and the sizes of the wall elements differed—ranging from smaller components to full room-sized panels—resulting in varying conditions for the reuse process.

4.2.1 Deconstructability

The deconstruction of the four donor buildings that enabled the seven receiver projects examined in this study demonstrates both partial deconstruction approaches, as seen in Norra Bergsjön and Navestad, and the complete dismantling of entire buildings, as undertaken in Hammarkullen and Lövgärdet.

In both Norra Bergsjön and Hammarkullen, the disassembly process proved more straightforward than initially anticipated. A key advantage of the Elementhus 65 system in Norra Bergsjön was that the elements were simply stacked without welding, secured using a hook-and-loop connection system. Although the system was not originally designed for disassembly, the original connectors effectively facilitated the lifting and separation of components during deconstruction (Tibblin, 1986). Similar to the experience in Norra Bergsjön, the actors involved in the deconstruction of the housing block in Hammarkullen expressed surprise at how well the precast concrete system could be disassembled (Bengtson, 1997). Because the system was based on the repetition of the

same configuration of elements on every floor, the disassembly procedure could be consistently repeated once the appropriate technique had been established (Lövkvist, 1996).
However, in Navestad, the deconstruction process turned out to be more complex and costly than anticipated. Many elements were damaged during dismantling and rendered unusable, necessitating more careful handling than initially expected. As a result, the costs associated with reusing entire concrete elements significantly exceeded those of crushing and recycling, ultimately leading Hyresbostäder to pursue the latter approach (Eklund et al.). Consequently, following the completion of the Nya Udden project, the ambitious plan to salvage concrete elements for the construction of an additional 500 student apartments in Linköping was abandoned. Although the design for these buildings had been developed to accommodate reused elements from Navestad, construction ultimately proceeded with newly produced concrete elements (P.O. Kelpe, personal communication, January 21, 2025). A published report on the Nya Udden project provides valuable insights into opportunities for improving the deconstruction and reuse processes, as identified by the actors involved (Eklund et al., 2003).
Organizational challenges were largely attributed to a compressed timeline, which hindered proper coordination between deconstruction and reconstruction. The lack of preparatory planning led to disassembly progressing ahead of new construction, requiring improvised and inefficient temporary storage. Additionally, assigning different subcontractors to each phase resulted in poor material handling due to limited understanding of the elements' intended reuse, causing damage and waste. The involved actors therefore recommended synchronizing deconstruction and reconstruction to reduce storage needs and involving the same personnel throughout to ensure careful handling and promote collaboration
At Norra Bergsjön, the partial deconstruction and conversion of the remaining housing blocks at Norra Bergsjön into rowhouses continued in subsequent stages, even in the absence of further government financial support. However, efforts to reuse the remaining disassembled elements gradually diminished, as matching them with new building projects proved more difficult than initially expected. Attempts to export the salvaged components abroad turned out to be a complex and prolonged process that ultimately failed to materialize (Ekelund, 1985). Even within the Gothenburg region, reuse efforts were problematic due to a tendency to 'cherry-pick' only the best elements, leaving many components unused (Beck-Friis, 1984). Reflecting on this issue, engineer Helmut Junker suggested that the ideal scenario would be to sell an entire building rather than individual elements (Beck-Friis, 1984). In their evaluation of the project, architect Bengt Forser and Jan Sundbom, the technical manager at the housing company, concluded that the refurbishment process carried out at the factory had been too costly. As a potential improvement, it was proposed that, in future projects, elements should be transported directly from the deconstruction site to the new construction site, with refurbishment undertaken only when necessary to minimize costs (Forser & Sundbom, 1986).
Similarly, at Lövgärdet and Hammarkullen, none of the ambitious plans to export the salvaged elements to countries experiencing housing shortages ultimately materialized. At Hammarkullen, only a small percentage of the salvaged elements were ultimately reused in the construction of semi-detached houses on the same site, while the remaining materials were eventually crushed (D. Chroneberg, personal communication, June 25, 2024).

4.2.2 Architectural Flexibility

In the process of deconstruction and reconstruction, it is unlikely that the spatial and functional requirements of the new building will align precisely with those of the original structure (Huuhka et al., 2015). This is evident in the seven recipient projects analysed in this study, all of which differ from their respective donor buildings in terms of building type, size, and floor plan. Consequently, a critical factor influencing the reuse potential of a precast concrete system is its inherent flexibility and the extent to which its constituent elements can be reassembled into new configurations.
The versatility and flexibility of precast concrete systems were already central concerns during the construction of the Million Programme. Every precast system faced the inherent challenge of balancing maximum production efficiency—achieved through the mass production of a limited number of element types—with the need to preserve a high level of spatial quality and flexibility in its application. Several studies were undertaken to evaluate the advantages and limitations of existing systems, with particular emphasis on assessing their architectural potential for generating flexible apartment layouts and accommodating a variety of building types (Anderson, 1967; 1968a; 1968b; 1968c; Wallinder, 1969).The most defining feature of precast concrete systems is the length of the floor slabs, which determines the positioning of the vertical load-bearing components within a system (Andersson, 1968a). Two main types can be distinguished: slabs that span the width of a single room and slabs that span longer distances, exceeding the length of one room. A vertical load-bearing element in a precast concrete system for residential buildings is typically a wall element, although some systems also incorporate columns. Depending on the length and orientation of the slabs, a wall-frame system can follow three different structural typologies (Hernández Vargas & Stenberg, 2024). In an integral wall system, room-sized slabs are supported by load-bearing walls positioned both longitudinally and transversely in relation to the building block. In a cross-wall system, floor slabs are supported by load-bearing walls positioned transversely relative to the building block. In a spine wall structure, the facade elements function as load-bearing components, allowing the floor slabs to span the full depth of the building.

https://doi.org/10.52202/080684-0001

A general advantage of the cross-wall typology, employed in all four donor buildings, is the potential flexibility it offers in floor plan design, as non-load-bearing partition walls can be freely positioned between the structural cross-walls. Another recognized benefit is the design freedom it affords for façades, since the exterior walls are non-load bearing, except for the gable walls (Wallinder, 1969). However, the range of possible floor plan configurations within a cross-wall system is largely determined by the maximum span length of the floor slabs used. In the *Elementhus 65* system, the maximum slab span of 4.8 meters significantly constrained the possible arrangement of room types within the floor plan (Wallinder, 1969). At the building scale, the system was designed for the construction of rectilinear buildings with a fixed depth of approximately 12 meters. Furthermore, building height was restricted to four storeys, as the system did not incorporate elevator shafts (Wallinder, 1969). In the development of the system into the later *Byggtema* version, the ability to accommodate other building types, such as tower blocks, was incorporated (Hjerten, 1969).

The *Göteborgs Stad Bostäder* system was the only system that incorporated volumetric elements. While these elements represented a high degree of prefabrication, they offered limited flexibility in the spatial layout of apartments. Moreover, all volumetric units were dimensioned for three-bedroom flats, resulting in oversized units when used for smaller apartments (Wallinder, 1969). Similar to the *Elementhus 65* system, the *Göteborgs Stad Bostäder* system was intended to be assembled into rectilinear housing blocks and featured little variation in possible building types (Wallinder, 1969). The Norrköping system used in Navestad is the only one of the four systems that was applied to various types of buildings, including tower blocks and curved structures (Wallinder, 1969). However, the flexibility of the system's spatial layout was considered limited due to the restricted span of the floor slabs and the fixed positions of the kitchen and bathroom, which were connected to special floor slabs containing installations (Wallinder, 1969).

Nevertheless, the recipient projects examined in this study demonstrate that the versatility of precast concrete systems can extend far beyond their original intended applications. At the building level, all receiver buildings differ significantly from their respective donor structures. In most cases, elements from large-scale mass housing blocks were repurposed into smaller-scale residential developments, including, low-rise multifamily buildings (Lerum and Backatorp), row houses (Ytterby), and detached houses (Hammarkullen). The most ambitious project—also unique in an international context—is the receiver building in Olivedal. In this case, the new structure features more storeys than the original donor building and is constructed on an irregular city block site, positioned between two existing buildings. This was made possible through the adaptation of existing elements combined with the addition of complementary new elements. In addition, the increased number of floors in the new building was facilitated by the incorporation of lift shafts positioned adjacent to the original staircase within the floor plan. Only in the case of the student housing in Ryd does the new building resemble the donor structure, as the salvaged elements were reassembled to follow the same circular layout as in Navestad, although the new building comprises only two sections. The smaller number of salvaged elements reused in the slab foundations of the row houses in Ytterby and in the basement of the warehouse in Lärjeån represents an alternative reuse scenario for salvaged concrete elements. These cases demonstrate that, rather than forming the complete structural frame, salvaged elements can also be successfully reused in combination with other structural materials, such as wood.

At the room level, the spatial layouts of all seven receiver buildings differ significantly from those of the original donor structures. At Lerum and Backatorp, all housing units were designed with private entrances, marking a distinct departure from the original floor plan at Norra Bergsjön, where apartments were arranged around a central stairwell in each section of the building blocks. In Olivedal, the arrangement of apartments around a central stairwell was retained, but the reused elements were reassembled in sometimes new combinations, allowing for new apartment layouts. These three different receiver projects involving reclaimed elements from Bergsjön highlight the system's versatility and flexibility, demonstrating that the components could be effectively repurposed across various new building types of differing sizes and spatial configurations. Architect Bengt Forser attributed this adaptability to the relatively small size of the elements, which allowed for numerous combinations in diverse applications (Forser & Sundbom, 1986).

In the case of the student housing building in Linköping, the original layout of two two-bedroom flats arranged around a stairwell in Navestad was adapted into five apartments in the new floor plan, varying in size from studios to one-bedroom apartments. At Hammarkullen, the reassembly of elements into row houses resulted in distinct layouts for each floor to accommodate the separation of functions across two storeys. This represents a significant departure from the conventional logic of the precast concrete system, which typically depends on identical configurations of elements on each floor of a building.

5 CONCLUSIONS

All seven early reuse projects that characterize the initial phase of precast concrete reuse in Sweden originated from the deconstruction of mass housing developments built under the so-called Million Homes Programme, a national housing initiative aimed at constructing one million new dwellings between 1965 and 1974. Closely tied to the public housing sector, these initiatives emerged as pragmatic responses to uneven patterns of housing demand. The rapid expansion of housing during the

Million Programme resulted in a surplus of apartments in large-scale, often less desirable residential areas, leading to widespread vacancies. Simultaneously, there remained a demand for smaller-scale residential developments in other parts of the country. Although interest in the reuse of precast concrete elements has grown in recent years, and several new projects have been initiated, contemporary efforts have yet to match the ambitious scope of the projects realized during this initial period.

Interestingly, the very characteristics that led to widespread criticism of many of the mass housing areas built under the Million Homes Programme—namely their systematization, standardization, and repetition—are the same attributes that made these precast concrete systems particularly well-suited for reuse. Especially during the deconstruction phase, the reuse process benefits from economies of scale, as upfront costs are reduced and standardized working methods can be systematically repeated throughout the disassembly process.

All four donor buildings involved in the reuse efforts analysed in this study were based on cross-wall structural systems, with load-bearing walls positioned transversely relative to the building block. However, the spans of the floor slabs varied between systems, and the sizes of the wall elements ranged from smaller components to full room-sized panels, resulting in differing levels of flexibility in design at both the building and room scales.

The most important finding of this study is that the recipient projects demonstrate that the architectural potential of precast concrete systems extends well beyond their original applications. Through relatively simple design interventions—such as reassembling elements into new configurations, adapting existing components, and integrating complementary new elements—these systems can be successfully repurposed to meet diverse site contexts, building types, and floor plan requirements.

Yet, despite the demonstrated reuse potential of precast concrete systems, both in terms of their deconstructability and their adaptability across a wide range of recipient buildings, these early initiatives did not lead to the widespread adoption of precast concrete reuse within the Swedish construction sector. With the exception of the Navestad project, it can be concluded that the primary reason most reuse initiatives during this initial phase failed to result in a greater incorporation of salvaged elements into new constructions was not a reflection of the reuse potential of these elements, but rather the difficulty in generating sufficient demand for salvaged components in new projects.

ACKNOWLEDGEMENTS

This work was supported by the European Union's Horizon 2020 research and innovation programme under the RECREATE project (grant agreement No. 958200). Special thanks are also extended to photographer Lars Mongs for his assistance with many of the photographs included in this paper, and to all interviewees who contributed valuable insights to the studied cases.

REFERENCES

Addis, B. (2006). *Building with Reclaimed Components and Materials: A Design Handbook for Reuse and Recycling.* Earthscan.

Alén, C., Roth, L., Dahlgren, S., Östberg, G., & Andersson, M. (1999). *Rapport Profilen-Ryd. A pilot project reusing concrete frames.*

Alonso, P., & Palmarola, H. (Eds). (2019). *Flying Panels.* DOM Publishers.

Andersson, G. (1967) Inventering av stomsystem för elementbyggda flerfamiljshus [Inventory of structural systems for prefabricated multi-family buildings]. *Byggforskningen.*

Andersson, G. (1968a). Flerfamiljshus med stomelement av betong [Multi-family buildings with concrete structural elements]. *Byggmästaren,* 6, 24–35.

Andersson, G. (1968b). Svenska elementbyggsystem för flerfamiljshus 1 [Swedish precast concrete systems for multi-family buildings 1]. *Byggforskningen.*

Andersson, G. (1968c). Svenska elementbyggsystem för flerfamiljshus 2 [Swedish precast concrete systems for multi-family buildings 2]. *Byggforskningen.*

Asam, C., Biele, H., Liebchen, J. (2005). *Untersuchung der Wiederverwendungsmöglichkeiten von demontierten Fertigteilelementen aus Wohnungsbautypen der ehemaligen DDR für den Einsatz im Wohnungsbau* [Investigation of the Reuse Possibilities of Dismantled Prefabricated Elements from Residential Building Types of the Former GDR for Use in Housing Construction]. Fraunhofer IRB Verlag.

Asam, C. (2007) Recycling prefabricated concrete components – a contribution to sustainable construction. *IEMB Info,* 3.

Axelsson, A. (1983). Förorternas hus flyttas till centrum [The suburban buildings are being moved to the city center]. *Arbetet Väst,* 8 November.

Beck-Friis, U. (1984). Från Bergsjön till Malaysia: Nedmontering av tomma hus blev lyckat projekt [From Bergsjön to Malaysia: The dismantling of empty buildings became a successful project]. *Arbetet Väst.* 9 October.

Bengtson, U. (1997). Ett elementhus blir flera – återbruk på hög nivå [One prefabricated building becomes several – high-level reuse]. *Byggindustrin,* 4, 24-26.

Blomquist, E. (1997). Fräcka parhus av rivningsrester i förort [Stylish semi-detached houses made from demolition leftovers in the suburbs]. *Svenska Dagbladet,* 5 October.

Ekenstam, C. (1997). Vill inte ha hjälp - Polen tackar nej till husen [Doesn't want help – Poland declines the buildings]. *GT,* 1 October.

Ekelund, J-O. (1985). Ny svensk exportvara: Förort i betong! [New Swedish export product: Suburb in concrete!]. *Offentliga Affärer,* 1.

Eklund, M., Dahlgren, S., Dagersten, A., Sundbaum, G. (2003). The conditions and constraints for using reused materials in building projects. In R. Chini (Ed.), *Deconstruction and Materials Reuse,* vol. 287. (pp. 248-259). CIB Publication.

Engelbrektson, L. (1997). Myllans mästare [Master of the soil]. *Arbetet Nyheterna,* 12 December.

Fischer, A., Huber, R.K., Asa, C., Winter, P. (2012). *Plattenvereinigung Berlin 2010/2011, Abschlussbericht* [Plate Association in Berlin 2010/2011, Final Report]. Deutsche Bundesstiftung Umwelt.

Forser, B., & Sundbom, J. (1986). Att vända utvecklingen i Norra Bergsjön [Reversing the trend in Northern Bergsjön]. *Plan,* 3, 140-143.

https://doi.org/10.52202/080684-0001

Framtiden Byggutveckling. (2024) *Gamla våningsplan lägger grunden för nya Kvarteret Återbruket* [Old floors lay the foundation for the new quarter Återbruket]. Availble at: https://framtiden.se/blog/2024/03/13/gamla-vaningsplan-lagger-grunden-for-nya-kvarteret-aterbruket.htm (Accessed 30 January 2025).

Hall, T. (Ed.). (1999) *Rekordåren: en epok i svenskt bostadsbyggande* [The Record Years: An Era in Swedish Housing Construction]. Boverket.

Hall, T., & Vidén, S. (2005). The Million Homes Programme: a review of the great Swedish planning project. *Planning Perspectives*, *20*(3), 301–328.

Hernández Vargas, J. & Stenberg, E. (2024). *Building, component, and connector typologies of precast concrete building systems.* The ReCreate Project.

Heyman, E. (2000). Grönt kafé slår ut i Lärjeåns Trädgårdar [Green café blossoms in Lärjeån's Garden]. *Göteborgs-Posten*, 31 July.

Hjertén, I. (1969) *Elementa.* Fastighets AB Göteborgsbostäder.

Huuhka, S. (2010). *Kierrätys arkkitehtuurissa. Betonielementtien ja muiden rakennusosien uudelleenkäyttö uudisrakentamisessa ja lähiöiden energiatehokkaassa korjaus- ja täydennysrakentamisessa* [Recycling in architecture. Reusing reclaimed prefabricated concrete elements and other second-hand structural components when building new and rehabilitating and infilling concrete neigbourhood units]. Tampere University of Technology.

Huuhka, S., Kaasalainen, T., Hakanen, J.H., Lahdensivu bet, J. (2015). Reusing concrete panels from buildings for building: Potential in Finnish 1970s mass housing', *Resources, Conservation and Recycling,* 101, 105-121.

Huuhka, S., Naber, N., Asam, C., & Caldenby, C. (2019). Architectural Potential of Deconstruction and Reuse in Declining Mass Housing Estates. *Nordic Journal of Architectural Research*, 1, 139-179.

Huuhka, S., Aarikka-Stenroos, L., Lahdensivu, J., Jonker-Hoffrén, P., Arnold, V., Stenberg, E., Blok, R., Kjartan Gudmundsson, Patrick Teuffel, and Angelika Mettke. (2023). Recreating the Construction Sector for Circularity: Catalysing the Reuse of Prefabricated Concrete Elements. *The Routledge Handbook of Catalysts for a Sustainable Circular* Economy (pp. 42-66). Routledge.

Ikano Bostad. (2023). *Samarbete för hållbart återbruk minskar klimatpåverkan i byggprojekt* [Collaboration for sustainable reuse reduces climate impact in construction projects]. Availble at: https://press.ikanobostad.se/posts/pressreleases/samarbete-for-hallbart-aterbruk-minskar-klima.html (Accessed 30 Jan 2025).

Isemo, A. (1989). Gamla fasader blir som nya [Old facades become like new]. *Göteborgs-Posten*, 31 August.

Johansson, B. (Ed.). (2012). *Miljonprogrammet– utveckla eller avveckla?* [The Million Program – to develop or dismantle?]. Forskningsrådet Formas.

Johansson, B. (1995). *Bygg- och rivningsmaterial i kretskloppet* [The Million Program – to develop or dismantle?]. Byggforskningsrådet.

Johnson, A. (1965). *Elementbyggnadssystem för bostadshus i Norrköping* [Precast concrete system for residential buildings in Norrköping].

Kraenzmer, W. (1999) *Selektiv rivning av bostadshus* [Selective demolition of residential buildings]. KTH Royal Institute of Technology.

Kubu, M. (1983). Så löses problemen med tomma lägenheter: Välfärdens betongklossar blir radhus [Solving the problem of vacant apartments: The concrete blocks of the welfare state become row houses]. *Dagens Nyheter*, 12 September.

Lövkvist, C. (1996). Nu ska hela rasket rivas...Nu ska hela rasket bort [Now the whole lot is being torn down...Now the whole lot must go]. *Göteborgs-Posten,* 17 August.

Malmqvist, T., Borgström, S., Brismark, J., & Erlandsson, M. (2023). *Referensvärden för klimatpåverkan vid uppförande av byggnader, Version 3* [Reference values for climate impact in the construction of buildings, Version 3], KTH Royal Institute of Technology.

Mettke, A. (1995). *Wiederverwendung von Bauelementen des Fertigteilbau* [Reuse of construction elements of precast buildings]. Werkstatt Verlagsauslieferung.

Nilsson, S., & Eliaeson, P. (1998). *Navestad 1998.* KTH Royal Institute of Technology.

Nyström, U. (1984). Gamla bostäder blir som nya...[Old homes become like new...]. *Göteborgs-Posten*, 20 December.

Küpfer, C., Bastien-Masse, M. & Fivet, C. (2023). Reuse of concrete components in new construction projects: Critical review of 77 circular precedents. *Journal of Cleaner Production*, 383, 135235.

AB Stångåstaden. (2002). *Nya Uudden.*

Stenberg, E. (2021). Restructuring Swedish Modernist Housing. *Nordic Journal of Architecture*, 3(2), 89–93.

Stenberg, E. (2016). Revisiting Sweden's Million Program Era: Contemporary Design Strategies for Prefabricated Structural Systems. In. W. Nägelil & N. Tajeri (Eds.), *Small Interventions: New Ways of living in Post-War modernism* (pp. 107-114). Birkhäuser.

Tibblin, H. (1986). Tomma lägenheter blev efterfrågade radhus [Vacant apartments became sought-after row houses]. *Fastighet'86,* 4.

Vidén, S., & Botta, M. (2006) *Bostadsförnyelse och miljöåtgärder med stöd av lokala investeringsprogram- Hållbar utveckling i 50-60-70-talens bostadsområden* [Housing renewal and environmental measures supported by local investment programs – Sustainable development in residential areas from the 1950s, 60s, and 70]. KTH Royal Institute of Technology.

Wallinder, J. (1969) *Utställningen Elementbyggda Flerfamiljshus* [The Exhibition: Prefabricated Multi-Family Buildings]. Chalmers University of Technology.

Sahlberg, A. (1997). Rivna höghuset blir låga parhus [Demolished high-rise becomes low semi-detached houses]. *Göteborgs-Posten*, 8 June.

Statistics Sweden. (2025). *Number of dwellings by region, type of building and type of ownership (including special housing, Year 2024.*

Stenberg, E (Ed.). (2013). *Structural Systems of the Million Program Era.* KTH Royal Institute of Technology.

Vidén, S., & Lundahl, G. (Eds.). (1992). *Miljonprogrammets bostäder: bevara – förnya förbättr*a [Million Program housing: preserve – renew – improve]. Statens råd för byggnadsforskning.

Återhus – att bygga hus av hus. (2023). *Bilaga 9: Arbetspaket 7-Pilottester* [Appendix 9: Work Package 7-Pilot Tests]. Availble at: https://aterhus.nu/rapportudi2.htm (Accessed 30 January 2025).

REUSE CHALLENGES IN CIRCULAR DECONSTRUCTION: LESSONS FROM EXPLORATIVE CASE STUDIES

Li Jiang[1*], Marc van den Berg[1], Hans Voordijk[1], and Andreas Hartmann[1]

[1*]University of Twente, Faculty of Engineering Technology, Department of Construction Management & Engineering, the Netherlands

*Corresponding Author: l.Jiang@utwente.nl

ABSTRACT

Background and aim. There is a lack of empirical insights into the challenges faced in deconstruction processes aimed at building element reuse, particularly from the perspective of demolishers. This study aims to address this gap by identifying the challenges that hinder the recovery of building elements for reuse in deconstruction processes.

Methods and Data. Using a multiple case study design, we examined deconstruction practices in two projects, an outpatient clinic and a brick factory. Qualitative data were gathered through ten semi-structured interviews, project documentation, and field visits. A combination of deductive and inductive approaches was applied to data analysis.

Findings. Our findings reveal several challenges that hinder reuse practices in deconstruction projects. We have categorised these into four key system elements: technology, people, processes, and policy. These challenges collectively impede the transition towards a more circular practice in the demolition industry.

Theoretical/ Practical / Societal implications. This study provides a holistic understanding of the challenges that demolishers encounter when attempting to reuse building elements. It also extends existing research by providing empirical insights into deconstruction practices.

KEYWORDS: Construction industry, Circular practices, Deconstruction, Reuse, Reverse Logistics

1 INTRODUCTION

The construction industry urgently requires insights into reuse processes to facilitate more sustainable waste management. The industry generates about one-third of the total amount of waste in Europe (European Commission, 2022), which is often destined to be recycled or landfilled (Chileshe et al., 2019). Specifically, the construction and demolition waste (CDW) generated from the end-of-life (EoL) of an asset, so-called demolition waste (DW), is the largest contributor to the CDW (Jiang et al., 2017; Wijewickrama et al., 2020). Reverse logistics supply chain (RLSC) management has emerged as a crucial part of sustainable practices (Mallick et al., 2023). In the construction sector, this refers to the process of moving building elements and components from the point of salvaged buildings to the point of new construction (Hosseini et al., 2015). This mainly involves recovered elements resulting from the processes of selective demolition or deconstruction (Elghaish et al., 2023; Ghobakhloo et al., 2013; Wibowo et al., 2022), with the aim of salvaging and recovering (a portion of) elements with reuse capability (Akbarieh et al., 2020). In a deconstruction process, a sequence of preferred circular actions is established. Among others, reuse is one of the preferred options (Parto et al., 2007). It represents using a building element again, either for its original purpose or for a similar intent (Van den Berg et al., 2020a).

Several challenges are yet to be overcome before the widespread adoption of reuse practices. In this regard, many studies provide valuable knowledge on the challenges associated with reuse in circular demolition processes. For example, Purchase et al. (2021) highlighted some challenges hindering circular practices in the construction and demolition sectors. In particular, the authors used a literature review to summarise several main barriers, including, policy and governance, permits and specifications, technological limitations, quality and performance and implementation costs. Similarly, Ferriz-Papi et al. (2024) reviewed and categorised the challenges

 https://doi.org/10.52202/080684-0002

that prevent improved CDW management. These challenges range from political, economic, social, and technological to environmental aspects. Wijewickrama et al. (2020) outlined challenges and future opportunities regarding information sharing in the RLSC on demolition waste, with a systematic literature review. However, most existing studies are primarily theoretical or conceptual, and empirical insights into the challenges of the deconstruction process aimed at reuse are lacking. Furthermore, in terms of target groups, researchers have mainly focused on project actors such as designers and clients (Eikelenboom et al., 2024). Little is known about the challenges faced by demolishers. For reuse to occur, demolishers must shift their focus from demolishing building parts to recovering them (Van den Berg et al., 2020b). Fini and Forsythe (2020) also identified demolishers, along with building owners, as the primary influencers in determining the extent of waste reduction sent to landfills. They need to execute several (new) tasks, including advising, redistributing, storing and supplying elements in circular construction projects (Eikelenboom et al., 2024). Despite their key roles in the decision-making processes for EoL scenarios of building elements, relatively few studies focus on demolishers (Van den Berg et al., 2020a).

This study aims to address these gaps by identifying the challenges that hinder the reuse of building elements in deconstruction processes, especially from the perspective of demolishers. An element refers to any physical part of a building that can be handled separately, such as façade elements and ceiling tiles (Van den Berg et al., 2020b). The succeeding sections are structured as follows. We first present the literature review on circular demolition and embedded challenges. This is followed by an explanation of the multiple-case studies methodology adopted. We then present our findings on the challenges faced in deconstruction projects intended for element reuse. A discussion is followed regarding its contributions and future work. The paper ends with a conclusion.

2 LITERATURE REVIEW

Construction practitioners have traditionally focused on the materials flow from the point of extraction to their consumption, while an RLSC keeps materials in a loop by harvesting them from buildings (Hosseini et al., 2015). A RLSC starts with dismantling existing buildings (Wijewickrama et al., 2020). Various dismantling techniques are available, including demolition and deconstruction. Specifically, demolition represents the conventional practice of removing a building without considering potential reuse possibilities. In contrast, deconstruction is sometimes understood as construction in reverse, serving an important role in buildings' circularity (Bertino et al., 2021). Those dismantled elements are then subjected to recovery through reuse or other strategies (Wijewickrama et al., 2020). In this study, the terms "circular demolition" and "deconstruction" are used interchangeably, both referring to the process of recovering elements with reuse potential.

The transition from demolition to deconstruction of buildings has gained traction. Allam and Nik-Bakht (2023) summarised three key areas of deconstruction-related research, each corresponding to a major phase of construction projects: (1) the design phase, with an emphasis on Design for Deconstruction; (2) the EoL phase, which focuses on deconstruction planning and waste management; and (3) the second-life phase, examining the performance of recovered construction elements. The authors further emphasised the need for EoL insights regarding the destination of the recovered elements and deconstruction processes. During the EoL phase, recycling – reprocessing components to produce new ones (Hosseini et al., 2015), has received significant attention. Several European Member States, such as the Netherlands, Germany and Finland, have already achieved a 70% recycling rate, meeting the target set by the European Commission in 2014 (Gálvez-Martos et al., 2018). To further improve resource efficiency, maximising reuse is regarded as one of the best practices in Europe (Gálvez-Martos et al., 2018). Compared to recycling, reuse minimises the consumption of additional materials, energy, and labour, making it a more circular option (Ellen MacArthur, 2013). Accordingly, the European Union has introduced the Waste Framework Directive, prioritising reuse over recycling (Huuhka et al., 2015). To support the reuse of building elements, a series of activities with a focus on reuse should be planned in circular demolition projects. Van den Berg (2024a) distinguished three main phases in a deconstruction process, namely, identifying, harvesting and distributing. Any deconstruction process is, accordingly, initialised by identifying building elements presenting a high reuse potential. The term "harvesting" represents the activity of reclaiming those valuable elements from the existing built environment, for further facilitating reuse in new projects (Jongert et al., 2011). Lastly, distributing presents the diverging movement of harvested elements away from a demolition site.

Wijewickrama et al. (2021) supposed that an integrated system of technology, people, process and policy is necessary in the RLSC of DW. Technology is required to provide reliable, accurate and sufficient information in digital form for building element reuse (Byers et al., 2023). Iyiola et al. (2024) showed that various digital technologies, such as Building Information Modelling (BIM) and blockchain, can support reuse practices. The second element, people, considers how demolishers and other stakeholders (e.g., contractors and clients) collaboratively work during buildings' EoL. For reuse to take place, the elements recovered by demolishers should be utilised in new projects by contractors and clients. In this context, Eikelenboom et al. (2024) studied the changing role of demolishers in circular construction projects, compared to their conventional roles regarding tasks, timing, position and image. Furthermore, the third element, process, refers to a set of interrelated activities designed to achieve a defined output (Cruz et al., 2015; Hammer & Champy, 2009; Ko, 2009). Information sharing among these processes is significant to ensure the effectiveness of demolition operations (Rameezdeen et al., 2016; Wu et al., 2022). The last element, policy, represents the rules, standards, and guidelines that provide demolishers with clear directions on what is expected of them (Wijewickrama et al., 2021). Wijewickrama et al. (2021) proposed that a RLSC, or a reuse process, should be achieved through the integration and coordination of technology, people, process and policy. The authors developed a conceptual integrated framework by integrating these four system elements while acknowledging the need for empirical validation and testing. Previous studies have explored the challenges associated with the deconstruction process but lack empirical insights into the challenges demolishers face when reusing building elements, particularly in relation to the interconnected system elements of technology, people, processes, and policy.

3 METHODS

This study employed a multiple-case study method to enhance theoretical knowledge by incorporating new empirical insights from real-life cases (Çetin et al., 2022). Two case studies were purposefully chosen: the deconstruction projects concerning an outpatient clinic and a brick factory (
Table 1). Both case projects were completely deconstructed at the moment of this study, providing the possibility of understanding the whole deconstruction process and enabling the observation of reuse practices in the target buildings. They are both located in the Netherlands, which is recognised as a global leader in the implementation of circularity (Marino & Pariso, 2020). Conducting the study in this context is ideal for generating valuable insights into the reuse-related challenges. These cases are also considered "unique" (Yin, 2014), since they share exceptional circularity ambitions: large quantities of old elements were planned to be reused. This circular approach is, also in Europe, rarely adopted in the demolition industry, where construction materials are often either disposed of or recycled (Gálvez-Martos et al., 2018).

Specifically, in the outpatient clinic project, a temporary outpatient clinic was dismantled in its entirety and rebuilt as a healthcare centre at a new location. The brick factory was composed of four halls. One obsolete hall was repurposed for use by a (local folklore) parade association, and many elements from some other halls were dismantled and planned to be sold through different channels. The project was awarded a Dutch certification for its circular approach. Both projects are considered circular projects with a high reuse percentage, supported by several favourable conditions. First, both projects were designed to be easy to disassemble. The outpatient clinic, a modular building with prefabricated components, was specifically designed for easy disassembly to support future reuse. Similarly, the brick factory employed a steel-based construction, which was notable because projects of that scale "were normally built by concrete (at that time) and then it is very difficult to demolish", introduced by the project manager. Second, both projects were driven by strong deconstruction ambitions. This focus ensured that stakeholders carefully considered the destinations of elements throughout the projects. These two similar deconstruction projects, hereby, offer an opportunity for an in-depth exploration of the common challenges faced by demolishers.

Table 1. Case studies overview

	Outpatient clinic	Brick factory
Characteristic	Rebuilding a modular building for a healthy purpose	Reusing elements from a steel-based construction
Gross floor area	1100 m2	3240 m2
Construction year	2016	1909
Demolition year	2023	2023
Exemplary element reuse	Facade elements, floor plates	Steel, roof plates

3.1 DATA COLLECTION

In line with the triangulation principle (Eisenhardt, 1989), multiple sources of evidence were used, including documents, field visits, and in-depth semi-structured interviews (see Table 2). Those information sources are intended to offer insight into the embedded challenges in hindering element reuse in deconstruction processes. Data was collected from June 2024 to November 2024. Ten semi-structured interviews were conducted with key informants in demolition teams from both projects, including two project managers, one project planner, one digital expert, one material harvester and two site managers. Incorporating the perspectives of diverse stakeholders can enhance the understanding of a complex problem or phenomenon being studied (Van de Ven, 2007). Semi-open questions were designed to understand challenges in the deconstruction process, from identifying, harvesting and distributing reusable elements from "donor" to "target" buildings (Van den Berg, 2024b). Each interview lasted between 60 and 90 minutes and was audio-/video-recorded and transcribed. The case studies were also informed by project documents, including, among others, construction drawings, materials inventory and demolition plans (see Table 2).

Table 2. Data collection

	Outpatient clinic	Brick factory
Semi-structured interviews	Project manager (1x) Project planner (2x) Digital expert (1x) Site manager (1x)	Project manager (1x) Project planner (2x) Harvester (1x) Site manager (1x)

https://doi.org/10.52202/080684-0002

Project documentation	Construction drawings, Demolition plans, Project Contracts, etc.	Construction drawings, Materials inventory, Project photos, etc.
Site visits	Visits to both the demolition site and the rebuilt location	-

3.2 DATA ANALYSIS

For analysing associated challenges in the process, this study used thematic analysis as a qualitative method, incorporating both deductive and inductive approaches.

Table 3. Data analysis example

.Example quotes	Inductive analysis	Deductive analysis
"There is not, something like QR code <in the sticker> that you can scan and know the information of the component…you could not track the component."	Limited digital technologies	Technology
"My colleagues said, we can, for instance, use a tablet on a construction site to number the elements. But I knew 100% that the construction workers outside could not use it on the right"	Lack of expertise and experience	People

The thematic analysis consists of generating emerging themes or analytical categories as a description of the phenomena within the data (Fereday & Muir-Cochrane, 2006). The deductive approach uses an organising framework derived from extant literature to perform the analysis. The inductive approach, on the other hand, involves working exclusively from raw data (Azungah, 2018). This technique is widely used in various studies (see Wijewickrama et al., 2021).

Specifically, all collected data, including transcribed interviews and project documents, were first thoroughly read to familiarise the researchers with the content. Next, Table 3 provides examples of how the data was analysed in a deductive and inductive manner. Altas.ti was used to conduct the analysis, which is software for structuring, retrieving and analysing qualitative data in a continuous and cyclical way (Ronzani et al., 2020). It also supports the analysis of "theme intensity", which represents the number of statements referring to a particular challenge across a total number of statements (Wao et al., 2011). Complementing the qualitative method, this quantitative analysis provided insights into the relative importance of different themes/codes (see Figure 1).

4 FINDINGS

This study identifies several challenges hindering the recovery of building elements for reuse in circular demolition projects, which are grouped into four key elements: technology, people, process, and policy.

4.1 TECHNOLOGY

The use of digital technology was limited in both case projects (Table 4), which is one of the biggest challenges (with a theme intensity of about 24%, see Figure 1). In the case of the brick factory, a company inventory application was used to take photos and record relevant information on reusable elements. The application integrates with two

an inductive approach was applied to identify initial codes representing data features relevant to the research questions centred on reuse-related challenges. Those codes were then collated into four themes or elements deductively: technology-, people-, process- and policy-related challenges, building on Wijewickrama et al. (2020). The final step involved refining each theme by reviewing it in relation to the extracted codes and the entire dataset. This process aligns with the phases of thematic analysis proposed by Braun and Clarke (2006). It offered a basis to start understanding challenges in deconstruction processes.

other digital technologies, Insert and an online marketplace, to facilitate information sharing between buildings slated for demolition and new projects. Specifically, Insert is an initiative founded by several demolition companies in the Netherlands. One important function is to collect and store reusable elements for potential future reuse among its partnering firms. It was used in the case of the brick factory to disseminate information of reusable elements. An online marketplace was used in both projects for selling recovered products in the Netherlands. Moreover, Lidar (Light Detection and Ranging technology) was utilised in the case of the outpatient clinic with the potential of developing a digitalised representation of the to-be-demolished building.

Table 4. Digital technologies use cases in case projects

	Identifying	Harvesting	Distributing
Brick factory	An inventory application, Insert, an online marketplace	-	-
Outpatient clinic	Lidar Scanner, an	-	-

	online marketplace

Those digital technologies were primarily used during the identifying phase, with limited application in subsequent activities (Table 4). The deconstruction process was perceived as labour-intensive and characterised by a lack of technologies. New technologies are hereby required to support the process, especially during the harvesting and distributing phases. For example, one project planner expected the application of Quick Response (QR) codes for tracking recovered elements in the harvesting and distributing phase. "You can know, for instance, that container A contains all the wall insulation when you scan the QR codes", he explained. However, "There is not something like QR code (in the project)...you could not track the component", he further added. Furthermore, existing technologies face limited implementation due to their inability to effectively support deconstruction activities. For example, although a scanned model of the outpatient clinic was generated using Lidar, the demolishers did not use it in practice. This is because "it doesn't provide the functionality we need... it is not user-friendly", as one digital specialist noted. The participant further explained several challenges with this technique. First, the scanning technology lacks visual context: "You want a nice picture (of building components), rather than (only) a point cloud...if we want to sell them". The technique also encounters problems in incorporating additional information into the scanned models and connecting with other platforms like Insert. Moreover, "(for Lidar technology), that's one million points to get one door...we want one BIM of the door, including the picture of that specific door", the participant said. It can be potentially improved with automatic recognition of standardised objects (e.g., windows and doors) with technology like artificial intelligence. Another example is the Insert, which is still implemented on a limited scale and is primarily viewed as a "promotional initiative". No elements were actually sold through the Insert in the brick factory case, and it ultimately served only for demonstration purposes. "It <Insert> is a nice platform and also a good initiative, but if you want to sell something, you must use other platforms", one project planner added.

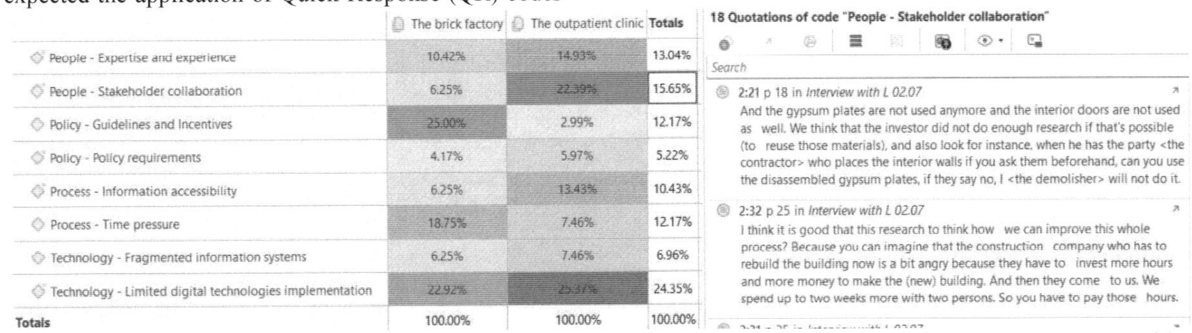

Figure 1. Theme density analysed by Atlas.ti

The use of different, disconnected information sources led to fragmented information management across the entire process. In the case projects, a common practice involves recording and exchanging information through documentation formats such as photos, digital spreadsheets, Portable Document Format (PDF) files, or other alternatives. The entire deconstruction process of case projects relied on these information sources, including annotated drawings detailing element dimensions and material inventory in spreadsheets. Furthermore, information was also available in the form of scanned models or within the inventory application. Those isolated and disconnected information sources led to information fragmentation across the deconstruction process. For instance, in the case of the brick factory, the inventory application was utilised during the identifying phase. However, as new information emerged during the harvesting phase (e.g., damaged products), updates were recorded in a separate Excel file due to its simplicity. "It is a bit difficult to add something here (in the application)", one project planner mentioned. During the phase of distributing, a purchase confirmation was manually prepared for each buyer, detailing the specific elements they purchased. However, this process was entirely paper-based, with no connection to the previous material inventory or other information sources.

Overall, digital technologies are used only to a limited extent in deconstruction projects, given some challenges. Furthermore, information is used in isolation, lacking an integrated information system.

4.2 PEOPLE

The extent of stakeholder collaboration—one of the most influential factors (with a theme intensity of about 15%) — can significantly impact the effectiveness of reuse practices in deconstruction projects. This was evident in the case of outpatient clinic, where demolishers were asked to label some building elements so that the (new) contractors could trace the origin of these components and reconstruct the structure in a nearby city. Two contractors were involved in both the deconstruction and reconstruction processes. During this process,

https://doi.org/10.52202/080684-0002

demolishers have discussed unexpected situations and new plans with contractors. For example, during the harvesting phase, actual labelling went differently from what was planned. The façade elements had been regarded as standard components, which could be repositioned on the façade of the new building rather than needing to be installed in their original locations. However, each façade element was found to have slight differences in its connection methods. The new plan was then discussed among demolishers and contractors, and those elements were labelled accordingly during the harvesting process. At the same time, the demolishers shared the 2D drawings with labelling and other documents (e.g., photos of containers) with the contractors. The early involvement of contractors facilitated the seamless information flow between the donor and the target building. However, their collaboration also faced some difficulties. For example, incorrect labelling was observed. The project planner explained that the construction company should "help us <demolishers>, but they did not spend enough hours during disassembly...now they have to spend more time (to rebuild due to wrong labelling done by the demolishers)". Moreover, the requirements of new contractors play a key role in determining the destination of elements. In the outpatient clinic, four containers of lamps, insulations and other loose elements were identified as reusables but were not installed in the new building as the new contractor "did not want to use them". The project planner also exemplified the gypsum plates, which were initially intended for reuse but ended up being disposed of. "If they <the new contractors> say no (to using those plates beforehand), I <demolishers> will not disassemble them...", he explained. Early involvement and information exchange are hereby important to make joint decisions in identifying, harvesting, and distributing reusable elements.

The lack of expertise and experience further hinders the technology implementation and circular practices. For example, although a scanned model of the outpatient clinic was generated, the project planner did not use it in practice. Instead of digital models, he relied on the physical drawings to label elements, "supposing you have 25 constructive façade elements, then I label them 1, 2, 3, 4, 5 to 25 and then I printed it out and came to the construction site to say (to frontman), that number 1 is left above". Except for the technological limitation of the scanned model, the project planner also discussed the challenges of conveying instructions to workers using the digital model. "My colleagues said we can, for instance, use a tablet on a construction site to number the elements. But yeah, I knew 100% that the construction workers outside could not use it on the right <way>, I know surely that when we did <use the tablets>, that we had more problems than not", the project planner explained. Despite the demolition company being one of the largest in the Netherlands, the project planner admitted, "This was the first time we've done a project like this (a circular rebuilt project)" and they lacked experience in tasks such as labelling building components and coordinating with different stakeholders.

In sum, the collaboration between demolishers and contractors affects the deconstruction processes. Additionally, demolishers lack the expertise and experience necessary for circular demolition practices.

4.3 PROCESS

The deconstruction process is guided through some standard procedures. Although there is no clear separation between identifying, harvesting and distributing building elements, the deconstruction process basically follows these three phases. It consists of an interconnected series of tasks, where the effectiveness of each step influences the others. In other words, (some) information generated from the early stages is needed for subsequent stages. For example, information on as-is building conditions gathered from previous owners/builders supports demolishers in identifying elements' reusability. With the information from identifying, demolition and separation plans can then be drawn up in the harvesting phase. "Based on what you want to do with the elements, you determine how you disassemble them", one project planner introduced. The deconstruction plan also supports subsequent transporting and storing in the distributing phase:" so you don't end up with elements you need first lying at the back", a project manager introduced. For certain common materials, demolishers have developed standardised recycling procedures based on established partnerships. "We have a partnership with a door manufacturer", one digital specialist mentioned. However, "very high-quality reuse is sometimes difficult", he added.

Information accessibility in deconstruction processes was one concern (with about a 10% theme intensity). It often stems from a reluctance to share data among stakeholders involved in different projects. In the case of the outpatient clinic, demolishers only received the basic drawings, which lacked technical details from previous contractors due to concerns over intellectual property. Despite this, demolishers were requested to label big building components (e.g., ceiling systems, floor elements, and façade plates). The project planner exemplified: "I had drawings for the façade (but without details of façade elements), then I made a red rectangle (to represent each façade element)". As acknowledged by the participant, this manual drawing process is prone to errors. Furthermore, to facilitate reuse, a key task in the identifying phase is to align the demand from target projects with the supply of reusable elements from donor projects. However, in the process, information of reusable elements is normally missing, given the information mismatch between demand and supply. Demolition companies largely depend on their own networks or past partnerships to find potential buyers. The manager of a brick factory explained that about 70% of the building elements were sold to their clients. Similarly, in the

https://doi.org/10.52202/080684-0002

project outpatient clinic, the project planner shared his concerns: "For this project, we had the opportunity to sell materials somewhere since we knew the clients, but that wasn't always this way". This implies that the potential for reusing elements is heavily influenced by the demolishers' ambition and network. Consequently, smaller demolition companies may face further limitations in this process due to their more restricted networks. Although two digital technologies were applied in case projects, their limited implementation hindered the information exchange between the supply and demand of reusable elements in the phase of identifying. As one project manager mentioned, "At first, most materials have a second life (from the identifying phase), but some materials didn't survive". The material inventory from the brick factory also documented some materials such as steel and boilers, which were considered to be reusable, while ultimately sent for recycling since "there was no demand". As a result, the reusable elements could not be collected and distributed in the subsequent phases due to the limited information available from the earlier phase.

Time pressure also affects the information flow appropriately in deconstruction processes. In the outpatient clinic, the phase of identifying was largely limited, while many unexpected situations and emerging information only appeared during the sequential activities (harvesting and distributing). The carpet was an example: "We thought it was something loose, but during the execution phase, we found it was glued to the wooden floor." When the first leading researcher asked if this information could be figured out during the identifying phase, with some pre-audit reclamations. The project manager introduced that this project suffered from a tight project schedule, "we have to start from week 40 and finish it before Christmas…but only in week 39, the building was empty (for demolition)…(because of this), sometimes you meet something which is not the same as planned". Similarly, several participants of the brick factory project highlighted the challenge of short timelines. The harvester explained that finding potential clients normally takes a lot of time, while demolition projects must be completed within strict deadlines. "We want to sell the recovered materials, but we have to move forward with the process", he noted. The site manager also explained that if some potentially reusable elements could not find destinations within the project timeframe, they would end up being recycled or disposed of. "That is why finding buyers quickly is crucial", he added.

Reuse practices in circular demolition projects are hindered by limited access to information during the deconstruction process. This is caused by the information mismatch between supply and demand and the unwillingness of information sharing among projects. Furthermore, reuse practices are also constrained by time pressure.

4.4 POLICY

Policy guidelines and incentives are key factors of element reuse in deconstruction processes (with about a 12% theme intensity). In the case of the brick factory, policy guidelines and incentives played a role, which received a certification for circular demolition. This certification encouraged demolishers to prepare an extensive material inventory including pictures of elements, material characteristics, optimal circular strategies (e.g., reuse and recycling), sale channels, appropriate demolition techniques, etc. As the harvester and project planner noted, "They <the certification organisation> are going to track what happens to every material we inventory", and "We have to sell these materials for the certification". It showed how policies encourage demolishers to choose the optimal destinations for salvaged elements. However, one harvester also mentioned that policy guidelines and incentives are still in the early stages and insufficient, "the government has to do something about it, to make it <using reused products> easier…the government have to stimulate the use of products getting free from the project".

Policy requirements also hinder reuse possibilities. "The construction company must provide a guarantee for the roof of the (new) building, and they could not use the old one", the project manager of the brick factory mentioned. Moreover, in the case of outpatient clinic, a project planner explained that old doors are often difficult to reuse because their height and fire safety features do not meet current regulations. As a result, these doors must undergo a remanufacturing process rather than reuse, such as being repurposed into a new door to meet the required height. However, the case of the outpatient clinic is an exception, where the new building was constructed with old doors, thanks to a special permit for temporary structures. In this regard, a site manager noted that building codes need to become more flexible to accommodate sustainable construction methods.

In sum, the policy can, on the one hand, facilitate the reuse of building elements by providing motivation and guidance; on the other hand, it can hinder reuse practices due to the policy requirements of building elements.

5 DISCUSSIONS

This study makes contributions to the understanding of circular practices in the demolition phase, an area that has been largely underexplored in existing literature. While much of the current research on reuse focuses on project actors such as designers and clients (Eikelenboom et al., 2024), it leaves the challenges faced by demolishers largely unaddressed. Given their critical role in decision-making processes regarding the fate of building elements at EoL, this study provides new empirical insights into the challenges to reuse, specifically from the perspective of demolishers. The study categorises these challenges into four key system elements—technology, people, process

https://doi.org/10.52202/080684-0002

and policy—offering a more holistic theoretical understanding of the challenges to reuse practices in demolition. This segmentation provides a comprehensive view of the multi-dimensional obstacles faced by demolishers, highlighting that the obstacles to circular demolition are not confined to any one dimension but are multi-faceted and interconnected (Wijewickrama et al., 2021).

Furthermore, this study extends existing research by providing empirical evidence that sheds light on these challenges. Specifically, technology-related challenges were frequently mentioned by the participants. The study reveals that the technologies currently applied in construction and demolition waste management are still in their early stages of development, consistent with the findings of Li et al. (2020). Additionally, deconstruction processes were constrained by fragmented information systems. To address this, the study recommends that future research focus on integrating emerging technologies (see Iyiola et al., 2024) and enhancing digital tools to enable seamless information exchange across the entire reuse-centric process, as an example provided by Kuzminykh et al. (2024). Regarding the element of people, the study highlights a key issue: limited stakeholder collaboration. As also highlighted by Küpfer et al. (2023), demolishers and contractors are typically responsible for separate tasks—deconstructing and reassembling reclaimed components, which disrupts the information flow in the process. In response, it is proposed that a new demolition process is needed, one that fosters stronger relationships and earlier collaboration among stakeholders, particularly between demolishers and contractors (Eikelenboom et al., 2024). This could involve the development of cross-disciplinary teams or closer working relationships to facilitate better knowledge sharing and information flow regarding both reclaimed elements and future construction projects. Additionally, the study underscores the need for demolishers to gain specialised expertise and experience in circular demolition practices, contributing empirical evidence to the existing research (see Wijewickrama et al., 2021). This implies that demolition companies should invest in training programs designed to upskill their workforce in resource efficiency (Sharma et al., 2022). Regarding process-related challenges, the study demonstrates that critical information is often not accessible throughout the deconstruction activities. This is partly due to mismatches between the supply and demand for reusable elements during the identifying phase, which is also highlighted in the study of Rakhshan et al. (2020). Because of this, demolition companies commonly rely on informal networks to find potential buyers, as demonstrated in the case studies. Instead, the development and implementation of digital platforms will be a potential solution (Köhler et al., 2024). Furthermore, limited reuse practices can also result from factors such as time pressure and a reluctance among stakeholders to share information. These gaps highlight the need for not only technological innovation but also cultural change within the demolition industry. Regarding the policy-related challenges, the study provides empirical insights into how regulations can influence element reuse in demolition. It suggests that policymakers should develop guidelines and incentives that encourage the reuse of building elements during demolition, such as establishing certification programs for element recovery (Wijewickrama et al., 2021). Furthermore, by supporting research on the regulation of reused element requirements, policymakers can further promote circular practices in construction projects.

However, this study acknowledges certain limitations, including its scope, which is restricted to two case studies. Further research incorporating additional case studies is recommended to strengthen and generalise the findings. Additionally, the focus on the demolishers' perspective may introduce bias, potentially overlooking the viewpoints of other stakeholders, such as contractors and clients. Similarly, the study primarily examines on-site demolition activities, without addressing upstream design decisions or downstream processes at new projects that influence reuse and circularity. Therefore, further studies are suggested to comprehensively understand the challenges in RLSC that influence reuse. Lastly, while this study offers valuable insights into deconstruction processes, there remains a limited understanding of how they differ from traditional demolition processes. Future studies are required to offer deeper insights into, for example, the specific tools, safety protocols, and workflows employed in deconstruction practices compared to demolition ones.

6 CONCLUSIONS

This study offers empirical insights into the challenges associated with deconstruction processes aimed at building element reuse, with a particular focus on the perspective of demolishers. Two case studies were selected to represent circular demolition practices. Qualitative data were gathered through ten semi-structured interviews, project documentation, and field visits. The study identifies and categorises the challenges into four key elements: technology, people, process and policy. In doing so, it provides valuable empirical evidence regarding the barriers demolishers encounter when attempting to divert waste from demolition sites for reuse. Based on these findings, the study suggests avenues for future research and practical strategies aimed at improving circular practices in demolition projects. Future research is needed to generalise these findings by incorporating additional case studies. Moreover, future studies should also explore upstream and downstream processes in RLSC that influence element reuse, from the perspective of different stakeholders.

ACKNOWLEDGEMENTS

The studies have received financial support from the Dutch Research Council (NWO) with the grant number NWA.1432.20.003.

REFERENCES

Akbarieh, A., Jayasinghe, L. B., Waldmann, D., & Teferle, F. N. (2020). BIM-based end-of-lifecycle decision making and digital deconstruction: Literature review. *Sustainability*, *12*(7), 2670.

Allam, A. S., & Nik-Bakht, M. (2023). From demolition to deconstruction of the built environment: A synthesis of the literature. *Journal of Building Engineering*, *64*, 105679.

Azungah, T. (2018). Qualitative research: deductive and inductive approaches to data analysis. *Qualitative research journal*, *18*(4), 383-400.

Bertino, G., Kisser, J., Zeilinger, J., Langergraber, G., Fischer, T., & Österreicher, D. (2021). Fundamentals of building deconstruction as a circular economy strategy for the reuse of construction materials. *Applied Sciences*, *11*(3), 939.

Braun, V., & Clarke, V. (2006). Using thematic analysis in psychology. *Qualitative research in psychology*, *3*(2), 77-101.

Byers, B. S., Raghu, D., Olumo, A., De Wolf, C., & Haas, C. (2023). From research to practice: A review on technologies for addressing the information gap for building material reuse in circular construction. *Sustainable Production and Consumption*.

Çetin, S., Gruis, V., & Straub, A. (2022). Digitalization for a circular economy in the building industry: Multiple-case study of Dutch social housing organizations. *Resources, Conservation & Recycling Advances*, *15*, 200110.

Chileshe, N., Jayasinghe, R. S., & Rameezdeen, R. (2019). Information flow-centric approach for reverse logistics supply chains. *Automation in Construction*, *106*, 102858.

Cruz, E. F., Machado, R. J., & Santos, M. Y. (2015). Deriving a data model from a set of interrelated business process models. International Conference on Enterprise Information Systems,

Eikelenboom, M., Oosterlee, M., & van Marrewijk, A. (2024). Demolishers or 'material experts'? Project actors negotiating changing roles in sustainable projects. *International Journal of Project Management*, 102605.

Eisenhardt, K. M. (1989). Building theories from case study research. *Academy of management review*, *14*(4), 532-550.

Elghaish, F., Hosseini, M. R., Kocaturk, T., Arashpour, M., & Ledari, M. B. (2023). Digitalised circular construction supply chain: An integrated BIM-Blockchain solution. *Automation in construction*, *148*, 104746.

Ellen MacArthur, F. (2013). *Towards The Circular Economy: Economic and business rationale for an accelerated transition*. https://www.ellenmacarthurfoundation.org/towards-the-circular-economy-vol-1-an-economic-and-business-rationale-for-an

European Commission. (2022). *Buildings and construction*. Retrieved 6/14 from https://ec-europa-eu.ezproxy2.utwente.nl/growth/industry/sustainability/buildings-and-construction_en

Fereday, J., & Muir-Cochrane, E. (2006). Demonstrating rigor using thematic analysis: A hybrid approach of inductive and deductive coding and theme development. *International journal of qualitative methods*, *5*(1), 80-92.

Ferriz-Papi, J. A., Lee, A., & Alhawamdeh, M. (2024). Examining the Challenges for Circular Economy Implementation in Construction and Demolition Waste Management: A Comprehensive Review Using Systematic Methods. *Buildings*, *14*(5), 1237.

Fini, A. A. F., & Forsythe, P. (2020). Barriers to reusing and recycling office fit-out: An exploratory analysis of demolition processes and product features. *Construction Economics and Building*, *20*(4), 42-62.

Gálvez-Martos, J.-L., Styles, D., Schoenberger, H., & Zeschmar-Lahl, B. (2018). Construction and demolition waste best management practice in Europe. *Resources, Conservation and Recycling*, *136*, 166-178.

Ghobakhloo, M., Tang, S. H., Zulkifli, N., & Ariffin, M. (2013). An integrated framework of green supply chain management implementation. *International Journal of Innovation, Management and Technology*, *4*(1), 86.

Hammer, M., & Champy, J. (2009). *Reengineering the corporation: Manifesto for business revolution, a*. Zondervan.

Hosseini, M. R., Rameezdeen, R., Chileshe, N., & Lehmann, S. (2015). Reverse logistics in the construction industry. *Waste Management & Research*, *33*(6), 499-514.

Huuhka, S., Kaasalainen, T., Hakanen, J., & Lahdensivu, J. (2015). Reusing concrete panels from buildings for building: Potential in Finnish 1970s mass housing. *Resources, Conservation and Recycling*, *101*, 105-121.

Iyiola, C. O., Shakantu, W., & Daniel, E. I. (2024). Digital Technologies for Promoting Construction and Demolition Waste Management: A Systematic Review. *Buildings*, *14*(10), 3234.

Jiang, H., Feng, W., & Ouyang, W. (2017). Suggestions on Implementation of Reverse Logistics for Construction Waste of Sustainable

https://doi.org/10.52202/080684-0002

Development. *Journal of Residuals Science & Technology*, *14*(3).
Jongert, J., Nelson, N., & Goosens, F. (2011). Recyclicity: A Toolbox for Resource-Based Design. *Architectural design*, *81*(6), 54-61.
Ko, R. K. (2009). A computer scientist's introductory guide to business process management (BPM). *XRDS: Crossroads, The ACM Magazine for Students*, *15*(4), 11-18.
Köhler, J., Nielsen, F. F., Verstermark, J., & Thuesen, C. (2024). Recirculation of construction and demolition Waste: A case study of Danish producers and demolishers. *Cleaner Materials*, *14*, 100276.
Küpfer, C., Bastien-Masse, M., & Fivet, C. (2023). Reuse of concrete components in new construction projects: Critical review of 77 circular precedents. *Journal of Cleaner Production*, *383*, 135235.
Kuzminykh, A., Granja, J., Parente, M., & Azenha, M. (2024). Promoting circularity of construction materials through demolition digitalisation at the preparation stage: Information requirements and openBIM-based technological implementation. *Advanced Engineering Informatics*, *62*, 102755.
Li, C. Z., Zhao, Y., Xiao, B., Yu, B., Tam, V. W., Chen, Z., & Ya, Y. (2020). Research trend of the application of information technologies in construction and demolition waste management. *Journal of Cleaner Production*, *263*, 121458.
Mallick, P. K., Salling, K. B., Pigosso, D. C., & McAloone, T. C. (2023). Closing the loop: Establishing reverse logistics for a circular economy, a systematic review. *Journal of environmental management*, *328*, 117017.
Marino, A., & Pariso, P. (2020). Comparing European countries' performances in the transition towards the Circular Economy. *Science of the Total Environment*, *729*, 138142.
Parto, S., Loorbach, D., Lansink, A., & Kemp, R. (2007). Transitions and Institutional Change: the case of the Dutch waste system. *Industrial innovation and environmental regulation*, 233-258.
Purchase, C. K., Al Zulayq, D. M., O'Brien, B. T., Kowalewski, M. J., Berenjian, A., Tarighaleslami, A. H., & Seifan, M. (2021). Circular economy of construction and demolition waste: A literature review on lessons, challenges, and benefits. *Materials*, *15*(1), 76.
Rakhshan, K., Morel, J.-C., Alaka, H., & Charef, R. (2020). Components reuse in the building sector–A systematic review. *Waste Management & Research*, *38*(4), 347-370.
Rameezdeen, R., Chileshe, N., Hosseini, M. R., & Lehmann, S. (2016). A qualitative examination of major barriers in implementation of reverse logistics within the South Australian construction sector. *International Journal of Construction Management*, *16*(3), 185-196.
Ronzani, C. M., da Costa, P. R., da Silva, L. F., Pigola, A., & de Paiva, E. M. (2020). Qualitative methods of analysis: an example of Atlas. TI™ Software usage. *Revista Gestão & Tecnologia*, *20*(4), 284-311.
Sharma, N., Kalbar, P. P., & Salman, M. (2022). Global review of circular economy and life cycle thinking in building Demolition Waste Management: A way ahead for India. *Building and Environment*, *222*, 109413.
Van de Ven, A. H. (2007). *Engaged scholarship: A guide for organizational and social research*. Oxford University Press, USA.
Van den Berg, M. (2024a). Digital Technology Use Cases for Deconstruction and Reverse Logistics. In *A Circular Built Environment in the Digital Age* (pp. 197-212). Springer International Publishing Cham.
Van den Berg, M. (2024b). An Ethnographic Account of Circular Demolition Practices: Realising reuse potentials. In *Embracing Ethnography* (pp. 267-280). Routledge.
Van den Berg, M., Voordijk, H., & Adriaanse, A. (2020a). Information processing for end-of-life coordination: a multiple-case study. *Construction Innovation*, *20*(4), 647-671.
Van den Berg, M., Voordijk, H., & Adriaanse, A. (2020b). Recovering building elements for reuse (or not)–ethnographic insights into selective demolition practices. *Journal of Cleaner Production*, *256*, 120332.
Wao, H. O., Dedrick, R. F., & Ferron, J. M. (2011). Quantitizing text: using theme frequency and theme intensity to describe factors influencing time-to-doctorate. *Quality & Quantity*, *45*, 923-934.
Wibowo, M. A., Handayani, N. U., Mustikasari, A., Wardani, S. A., & Tjahjono, B. (2022). Reverse logistics performance indicators for the construction sector: a building project case. *Sustainability*, *14*(2), 963.
Wijewickrama, M., Chileshe, N., Rameezdeen, R., & Ochoa, J. J. (2020). Information Sharing in Reverse Logistics Supply Chain of Demolition Waste: A Systematic Literature Review. *Journal of Cleaner Production*, 124359.
Wijewickrama, M., Chileshe, N., Rameezdeen, R., & Ochoa, J. J. (2021). Quality assurance in reverse logistics supply chain of demolition waste: A systematic literature review. *Waste Management & Research*, *39*(1), 3-24.
Wu, Z., Yang, K., Xue, H., Zuo, J., & Li, S. (2022). Major barriers to information sharing in reverse logistics of construction and demolition waste. *Journal of Cleaner Production*, *350*, 131331.
Yin, R. K. (2014). *Case study research: design and methods*.

LUND UNIVERSITY | LTH FACULTY OF ENGINEERING

CIRCULAR ECONOMY PRINCIPLES IN INNOVATIVE FACADE APPLICATIONS: A REVIEW OF FIBRE-REINFORCED PULTRUDED PROFILES APPLICATIONS FOR ENERGY-EFFICIENT BUILDINGS

Paolo Giussani[1*], Alberto D'Occhio[2], Enrico Sergio Mazzucchelli[1], and Paolo Rigone[1]

[1*]Department of Architecture Built environment and Construction Engineering, Politecnico di Milano, Italy

[2]Gualini SpA, Costa di Mezzate, Italy

*Corresponding Author: paolo1.giussani@polimi.it

ABSTRACT

Background and aim. Addressing resources and energy efficiency within the construction sector is the key to achieve the EU's ambition of climate neutrality and fully decarbonised of building stock by 2050. This paper explores the integration of circular economy principles, with a specific focus on fibre reinforced pultruded profiles as a sustainable material. It provides an overview of the characteristics and manufacturing processes of continuous pultruded profiles, exploring their potential implementation in facade components, and conducts a theoretical comprehensive sustainability assessment of their environmental impact. These materials contribute to potentially increase the environmental sustainability of the construction sector, reducing the overall lifecycle expenses, and boosting the energy performance of buildings.

Methods and data. A mixed-methods research design was employed, combining a comprehensive literature review and analysis of case studies. These methods evaluated the characteristics, manufacturing processes, and environmental performance of fibre-reinforced pultruded profiles in facade applications.

Findings. The research highlights key technologies that can increase resource efficiency and reduce waste in the fibre reinforced polymer industry. Prospects for technological advances in pultrusion processes are discussed. The findings reveal that pultruded composite materials offer significant advantages, including resource efficiency, waste reduction, and improved energy performance of building skins for durable and low-maintenance facade systems.

Theoretical / Practical / Societal implications. Practically, this research highlights the potential of pultruded profiles for innovative facade design by incorporating circular economy principles. Societally, the findings support the transition to sustainable building practices, contributing to climate goals and resource conservation. This theoretical interdisciplinary approach addresses the challenges of modern façades systems and lays the groundwork for sustainable, energy-efficient buildings.

KEYWORDS: Circular economy, Facade application, Glass fibres, Pultrusion, Recyclable, Sustainability

1 INTRODUCTION

In March 2020, the European Commission presented the circular economy action plan, which aims to promote more sustainable product design, reduce waste and empower consumers (Circular Economy Action Plan For a cleaner and more competitive Europe, European Commission 2020_98_final report). Energy efficiency and sustainability in the building sector are necessary to achieve the 2030 Agenda for Sustainable Development Goals (General Assembly, 2015, the 2030 Agenda for Sustainable Development). The construction industry is one of the major responsible of climate change and global waste production (Pastori et al., 2021).

Heating and cooling in buildings and industry account for 50% of the European Union's energy consumption (Towards a smart, efficient and sustainable heating and cooling sector, European Commission. 2016_final report). Governments, companies, and consumers each have a crucial role when transitioning from a linear to a circular model of production and consumption. However, the circular economy is not only focused on technical

https://doi.org/10.52202/080684-0003

aspects. The approach involves the complete value chain, from product design and production processes to consumption and waste management. The regeneration process become a valuable resource. To fulfil this ambition, the EU needs to accelerate the transition towards a regenerative growth model that gives back to the planet more than it takes, advance towards keeping its resource consumption within planetary boundaries, and therefore strive to reduce its consumption footprint and double its circular material use rate in the coming decade (Circular Economy Action Plan For a cleaner and more competitive Europe, European Commission 2020_98_Annex). The transition to a circular economy goes hand in hand with new legislative framework, new business models, standardisation, green public procurement, and a new design thinking that takes into account reparability, durability, and recyclability (Mrotzek-Blöß et al., 2019). The key point is replacing virgin raw materials with recycled raw materials.

Literature suggests that implementing circular economy strategies can help mitigate impact reduce carbon emissions and improve sustainability in construction practices. However, while existing studies have addressed the potential of circular practices, there is a gap in understanding how these strategies can be specifically applied to the lifecycle of Fibre-Reinforced Polymer (FRP) materials in building façade applications. In the following paragraph the research focuses on possible strategies for new application of FRP in building façades components represent one of the key to the building systems integration necessary to realise critical health, carbon, resilience, and sustainability goals in buildings and urban habitats (Boswell et al., 2021).

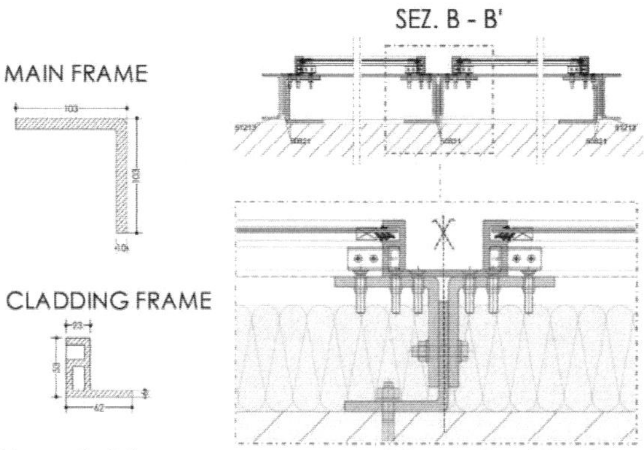

Figure 1: Schematic representation of the ventilated façade system and its components – system design

The components of the innovative façade system (see Figure 1) include a combination of mullion profiles and brackets, both made of pultruded composite material. The substructure consists solely of vertical elements, which are connected to the support using a bracket system that allows for the adjustment of installation tolerances on-site. The cladding consists of cassette panels made of composite material as well, with an infill photovoltaic glass unit. The use of fibre-reinforced polymer (FRP) leverages its advantageous properties, including high mechanical performance, favourable thermal behaviour, fire and weather resistance, and durability. To ensure that buildings are fit for the EU's enhanced climate ambition under the European Green Deal, the revised directive will contribute to the objective of achieving climate neutrality by 2050 (European Commission, Fact Sheet, 2016. *Energy Performance of Buildings Directive, European Commission*). The drivers of climate change and biodiversity loss are global and are not limited by national borders (The European Green Deal, European Commission, 2019_640_final report). The paper is divided into five sections. Section two explores the phases of the pultrusion process and discusses material characteristics in the case study. Section three is related to the methods and provides an overview of the main sustainability regulations and guidelines. Section four addresses the main strategies and current approach for disposing of composite materials. The final section five, highlights the potential of pultrusion technology and outlines future prospects related to a circular economy framework concerning FRP composite materials.

2 CASE STUDY: PULTRUDED PROFILES DEFINITION, MANUFACTURING PROCESS AND CHARACTERISTICS

Compared to traditional manufacturing processes that may involve energy-intensive procedures, pultrusion stands out as a sustainable alternative. The process allows for precise control over material composition and results in profiles with consistent quality and performance. Pultruded profiles offer an innovative solution that surpasses conventional materials in terms of durability and design flexibility.

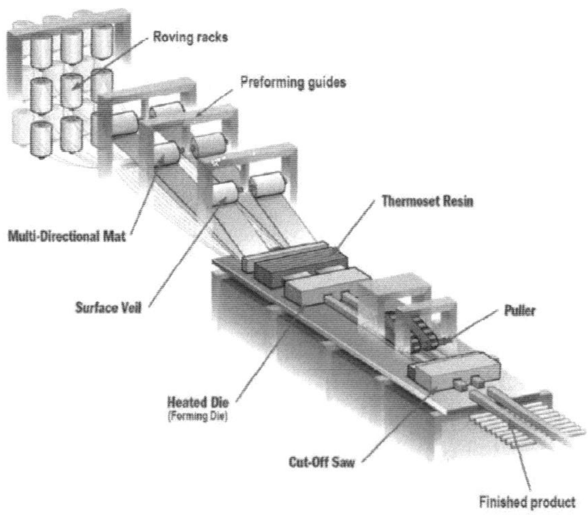

Figure 1: Right: Manufacturing process scheme - source adapted by the author from [Bedford reinforced plastics, Inc.]

https://doi.org/10.52202/080684-0003

FRP components are characterized by limited maintenance than traditional building materials such as wood, steel or aluminium. As a consequence, pultruded profiles align with circular economy principles by optimizing resource use, minimizing waste, and reducing the environmental impact associated with material production (figure 2). Specifically, the process uses matrix and fibres for the manufacturing of composite standard profile or custom parts for very specific needs (figure 3). Common materials used in the reinforced phase are glass, polymers, metals, ceramics, and graphite, which provides easy compatibility with matrix polymers at low cost. Thermosets and thermoplastics are the most commonly used polymers in the manufacturing of composite materials. This method involves the pulling of reinforcing fibres through a resin bath, ensuring uniform distribution, and subsequent curing to form a robust and versatile composite material (figure 4).

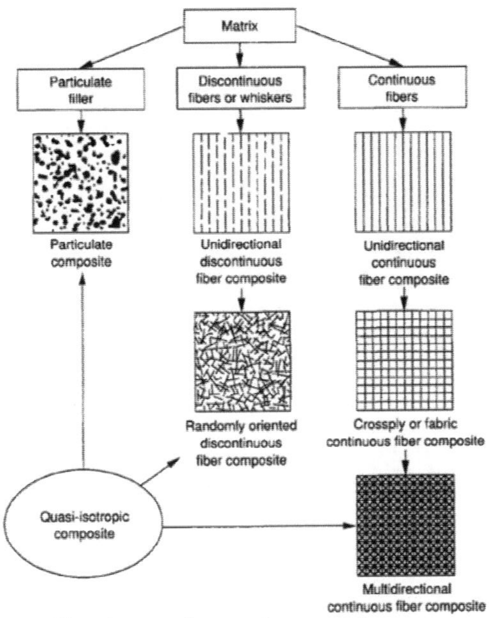

Figure 2: Classification of composite material systems – source [Ishai O, 2018]

The key properties responsible for the rising demand for pultruded composites include excellent tensile strength, high chemical resistance, non-magnetic properties, low thermal expansion and low maintenance. From a technical point of view, they are not comparable to other traditional materials but these materials are still little used in construction field due to a lack of knowledge on the part of designers. Pultrusion is an automated technological process that creates high-quality composite profiles with a consistent cross-section. This is achieved through the impregnation of fibres with resin and their passage through a heated die (Nguyen et al., 2013). Subsequent to impregnation, the profile is shaped using a heated die, ensuring proper curing and solidification.

Figure 3: Right: Pultrusion process, roving racks, preforming guides, roving, surfacing mat – source [Italcomposites Doo]

The characterization in the other two directions is mainly given by the thermoset resin selected. Over time the process evolved and, as first improvement, it was possible to die along with the longitudinal fibre also layers of mat or fabric in 1960s; later in 1970s (Barkanov et al., 2022). Fibre-Reinforced Polymer profiles, engineered through pultrusion, have proven their versatility in diverse architectural elements, ranging from cladding panels to load-bearing structures. The inherent characteristics of pultruded profiles make them particularly well-suited for sustainable construction practices. Notably, their high strength-to-weight ratio imparts structural integrity without excessive material consumption.

The material's non-conductive properties contribute to improved energy efficiency in building applications. Moreover, the application of FRP pultruded profiles in façade components contribute in a reduction in the overall weight of the facade, contributing to a more efficient and cost-effective construction process. These positive outcomes underscore the practical advantages of pultruded profiles in real-world construction scenarios, aligning with the principles of circular economy by promoting resource efficiency and waste reduction. FRP composites have also a good ratio between price and low level of embodied energy which also means that the energy required to manufacture the component is markedly lower than in the case of a metal assembly (Knippers et al., 2011). Additionally, the versatility of pultruded profiles allows for customization to diverse design requirements in facade geometry.

Figure 5 utilizes different colors to distinguish between various materials. Gray circles represent the baseline or reference group, while colored circles correspond to specific materials such as pultruded FRP, extruded aluminum, etc. The comparison shown in the graph represents the price-to-embodied energy ratio for standard profile dimensions up to a maximum of 150 x 100 mm (this size can generally be considered a good average dimension for the type of façade application discussed in the paper). The analysis presented in the graph takes into consideration the embodied energy for the primary production of metal-based and composite-based materials (X-axis), with the prices per unit mass shown on the Y-axis.

https://doi.org/10.52202/080684-0003

3 METHODS: OVERVIEW OF THE MAIN SUSTAINABILITY ASPECTS AND REGULATIONS

A United Nations projection (A/RES/70/1) estimates that the world's population will reach 9.7 billion by 2050 and 10.4 billion in 2100. An estimated 70% of these people will live in cities, up from 54% today. This growth involves gigantic challenges in term of infrastructure, energy and the environment. Huge new investments will be required and the use of sustainable building techniques and new building materials will be essential. The Ance Study Centre (Italian Association of Building Manufacturers) estimated that approximately 430,000 energy efficiency interventions will be necessary (JEC composites magazine, 2024, Composite material in civil engineering and architecture, n. 154).

The European market (figure 6) for glass-reinforced polymers has experienced a steady yearly growth of + 2% since 2009. Pultrusion was the fastest growing, with a pultrusion market valued at €1.87 billion in 2020, which is expected to grow by €2.98 billion by 2028 (Elmar et al., 2018). Current practices, especially in the construction sector, are mainly oriented towards waste management and recycling, which is the least optimized solution in the hierarchy of circular actions, but also the most promoted by the current European legislative framework.

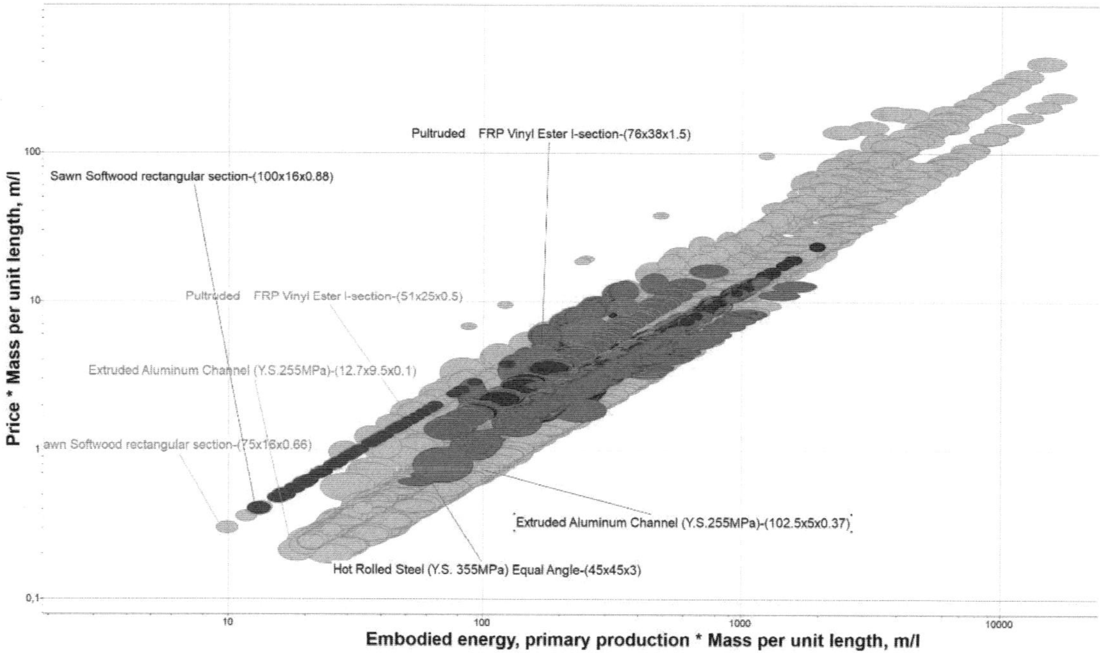

Figure 5: Comparison of Price vs. Embodied Energy: FRP vs. Traditional Materials (Profiles up to 150 x 100 mm) – Analysis calculated using Granta ANSYS Software.

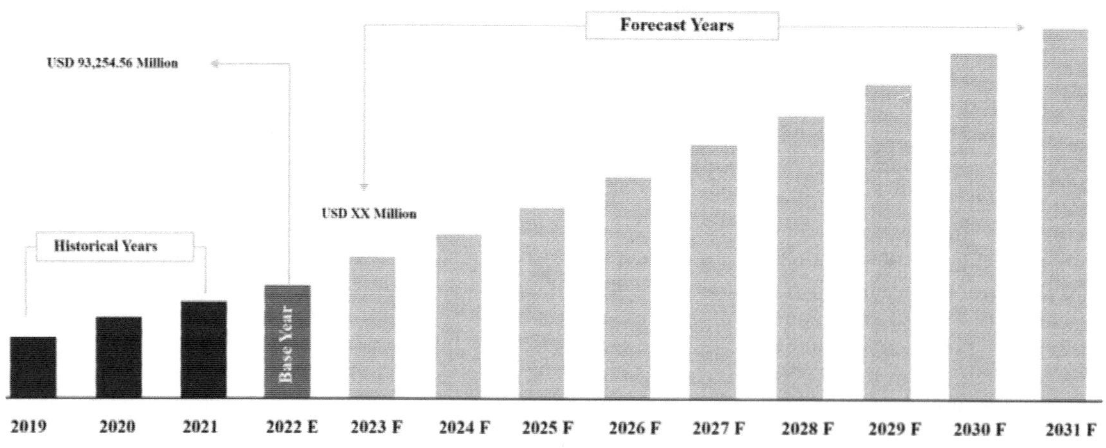

Figure 6: Fibre-Reinforced Composites market size, demand over the years - Study Period 2018-30 (Report Code SRCH1729DR) - source adapted by the author from [Straits, 2024]

Moreover, downcycling activities, such as the reuse of aggregates for the construction of road foundations, deriving from the need to solve the problem of managing construction and demolition waste at the end of the building service life, are the most practiced (Lavagna et al., 2022). A detailed analysis of the current regulatory framework was conducted to understand how existing sustainability strategies align with European and global policies. the study examined how the circular economy principles, waste management regulations, and recycling technologies impact the use of pultruded fibre-reinforced polymer materials. The research also evaluated the implementation of the waste hierarchy in FRP production and end-of-life scenarios, highlighting the need for more efficient and sustainable practices.

Circular economy (figure 7) means an economic system whereby the value of products, materials and other resources in the economy is maintained for as long as possible, enhancing their efficient use in production and consumption, thereby reducing the environmental impact of their use, minimising waste and the release of hazardous substances at all stages of their life cycle, including through the application of the waste hierarchy (Regulation (EU) 2020/852). A more circular economy increases the life-span of products increasing reuse, reparability, durability, upgradability and promoting innovative forms of consumption such as the collaborative economy (European Commission, SWD 306 final, 2023).

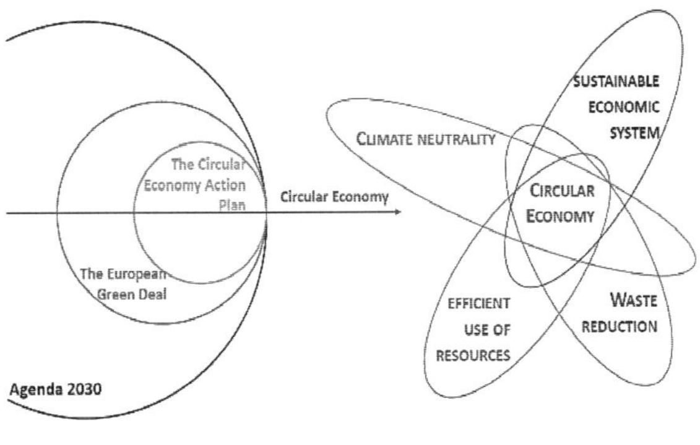

Figure 7: The multiple values of circular economy – source adapted by the author from [Dalla Valle et al., 2022]

Current trends reveal how remanufacturing is an activity implemented not only as an End-of-Life sustainable strategy but also during the whole life cycle of a product. In this scenario, the waste hierarchy is a framework that outlines the preferred approaches for managing waste in order of priority, with the goal of promoting sustainability and minimizing environmental impact.

Applying the waste hierarchy to pultruded profiles emphasizes the importance of sustainable practices throughout their life cycle, from production to end-of-life management. Incorporating specific sustainability metrics, such as energy consumption, CO_2 emissions (manufacturing, installation, etc), and material waste (quality), would provide measurable indicators to evaluate the effectiveness of these practices. By prioritizing prevention (minimize the generation of waste), promote recovery and recycling techniques, reduce pressure on resources and boost the transition to a circular economy (figure 8).

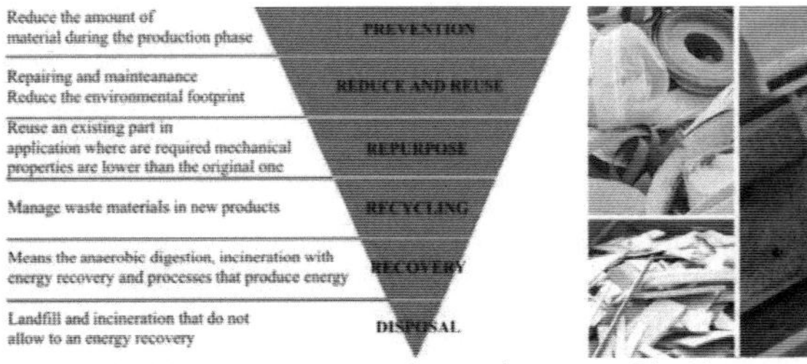

Figure 8: Waste management hierarchy as per The European Union's 2008/98/EC directive - source by the author from [De Fazio et al, 2023]

The waste management principles reduce the environmental footprint and contribute to a more circular and sustainable approach to the material use (minimises the incineration of waste and avoids the disposal of waste, including landfilling, in accordance with the principles of the waste hierarchy). FRP waste material negatively impacts the local environment by contaminating the soil, air, and groundwater. Landfill and incineration are not recycling methods. The incineration route still leaves behind 50% of the waste material as ash, which still needs to be landfilled (Jacob, 2011). Most FRP production waste ends up in a landfill. Dumping this waste in a landfill may not be the best sustainable waste disposal solution and commonly the space available in landfill sites may be limited in some countries.

The transformation, however, will not happen by itself, but important promotion actions need to be taken by various stakeholders, like policy-makers, industrial companies, research and academic community as well as the general public (Karvonen et al., 2017). Circular practices in the Fibre Reinforced Plastic (FRP) profiles industry are still emerging but gaining traction due to the growing awareness of sustainability and environmental concerns. Potential circular practices take in consideration the closed-loop manufacturing processes where waste materials generated during production are collected, recycled, and reintegrated into the manufacturing process.

This aspect reduces the reliance on new raw materials and minimizes the environmental impact of production (in Figure 9, the recycling loop illustrates the process from raw material to the machining phase, leading to the finished components. The loop also presents the potential end-of-use phase after utilization and disassembly. On the right side of Figure 9, the pultruded profiles are categorized under the group of continuous fibres with different orientations in a grid).

https://doi.org/10.52202/080684-0003

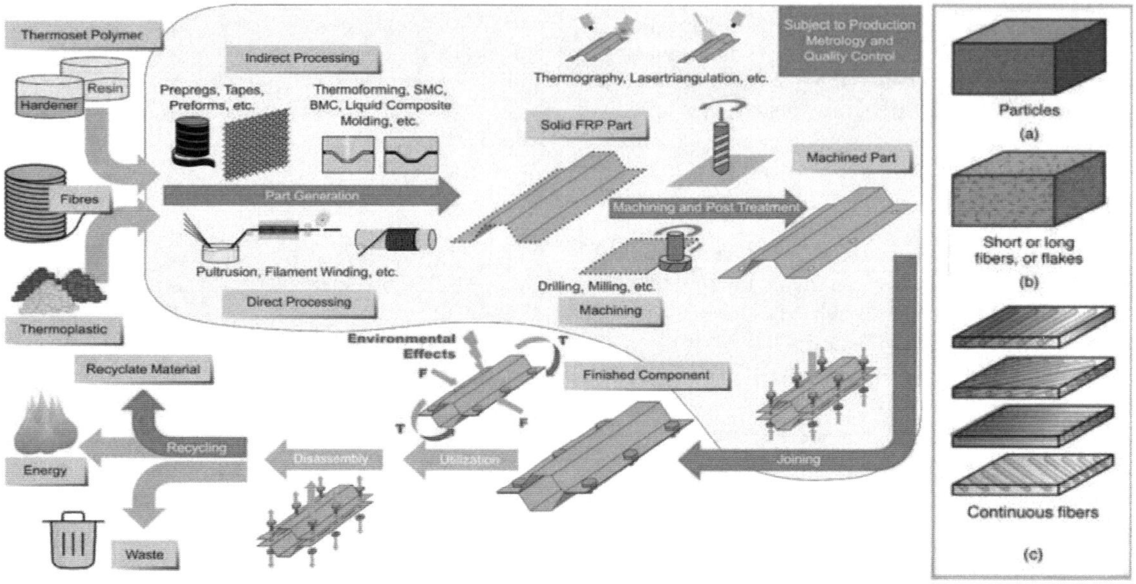

Figure 9: Methods of reinforcing plastics in particles (a), short fibres (b) and continuous fibres (c) – source adapted by the author from [UCIMU Association]

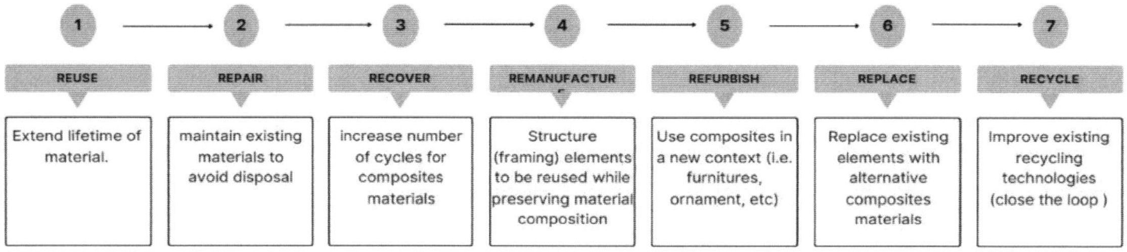

Figure 10: Circular thinking and the use of the 'R' strategies for composite material

The main critical and controversial Life Cycle Assessment issues concern the definition, on one hand, of the service life and, on the other, of the allocation procedures. Incorporating LCA parameters such as environmental impact, material flow analysis, and energy consumption throughout the life cycle of FRP components would provide a more rigorous framework for evaluating sustainability practices. Indeed, the potential for multiple life cycle is limited by the products reversibility and the number of use cycles strictly depends by the associated service life (Dalla Valle et al., 2022).

There are many options for extending the lifetime of composite components. For example, the product might be used as it is for the same purpose in a different application (reuse), or its use can be extended by applying conventional maintenance techniques both in situ (repair) and through industrial processes (refurbish and remanufacture). If repair is not possible, the product can often be used for a different function (repurpose). When these options are finally exhausted, recycling is possible through both closed loop and open loop processes (EuCIA Association, 2024). The EU policy on Construction Product Regulation (CPR) and its Basic Requirement of Construction Works (BRCW) 7 Sustainable use of natural resource could provide a good basis for optimizing resources, including reuse (Hobbs et al., 2017).

It is challenging but potentially very important to resolve Circular Economy in facades with a view to transitioning from the take–make–waste mindset to one of reduce–reuse–recycle. In other words, in order to approach a circular thinking regarding the difficult end-of-life opportunities of composite materials, many R-strategies (Chatziparaskeva et al., 2022) of circularity (e.g., reuse, reduce, recycle, refurbish, remanufacture, etc.) should be taken into account (figure 10). The wind industry in Europe is projected to generate approximately 15,000 tonnes of blade waste annually between 2020 and 2023 (data from Association WindEurope). This amount is expected to rise in the next years and exceed 60,000 tonnes annually by 2030 (figure 11). Following this information, the wind energy sector has more precise knowledge about the amounts of composite materials to be decommissioned each year (compared to the construction sector).

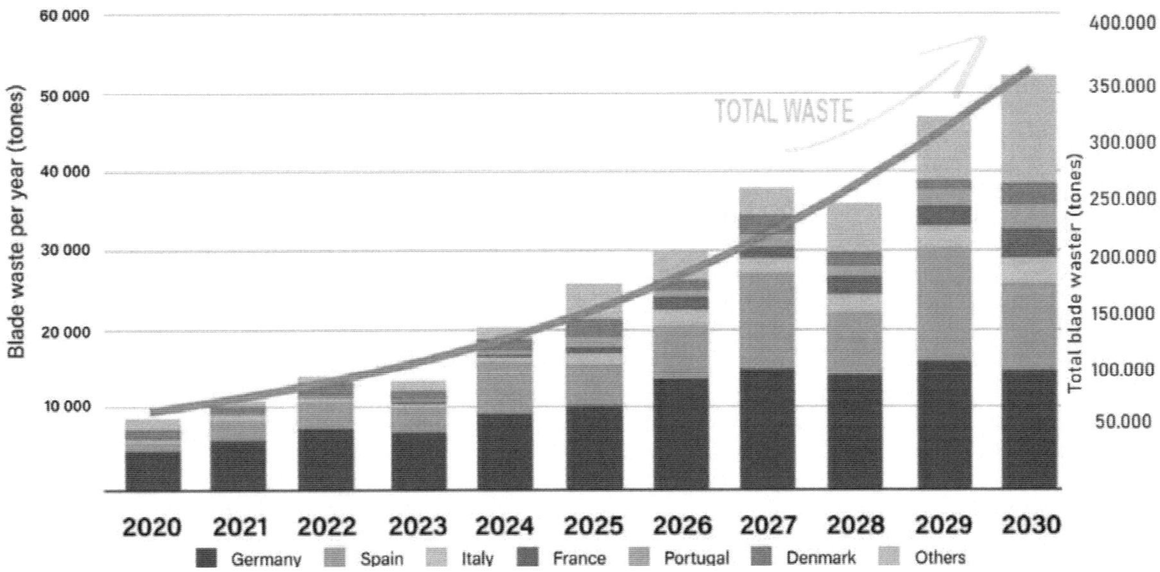

Figure 11: Decommissioned blade weight (including repowering) - source adapted by the author from [European Boating Industry AISBL, 2023]

This visibility on composite waste volumes makes the wind sector a prime mover in supporting the establishment of a business plan for the industrialisation of composite recycling/upcycling. Cement kiln co-processing, while costly, is the treatment technology that is already available and that could be increasingly used within the next years to transition to the circular economy approach. Glass fibre can be a source of silica that is needed in the cement production.

One ton of composite material leads to the saving of 460 kilograms of primary raw material (such as sand). Polymers can be used to produce energy. The high efficiency and lower CO_2 emission factor, reduces the total CO_2 impact of the cement production process. One ton of composite can save approximately 110 kilograms of CO_2 compared to fossil fuels (European Boating Industry AISBL, 2023).

4 RESULTS: STRATEGIES BASED ON PULTRUDED FIBRE-REINFORCED POLYMER MATERIAL

The circular economy model aims to optimize resource use, minimize waste, and promote sustainable practices throughout a product's life cycle. Applying this model to pultruded profiles involves understanding and managing material flows to enhance their environmental and economic sustainability.

Current methods of disposing composite materials involve mostly landfilling, cement clinker co-processing and recycling by matrix degradation. Co-Processing technique is the use of waste as raw material, as a source of energy, in industrial processes, such as cement, lime, steel, glass, and power generation (Vijay et al., 2016). As it is known, landfilling, especially of composite materials, represents a high loss of high-value materials and energy input for their production, as well as loss of opportunities for reusing composites in other investments. There is a need for innovation regarding composite end-of-life disposal methods in order to integrate the materials into a circular economy mindset and increase value chain following organizational models.

The findings indicate that while significant progress has been made in recycling and reusing pultruded FRP materials, real-world challenges remain in scaling up these technologies. the study shows that co-processing in cement kilns and advanced mechanical recycling methods can contribute to a more circular economy. However, the adoption of these technologies requires stronger policy support, industrial investments, and significant advancements in scaling these processes. Without adequate infrastructure and government incentives, the widespread adoption of these methods may be delayed. The high costs associated with these processes and the need for specialized facilities further complicate their implementation on a global scale. Additionally, the research reveals that product remanufacturing and reuse strategies can significantly extend the life cycle of FRP components, reducing waste generation and lowering carbon emissions. However, a critical barrier to the successful implementation of these strategies is the lack of standardized remanufacturing processes and quality assurance mechanisms across the industry. In the hypothetical scenario (figure 12) discussed in this paragraph, which investigate the use of FRP for innovative ventilated facade system, the value chain operates in several distinct phases. Initially framing components are sold to facade contractors. After the components have been used, the FRP suppliers then

https://doi.org/10.52202/080684-0003

repurchase the materials (either from other existing projects or from damaged work). Mostly re-manufacturer acquires old products from contractors. The next phase involves the transportation of these materials to a designated plant for remanufacturing. (physical decomposition). This involves sorting, cleaning, and grinding (mechanical recycling). During this phase, the quality of the secondary raw material (resource) is verified by a third-party lab (any contamination). Once processed, the materials are available for new reprocessing, such as re-manufacturing FRP cladding panels.

Once the remanufactured materials are ready, FRP suppliers retrieve them and stock them according to the established technical requirements for future use. These materials are not necessarily downcycled but are kept in reserve for re-manufacturing when demand arises. This system could face significant challenges related to logistical costs and inventory management**,** especially if remanufactured materials do not always align with market demand. Following the next step, the concept of ownership could change from the first original loop (virgin).

Finally, the remanufactured products, like FRP cladding panels, are sold to third parties or directly back to facade suppliers. This gives the FRP suppliers two valuable options: they can provide both the main framing for the ventilated facade system and the remanufactured cladding panels made from the same base material. By extending the useful life of these materials, the system helps to prevent waste generation. Products are sold through traditional channels, often at a reduced price, due to the value derived from the remanufactured materials. The specific path and options in steps four and five can vary depending on the type of product being remanufactured, such as shifting from facade framing to indoor or outdoor furniture, as an example. Despite these opportunities, market acceptance of remanufactured FRP products may be limited by consumer perceptions and regulatory barriers. FRP waste comes from production, usage, and end-of-life deconstruction. To minimize production waste, it's essential to evaluate the manufacturing process and pinpoint the most efficient methods.

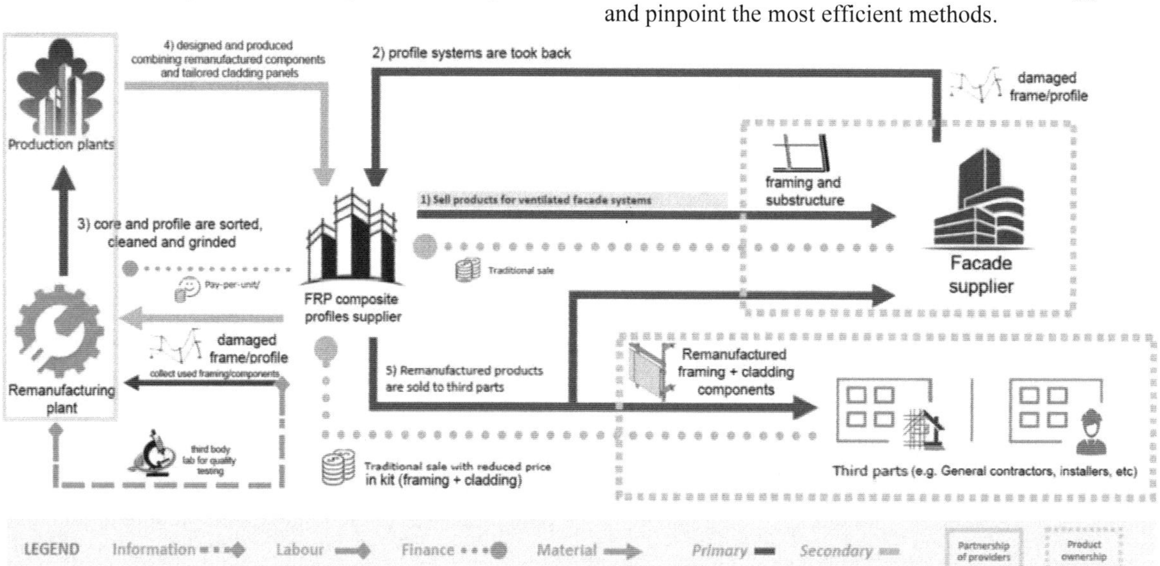

Figure 12: Exploring potential circular practices in FRP profiles industry: models and system map

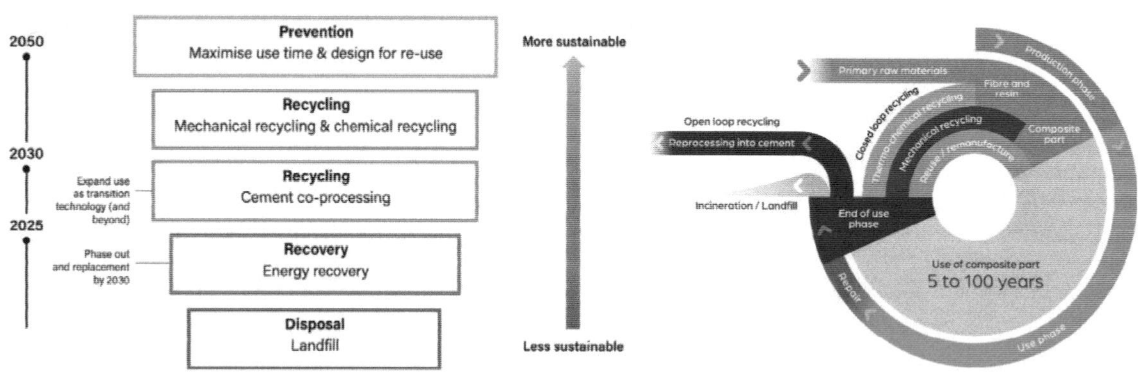

Figure 13: Proposed recycling pathway until 2030 and 2050 and composites circularity model – source adapted by the author from [UCIMU Association]

The European Commission (2018), in the document entitled *European Strategy for Plastics in a Circular Economy*, emphasizes that the low reuse and recycling rates of plastics at the end of their life cycle is a key challenge that must be addressed (Chatziparaskeva et al., 2022). Another black spot to recycling FRPs is the lack of effective separation techniques for composites during demolition, which could complicates their reuse. Deconstruction and reuse must be kept in mind when designing, constructing, and using FRP components for intended future reuse. Generally, automatic processes result in less waste than manual methods, but the initial cost and technological complexity of automated systems remain another black point to investigate. The use of different pre-impregnated recycled materials, different in their viscosity and stream characterized by mechanical testing and microscopy analysis - represent another possible option to transform a waste into a product with high added value, reducing the carbon footprint (Asensio et al., 2020). The aim for recycling/up cycling should be the conversion into new materials and products used in the manufacture of new composite products, enabling a circular approach within the composite sector. However, achieving this level of circularity requires overcoming challenges related to the integration of recycling systems and product design that anticipates end-of-life considerations. Assessing the carbon footprint of materials involves considering the emissions associated with their entire life cycle, including raw material extraction, manufacturing, transportation, installation, use, and end-of-life management.

It is important to note that specific values may vary based on factors such as production methods, energy sources, and transportation distances that must be weighted on the base of a detailed holistic evaluation. The approach for new recycling solutions has to be technology-open to identify the most suitable approach for all composite use industries. Generally speaking the recycling/disposal options about FRP products can be classified into five types: incineration, landfilling, thermal recycling, mechanical recycling, microwave and chemical recycling (figure 13). Incineration causes air pollution, CO_2 emission, and acidification of the disposal of composite products waste. Although landfills tend to have low air pollution, they can cause soil and groundwater pollution. Each of these methods comes with its own environmental and economic implications, and a careful, case-by-case evaluation is needed to determine the most sustainable approach for different types of FRP waste (also according to the waste material origin and possible contamination). Noted that a significant proportion of FRP products are thermally recycled or used as fillers in the cement industry. Microwave-assisted pyrolysis emerges as a technology with high potential to deal with the problem of recycling composite materials, since it solves traditional limitations of the pyrolysis process. The solvolysis process, or thermo-chemical recycling process, consists of the decomposition of the polymer matrix by a solution of acids, bases and/or solvent. Finally the mechanical recycling involves collection/segregation, cleaning and drying, chipping/sizing, colouring/agglomeration, palletisation/extrusion, and manufacturing the end product (Julian et al., 2022). During this process, the final new mix is composed of the same particles but in a different mixed recipe and a varying percentage of glass fibres. It is important to consider the potential fibre length reduction after the grinding process, which will affect the final properties of the recovered fibre. The new mix must be weighed, and recipe eventually modified with a minimum percentage of virgin raw materials and additives (e.g., glue primers). The variation in colour represent an opportunity (figure 14).

5 CONCLUSIONS AND FUTURE PROSPECTS

In general, pultrusion technology is innovating at a rapid pace and driving growth in the global market. One of the key driving factors for the global market growth for pultruded profiles is the rising demand for lightweight structural composite with high performance characteristics. The exploration of pultruded profiles revealed their unique characteristics, including high strength-to-weight ratio, corrosion resistance, and design flexibility, positioning them as a sustainable alternative to traditional facade materials.

Even if composite materials offer great engineering opportunities, their integration in the circular economy remains challenging. By systematically managing material flows in the circular economy framework, the application of pultruded profiles can contribute to resource efficiency, waste reduction, and sustainable practices in the construction industry. This approach aligns with the broader goals of promoting circularity, reducing environmental impact, and fostering a more sustainable and resilient economy.

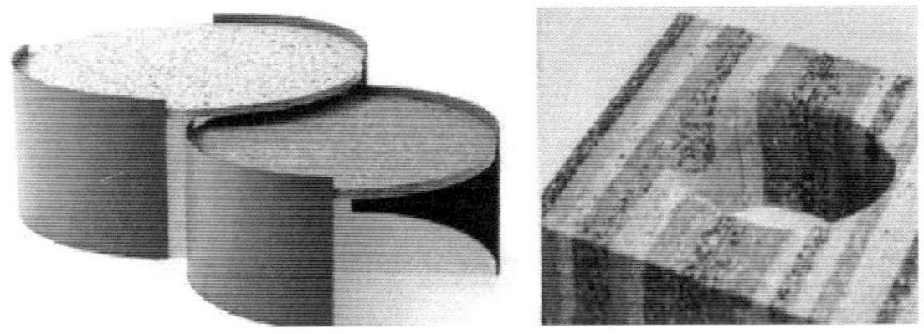

Figure 14: Panel and furniture obtained from the recycling of fibreglass and rigid expanded thermosets – source [Gees Recycling Srl]

Exploring innovative building materials has become a priority to reduce the overall footprint of construction activities. Nowadays, the adoption of circular economy principles in the construction sector, has garnered attention from researchers, designer and practitioners worldwide.

https://doi.org/10.52202/080684-0003

Regarding the composites, the development of a general circular business model is necessary, but also quality protocols concerning end-of-waste strategies of each individual mix of materials are needed. Ongoing research and development efforts aim to enhance the efficiency and precision of pultrusion, reducing energy consumption and expanding the range of design possibilities. Innovations may include new process to pursue the R-Strategy in composite materials. Furthermore, the LCA analysis is currently in progress to assess the environmental impacts of specific pultruded profiles, with additional results to be incorporated in future work. Other suggestions for future work could include research into the long-term durability and performance of pultruded profiles under diverse environmental conditions, exploration of hybrid composite systems to unlock new opportunities for optimizing performance and sustainability, and the integration of smart technologies into pultrusion processes, which could lead to improvements in efficiency, precision, and product quality. In conclusion, the transition to a circular economy for FRP materials requires overcoming significant technological and economic challenges, but there are viable strategies and growing opportunities that can help mitigate these barriers. Policymakers can play a crucial role in facilitating a supportive regulatory environment by offering incentives, subsidies, or tax breaks for projects that incorporate innovative and environmentally friendly materials like FRP profiles. As the construction industry continues to evolve, embracing innovative materials becomes imperative for achieving an harmonious balance between environmental responsibility, economic viability, and social progress.

ACKNOWLEDGEMENTS

The author would like to acknowledge Gualini SpA and Italcomposites Doo for the know-how in the manufacturing process of pultrusion, and Gees Recycling for sharing their dedication and expertise in recycle composite materials.

REFERENCES

Asensio, M., Esfandiari, P., Núñez, K., Silva, J.F., Marques, A., Merino, J.C., Pastor, J.M. (2020). Processing of pre-impregnated thermoplastic towpreg reinforced by continuous glass fibre and recycled PET by pultrusion, *Composites Part B: Engineering*, volume 200, pp. 108-365, https://doi.org/10.1016/j compositesb.20 20.108365

Barkanov, E., Akishin, P., Namsone-Sile, E. (2022). Effectiveness and Productivity Improvement of Conventional Pultrusion Processes. *Polymers*, 14, 841. https:// doi.org/10.3390/polym14040841

Boswell, K., Hoffman, S., Selkowitz, S., Patterson, M. (2021). Skin Metrics: The Wicked Problem of Façade System Assessment, *Journal of Facade Design and Engineering*, Vol. 9(1), pp. 131–146, https://doi.org/10.7480/jfde.2021.1.5538

Chatziparaskeva, G., Papamichael, I., Voukkali, I., Loizia, P., Sourkouni, G., Argirusis, C., Zorpas, A.A. (2022). End-of-life of composite materials in the framework of the circular economy, *Microplastics*, 1, no. 3, pp. 377-392, https://doi.org/10.3390/micro plastics1030028.

Dalla Valle, A., Campioli, A. (2022). The environmental assessment of re-manufacturing, in Talamo, C.M.L., *Re-manufacturing networks for tertiary architectures, Innovative organizational models towards circularity*, pp. 212-226, FrancoAngeli, Milano.

De Fazio, D., Boccarusso, L., Formisano, A., Viscusi, A., Durante, M. (2023). A Review on the Recycling Technologies of Fibre-Reinforced Plastic (FRP) *Materials Used in Industrial Fields*, Journal of Marine Science and Engineering, 11, no. 4: 851, https://doi.org/10.3390/jmse11040851

Elmar, W., Michael, S., Michael, K. (2018). Composites Market Report, Market Developments, *Trends, Outlook and Challenges* (accessed on 22 April 2024).

EuCIA. (2024). why composites are genuinely circular, *European national composites associations and industry-specific sector groups,* Brussels

European Boating Industry. (2023). *AISBL*, A roadmap on the implementation of the circular economy for end-of-life recreational boats, Brussels.

European Commission. (2023). Measuring progress towards circular economy in the European Union – *Key indicators for a revised monitoring framework, SWD 306 final*, Brussels. European document.

European Commission. (2020). *A new Circular Economy Action Plan For a cleaner and more competitive Europe.* Brussels. European document. https://www.eea.europa.eu/policy-documents/com-2020-98-final report (accessed on 13 December 2024).

European Commission. (2019). *The European Green Deal* (COM(2019) 640 final report). Brussels. European document. https://eur-lex.europa.eu/legal-content/EN/TXT/?uri=CELEX:52019DC0640 (accessed on 14 December 2024).

European Commission. (2018). *European Strategy for Plastics in a Circular Economy*, Brussels. SWD(2018) 16 final report. European document.

European Commission. (2016). *Towards a smart, efficient and sustainable heating and cooling sector*, Brussels. European document. https://ec.europa.eu/commission/presscorner/api/files /document/print/en/memo_16_311/memo_16_311_e n.pdf (accessed 16 December 2024).

European Commission. Regulation (EU) No 305/2011, Construction Products Regulation (CPR), https://single-market-economy.ec.europa.eu/sectors/c onstruction/construction-products-regulation-cpr_en (accessed on 14 December 2024). European document.

European Parliament Research Service, the circular economy: find out what it means, how it benefits you, the environment and our economy. https://www.europarl.europa.eu/topics/en/article/20151201STO05603/circular-economy-definition-importance -and-benefits (accessed 08 January 2025). European document.

Hobbs, G., Bre, K.A. (2017). Reuse of building products and materials – barriers and opportunities, *Proceedings of the International Hiser Conference on Advances in Recycling and Management of Construction and Demolition Waste*, Delft, 109-113.

Ishai, O. (2018). Engineering mechanics of composite materials, *Advanced Materials*, New York.

Jacob, A. (2011). Composites Can Be Recycled. *Reinforced Plastics*, Volume 55 (3), pp. 45-46, https://doi.org/10.1016/S0034-3617(11)70079-0.

JEC composites magazine. (2024). *Composite material in civil engineering and architecture*, n. 154. https://digital-magazine.jeccomposites.com/jec-composites-magazine/jec-composites-magazine/n154-2024 (accessed 3 January 2025).

Julian, I., García-Jiménez, A., Aguado, A., Arenal, C., Calero, A., Campos, V., Escobar, G., López-Buendía, A.M., Romero, D., Verdejo, E., García-Polanco, N. (2022). Advances in the circularity of end-of-life fibre-reinforced polymers by microwave intensification, *Chemical Engineering and Processing* - Process Intensification, Volume 178, pp. 109-015, https://doi.org/10.1016/j.cep.2022.109015.

Karvonen, I., Jansson, K., Behm, K., Vatanen, S., Parker, D. (2017). Identifying recommendations to promote remanufacturing in Europe. Journal of Remanufacturing, Springer, Volume 7(no.2-3), pp. 159-179, https://doi.org/10.1007/s13243-017-0038-2.

Knippers, J., Cremers, J., Gabler, M., Lienhard, J. (2011). Construction manual for polymers and membranes, *Edition detail*, Munich.

Lavagna, M., Monticelli, C., Zanelli, A. (2022). Circular economy and tertiary architecture, in Talamo, C.M.L, *Re-manufacturing networks for tertiary architectures, Innovative organizational models towards circularity*, FrancoAngeli, Milano, pp. 29-28.

Mrotzek-Blöß, A., Haas, A., Goldmann, D., Nuortila-Jokinen, J., Kulczycka, J., Smol, M., Petranikova, M., Ebin, B. (2019). On the way to a circular economy: Obstacles and challenges in the metal value chain, *Proceedings of the 17th International Waste Management and Landfill Symposium,* IFAD, Cagliari

Nguyen, Q., Ngo, T., Mendis, P., Tran, P. (2013). Composite Materials for Next Generation Building Façade Systems, *Civil Engineering and Architecture*, 1, no.3, pp. 88-95, https://doi.org/10.13189/cea.2013.010305.

Pastori, S., Mereu, R., Mazzucchelli, E.S., Passoni, S., Dotelli, G. (2021). Energy Performance Evaluation of a Ventilated Façade System through CFD Modeling and Comparison with International Standards, *Energies*, Volume 14, no.4, pp. 193, https://doi.org/10.3390/en14010193.

Regulation 2020/852 of the European parliament and of the council of 18 June 2020 on the establishment of a framework to facilitate sustainable investment and amending Regulation (EU) 2019/2088. European document.

Straits. (2024). *Fibre-Reinforced Composites market size, demand over the years - Study Period 2018-30* (Report Code SRCH1729DR). https://straitsresearch.com/report/fiber-reinforced-composites-market (accessed on 4 January 2025).

United Nations general assembly. (2015). Transforming our World: The 2030 Agenda for Sustainable Development, *outcome document of the United Nations summit A/RES/70/1*, New York. https://sdgs.un.org/2030agenda (accessed on 10 December 2024).

Vijay, N., Rajkumara, V., Bhattacharjee, P. (2016). Assessment of Composite Waste Disposal in Aerospace Industries, *Procedia Environmental Sciences.* Volume 35, pp. 563-570, https://doi.org/10.1016/j.proenv.2016.07.041.

https://doi.org/10.52202/080684-0003

DEVELOPMENT OF A PRELIMINARY, USAGE SPECIFIC PRODUCT DOCUMENTATION FOR RECLAIMED GLT – A NORWEGIAN CASE STUDY

Mahnert, Karl-Christian[1*], Altgen, Daniela[1], Bergset, Knut Helge[1], and Øvrum, Audun[1]

[1]NTI (Norwegian Institue of Wood Technology), Norway

*Corresponding Author: kcm@treteknisk.no

ABSTRACT

Background and aim. The paper aims to collect background information about the production of glued laminated timber (GLT) in Norway and to suggest content for usage specific product documentation for GLT intended for reuse based on laboratory work combined with practical experience from two case studies.

Methods and Data. Bond line quality in reclaimed GLT was assessed by testing the resistance of delamination. Adhesive systems were tentatively identified using stained microscopy sections and FTIR spectroscopy.

Findings. The paper illustrates the consequences of the youngest industrial history from the wood working industry in Norway for the reuse of glued laminated timber in load bearing applications. It shows difficulties with on-site evaluation of building products and points out the conflict between desired comprehensive knowledge of properties of reclaimed building elements and the need to keep destructive testing at a minimum.

Theoretical / Practical / Societal implications. The authors consider the findings of the paper practically relevant as they show the complexity of evaluating the reuse potential of a relatively simple building product. At the same time, they propose a solution for how this complexity can be overcome by suggesting test methods and deriving categories for usage specific product documentation.

KEYWORDS: Glued laminated timber, delamination, adhesives, casein, documentation

1 INTRODUCTION

Glued-laminated timber (GLT) is a type of engineered wood made by bonding together finger-jointed lamellae under pressure to form a large structural component. GLT is widely used in building applications such as columns, beams, and arches in mid to high-rise public, private and commercial structures. Due to the high added value compared to structural timber, sophisticated design, the high adhesive costs and the amount of stored carbon, GLT is considered highly relevant for reuse. However, architects and planners will need product documentation to include reused GLT in new structures. This paper describes specific criteria and methods to investigate reclaimed GLT with the aim of collecting information as a basis for issuing usage specific product documentation. A comprehensive outline of the history of adhesive utilization in Norway is compiled as framework for the investigations, which include state-of-the-art laboratory methods and practical experience. Compared to the mind map based holistic approach described by Yahmi et al. (2023), the current study breaks down the considerable list of material-related barriers for reuse into a clear list of criteria tailored to the specific situation in Norway,

1.1 BACKGROUND

GLT consists of strength-graded according to NS-EN 14081, finger-jointed lamellae, typically made of spruce in the Nordic countries, and can be produced in various shapes and sizes. It can be adapted to all types of load-bearing structures due to its variable cross-section and good formability. The requirements to timber used in the production of GLT, to adhesives and overall quality of GLT are defined in NS-EN 14080.

From a production standpoint, it is easier to manufacture GLT from spruce than from pine, which has a lower resin content. Impregnated pine lamellae are used in structures expected to be exposed to significant moisture variations. These lamellae undergo a moisture increase during the

https://doi.org/10.52202/080684-0004

impregnation process, followed by drying, which can affect the bonding quality. A distinction is made between homogeneous GLT (where all lamellae have the same strength class) and combined GLT (where the best-quality lamellae are placed in the outer parts of the cross-section). Combined GLT optimizes the yield from the timber resource and is therefore the most common type. The product standard NS-EN 14080 requires that the moisture content of all lamellae in a GLT cross-section be within 6-15% at the time of production, with a maximum 5% moisture variation between lamellae being bonded together. This ensures optimal bonding conditions and minimizes stresses that could lead to cracks. Some small cracks are expected, but they rarely affect the load-bearing capacity of GLT.

Reusing GLT can be expected to significantly contribute to both environmental sustainability and societal well-being by reducing the demand for virgin materials and minimizing construction waste. While research specifically focused on glulam reuse is not available yet, more general studies on timber reuse list environmental benefits—such as lower greenhouse gas emissions, reduced energy consumption, and prolonged material life cycles. For example, the deconstruction and reuse of timber elements in buildings can result in nearly zero CO_2 emissions, largely due to wood's carbon sequestration capabilities and optimized end-of-life strategies (Di Ruocco et al., 2023).

1.2 ASSESSMENT OF GLT PRIOR TO DISASSEMBLY OF CONSTRUCTION

1.2.1 Lamellae in GLT

Lamellae used in the production of GLT are usually strength graded. It is not possible to assess the strength class of the lamellae in a beam because only the sides of all lamellae are visible. If there is uncertainty regarding the strength classification of the lamellae, it is advisable to grade conservatively and assume a lower quality than what was standard at the time of production if the visual impression implies this.

Cracks in lamellae due to drying or internal stresses are not considered to affect the load-bearing capacity of the GLT. However, mechanical damage of lamellae should be considered as reducing the GLT's capacity and must be deducted from the cross-section used as a basis for evaluating its strength class.

1.2.2 Finger joints

A comprehensive inspection of finger joints is not practically possible. However, the finger joints in the top and bottom lamellae of the GLT can provide an indication of the quality of the finger joints throughout the entire GLT structure. General requirements to finger joints according to NS-EN 14080 are:

- Finger joints must not contain knots with a diameter greater than 6 mm or grain deviation
- The distance between knots larger than 6 mm in diameter and the finger joint must be at least three times the knot diameter.
- There should be no gaps between the fingers that are not filled with adhesive.

1.2.3 Bond lines

The bond lines must be tight to ensure the proper transfer of stress between the lamellae. Open bond lines indicate ageing of the adhesive and/or high internal stresses in the GLT. Open bond lines in newly produced GLT are obvious production failures, and a GLT beam with open bond lines would not pass the producers' quality control. Therefore, open bond lines disqualify reclaimed GLT for the use in load-bearing constructions. Thus, they must be repaired before the GLT can be used again.

1.2.4 Other criteria

For surface-treated GLT, the ability to inspect the material before and after disassembly is reduced. Still, it is unlikely that the surface treatment itself would prevent the reuse of GLT. Lead paint is the only surface treatment hazardous to health that has been used in Norway. However, it has been banned in 1929 (Lovdata, 2025) – about 29 years before the first production of GLT in Norway in 1958 (NLF, 2015). Therefore, this type of paint is considered unlikely to be found on GLT potentially available for reuse in Norway today.

If the GLT is made from impregnated wood, a chemical analysis of the lamellae must be performed to determine whether they were treated with a preservative containing chromium or arsenic.

The emissions from cured adhesive, regardless of adhesive type, are not harmful to health.

1.3 EVALUATION FOR RE-USE

GLT consists of lamellae made from finger-jointed lumber. Wood is known to react to moisture through dimensional changes. Such changes can create a dynamic stress pattern in the GLT, where the extent of the stress depends on indoor climate conditions and how they fluctuate throughout the year. If the movements become too large, joints, connections, and bond lines may be affected over time.

Climatic stresses such as temperature, rain, wind, and snow will influence the GLT and can reduce its capacity over time. Additionally, design rules and snow load requirements might have changed compared to those valid when the original structure was designed. This must be considered when assessing GLT for reuse. Thus, it should be assumed that the product would need reinforcement to fully utilize its span, even if it is in good condition.

For untreated GLT, fire resistance can be assumed to be equivalent to untreated wood (D-s2, d0). If the GLT is impregnated or surface-treated, the type of impregnation and/or treatment must be identified to obtain information on its fire resistance.

Cracks that weaken the GLT's capacity can be repaired using approved adhesives. Currently, epoxy or

https://doi.org/10.52202/080684-0004

polyurethane adhesives are commonly used for on-site repair of timber structures. Requirements for these adhesive types are specified in NS-EN 17418.

1.3.1 Original documentation

Original labelling facilitates the most efficient reuse since the properties of the specific GLT are either included in the label or can be relatively easily obtained by contacting the manufacturer or institutions involved in the relevant control scheme during the production period.

1.3.2 Assessment of load history

The load history of a structure is important to determine the probability that the construction has been exposed to loads exceeding its designed capacity. Potential overloading, such as heavy snow loads, may weaken joints or the GLT itself, which could reduce the residual capacity compared to the originally designed capacity. This assessment is the least reliable, as it is difficult to get a complete picture of the structure's load exposure over time.

1.3.3 Identification of the adhesive

The type of adhesive used in the production of a GLT beam is crucial to determine whether the beam can be reused in a load-bearing structure. This is because some adhesives that were commonly used in Norway in the early days of GLT production are no longer permitted for use in modern GLT manufacturing.

If this information is not part of the labelling or available from other sources, the color of the bond line becomes an important criterion.

Dark brown/black glue lines indicate the use of phenol-resorcinol (PF) or phenol-formaldehyde-resorcinol (PRF) adhesives (Hunt et al. 2018), which are known for their durable bond lines and thus high value for the reusability of GLT. This type of adhesive is approved for the production of load-bearing GLT according to current standards.

Light-colored bond lines may indicate the use of casein adhesive, which is protein-based and derived from milk. Casein adhesive was the only adhesive system available for GLT production in the Nordic region until World War II. Since casein adhesive is not moisture-resistant, it was only used in GLT constructions for indoor applications. It is not approved for the production of load-bearing GLT structures today.

Urea-formaldehyde (UF) adhesive also results in light-colored bond lines and is covered by NS-EN 14080. Currently, there is no UF adhesive approved for load-bearing GLT structures, but UF adhesive was previously commonly used in GLT for climate classes I and II.

Other adhesives that produce light-colored glue lines, such as MF (melamine-formaldehyde), MUF (melamine-urea-formaldehyde), polyurethane (PUR), and emulsion-polymer isocyanate (EPI), are approved under NS-EN 14080 and can be assumed to have met the requirements for adhesives approved for GLT at the time of production.

1.3.4 Suggestions for characterizing reclaimed GLT by testing

The testing requirements should depend on the available documentation for the product. CE-certified GLT is assumed applicable for load-bearing structures without further testing if the load history assessment does not indicate that the GLT's capacity was overutilized in previous use.

For older GLT or GLT that is believed to have been overutilized in a previous application, testing of the bond lines should be mandatory.

Testing the capacity of finger joints requires the removal of large sections from a beam, making it impractical. Instead, a visual assessment of the visible finger joints as outlined above should be decisive.

The capacity of the glue joints between lamellae, however, can be tested with a relatively small sample extraction. NS-EN 14080 specifies different testing requirements depending on the service class in which the GLT beam will be used:

- For service classes 1 and 2, the shear strength of the glue joint under compression is required.
- For service class 3, the resistance to delamination must be tested within the limits defined by the standard.

A scaled test program consisting of shear strength testing and assessing the resistance to delamination, depending on intended future use of the GLT, is assumed to provide necessary information for the classification of GLT for reuse.

2 MATERIAL AND METHODS

Two sets of sections from glued-laminated timber (GLT) beams, I and II, both recovered during the deconstruction of public buildings in the vicinity of Oslo, were used in this work. All samples were cut from full sized GLT beams which had been transported to intermediate storage locations in the Oslo region.

2.1 MATERIAL

The first set of sections (I) consisted of 19 GLT samples (A-S) from beams presumably produced in 1967-1969 used in the roof of the old Aker hospital. The beams had cross sections of GLT 450-650 mm x 120-200 mm (h x w) consisting of 14, 21 or 18 lamellae with corresponding 13, 20 or 17 light-coloured bond lines.

The second set of sections (II) consisted of four GLT samples (A-D) from beams produced in 1963, used in the roof of the old gymnasium "Rykkinnhallen". The beams had a cross-section of 90 x 633 mm (h x w), each consisting of 19 lamellae and corresponding 18 dark-coloured bond lines. The two upper lamellae in section A had been damaged during the deconstruction of the building, the corresponding bond lines were therefore excluded from the investigations.

https://doi.org/10.52202/080684-0004

2.2 METHODS

2.2.1 Assessment of bond line quality

The bond line quality of the samples was investigated by assessing the resistance to delamination of the bond lines according to NS-EN 14080, method B. For this, a 75 mm wide sample was taken from each section. The number and total length of the bond lines per sample was recorded before the samples were impregnated with water and subsequently dried to approximately their original mass. Immediately after reaching the final mass, the length of the openings per bond line was recorded and the delamination calculated as percentage of the total bond line length.

2.2.2 Identification of adhesives

The identification of the light-coloured adhesive present in set I was important to find out if the GLT was bonded with UF or casein adhesive. MUF would have also resulted in a light-coloured bond line but this adhesive type was first used in the production of GLT in Norway in the 1980's (Treteknisk, 1999) and is therefore not an alternative for the beams of set I produced before 1970. UF adhesives are covered by NS-EN 14080 and would potentially allow the re-use of the GLT in load-bearing application in service class 1 and 2 (NS-EN 1995-1-1), casein has never been covered by EN 14080 and would exclude the GLT for any load-bearing application. FTIR-spectroscopy was applied to identify the adhesive type used.

Samples from the hardened adhesives from set I and II, a UF- and a PRF reference were ground with mortar and pestle. The samples were analyzed with a FTIR spectrometer (Cary 630 FTIR spectrometer (Agilent Technologies, Inc., USA)) with a diamond ATR (Agilent Technologies, Inc., USA). The results were analysed with the software MicrolabExpert (Agilent Technologies, Inc., USA). An ATR correction was conducted prior to a 2-point base line correction for all spectra.

FTIR is a widely available technology frequently applied for the identification and investigation of wood adhesives and their curing reactions. However, the results of the FTIR-analysis did not allow a clear identification of the adhesive sample from the light bond line as either UF or casein. Therefore, ninhydrin was applied to indicate amino acid components (Lennart, 2005) present in casein adhesives but absent in UF-adhesives. A 70 µm thick section of a light-coloured bond line from section set I was prepared on a sliding microtome, stained with an aqueous solution of ninhydrin and dried at room climate for 16 hours. Light microscopy at 10x magnification was conducted on an Olympus BX60 (Olympus Europa SE & Co. KG, Germany).

Given the background information in this study, ninhydrin staining would have been sufficient to distinguish between the two relevant types of adhesive.

3 RESULTS

3.1 BOND LINE QUALITY

The bond line quality of set I (average delamination of 14.8 %) was significantly lower than the bond line quality of set II (average delamination of 3.1 %) (Table 1).

For set I, only samples J, R and S fulfilled the delamination requirement of maximum 4 % delamination after one test cycle, and maximum of 8 % delamination after two test cycles given in NS-EN 14080. The resistance to delamination for all samples from set II was better than the requirement of maximum 4 % delamination after one test cycle given in NS-EN 14080.

Table 1: Results from testing the bond lines' resistance to delamination

Set	Sample	Number of bond lines	Total delamination	
			[mm]	[%]
I	A	13	793	24.4
	B	13	234	7.2
	C	13	421	13.0
	D	13	356	11.0
	E	13	229	7.0
	F	13	277	8.5
	G	13	281	8.6
	H	13	528	16.2
	I	13	503	15.5
	J	13	106	3.3
	K	20	496	16.5
	L	20	1101	36.7
	M	20	714	23.8
	N	20	882	29.4
	O	20	263	8.8
	P	20	722	24.1
	Q	20	582	19.4
	R	17	267	3.9
	S*	17	445	6.5
II	A	16	50	3.0
	B	18	63	3.9
	C	18	38	2.3
	D	18	51	3.1

3.2 IDENTIFICATION OF ADHESIVE

3.2.1 FTIR spectroscopy

The FTIR-spectra for the light-coloured adhesive from set I and a reference UF-sample are shown in Figure 1. The black line for the spectra of set I shows the characteristic vibrational bands and 2920 and 2850 cm^{-1} linked to the higher concentrations of CH_2-groups in amino acids (Ptiček and Siročić, 2017). As expected the vibration peak of carbonyl groups was found between 1300 and 1100 cm^{-1} and at 1652 cm^{-1} (Ptiček and Siročić, 2017). Typical casein peaks, according to Ptiček and Siročić (2017), that were absent in our spectra are those

https://doi.org/10.52202/080684-0004

linked to amide stretching at 1585 cm^{-1} and the ones resulting from carbonyl groups (C=O) in the range of 1725-1750 cm^{-1}.

The spectra of the UF-adhesive show the band of N-H stretching of secondary amines around 3300 to 3350 cm^{-1}. The stretching of carbonyl groups (C=O) and C-N-stretching of secondary amines are represented by the peaks at 1632 and 1550 cm^{-1}, respectively (Liu, 2017). The band at 1380 to 1330 cm^{-1} is assigned to -CH_2OH groups in UF resins, the peak at 1130 cm^{-1} illustrates the C-O aliphatic ether (Singh et al., 2014).

The spectrum of the adhesive from set I suggests that the adhesive is based on casein rather than UF. To invalidate the uncertainties linked to the described absences of some typical casein peaks, the bond line will be investigated using a staining solution and light microscopy.

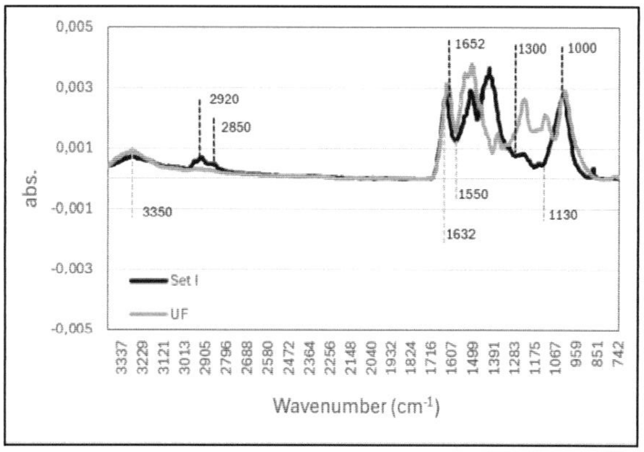

Figure 1: FTIR-spectra of adhesive sample from set I (assumed casein) and of reference UF-sample.

The dark-coloured bond lines in the samples from set II indicate a phenolic adhesive. The FTIR-spectra from the adhesive found in the bond line from set II (black line) and the spectra of the PRF-reference are shown in Figure 2. Both profiles show peaks at the characteristic bandwidths of 1595 cm^{-1} and 1500 cm^{-1} which are assigned to the C=C aromatic rings that are embodied into the cured adhesive (Özparpucu et al., 2022). The peaks in the spectral range between 1500 to 1310 cm^{-1} are related to the methylene and methyl groups of the adhesive (Alpert et al., 2012) which are a part of the methylene bridges between phenol-resorcinols (Poljansek and Krajnc,2005). The peak at 1085 cm^{-1} has been described by Poljansek and Krajnc (2005) and Bobrowski and Grabowska (2015) as linked to the ether bridges between methylol groups developed during condensation reactions. Thus, the adhesive used in set II is identified as PRF.

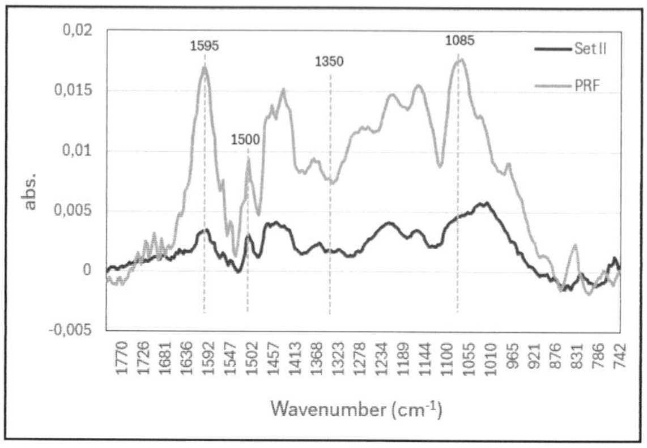

Figure 2: FTIR-spectra of adhesive sample from set II (assumed PRF) and of reference PRF-sample.

3.2.2 Light microscopy

The light-coloured bond-lines in the samples from set I indicate a casein or UF type adhesives. The color reaction of the bond line stained with ninhydrin (Figure 3) proved the presence of proteins, confirming the use of a protein-based adhesive, in this case casein, and excludes an UF adhesive which does not contain proteins.

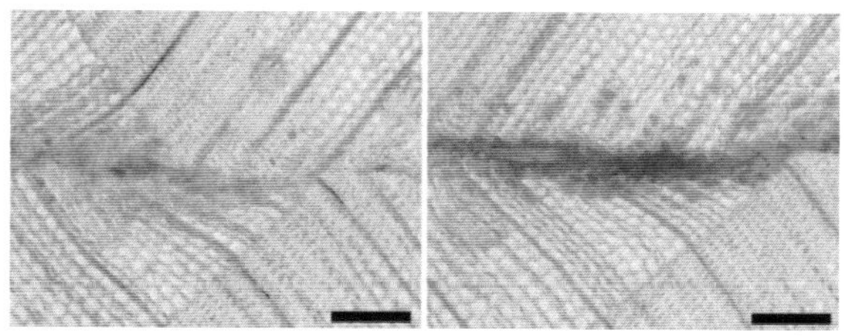

Figure 3: Microscopy images (10x magnification) of a light-coloured bond line in a sample from set I before (left) and after staining with ninhydrin (right). Scale bar 200 μm.

Taking into consideration the widespread application of casein adhesives in the GLT during the relevant period (Treteknisk, 1999), we assume that the GLT beams in set I are bonded with a casein adhesive.

3.3 DISCUSSION

The investigations of samples from set I showed low resistance to delamination for the samples from set I and proved that the GLT was bonded with a casein adhesive. Samples from set II showed high resistance to delamination, the relevant adhesive was identified as PRF.

Casein adhesives are also known to fail in the presence of water. Thus, the high delamination values for set I can most likely be explained by the use of a casein adhesive. PRF adhesives on the contrary are known for their excellent water resistance. Thus, the low delamination values for the samples from set II is characteristic for this type of adhesive.

Besides the diverging moisture resistance of the two adhesive types, intense tensions in one of the sample sets due to pronounced changes in moisture content during the service life or weakening of the timber close to the bond line by wood destroying organisms could explain the difference in resistance to delamination (Yahmi et al., 2023). The former would manifest itself in widespread cracking in the lamellae which would have been detected during preparation of the samples. The latter requires wetting of the GLT over elongated periods of time, accompanied by obvious discoloration of the lamellae and macroscopic alterations in the wood structure. Also these signs of damage would have been visible during sample preparation.

Another reason could be ageing of the bond lines. Still, both adhesive types have shown high reliability in application and long-term investigations (Deppe and Schmidt, 1994, Raknes, 1997).

Therefore, the differences in resistance to delamination between set I and II can be explained by the inherent difference in moisture resistance of casein and PRF adhesives.

3.4 SUGGESTIONS FOR A USAGE SPECIFIC PRODUCT DOCUMENTATION FOR RECLAIMED GLT

The reuse of GLT beams in load bearing applications requires the structural integrity of the beams, reliable bonding and sufficient capacity to carry the loads in the future applications.

A usage specific product documentation should show relevant national building codes and product standards applicable to the actual material, e.g. NS-EN 14080 for GLT beams. Furthermore, essential conditions and properties for the intended use should be listed (service class, cross-section, strength class and outer appearance).

The fulfillment of these usage-specific requirements should be documented by the seller, based on test reports from experts.
In this context it is important that destructive testing should be reduced to a minimum, of course within the limits of responsibility.
As mechanical testing will lead to the destruction of entire beams, the focus of an assessment should be on

- visual evaluation of the general condition of the beam and the quality of the lamellae and finger joints.
- conservative reduction of load bearing capacity in case of mechanical damage.
- scaled extend of service class specific testing of bond lines according to standard tests defined in NS-EN 14080.

Since the beams of sets I and II were produced before the product standard for GLT, NS-EN 14080 was in place, all three requirements need to be fulfilled. The investigations in the current paper address the aspect of reliable bonding only. The bonding of beams of set I is found not to fulfil today's requirements, the bond lines of samples from beams of set II, however, yielded sufficient resistance against delamination to allow the reuse of the GLT in load bearing applications in all service classes.

4 CONCLUSIONS

This paper compiles background information relevant for the assessment of reclaimed glued laminated timber (GLT) for reuse. It gives an overview over adhesive systems applied in the production of GLT in Norway, gives examples for the analysis of bond line quality and identification of relevant adhesive systems based on two national case studies and concludes with recommendations for a usage specific product documentation for the reuse of GLT.

ACKNOWLEDGEMENTS

The authors acknowledge the Norwegian Green Plattform Project "circWOOD", funded by the Norwegian Research Council (Project number 328698), for the funding of the activities covert in this paper.

REFERENCES

Alpert NL, Keiser WE, Szymanski HA (1970) *IR: theory and practice of infrared spectroscopy*. Springer Science & Business Media

Bobrowski A, Grabowska B (2015) FTIR method in studies of resol type phenol resin structure in the air atmosphere in some time intervals. *Metallurgy and Foundry Engineering* 41(3):107-113.

Deppe, H.J. and Schmidt, K. (1994) Ergänzende Untersuchungen zum Alterungsverhalten gestreckter Holzmontageverleimungen. *Holz als Roh- und Werkstoff.* 52: 355-360.

Di Ruocco, G., Melella, R., and Sabatano, L. (2023). Timber Buildings Deconstruction as a Design Solution toward Near Zero CO2e Emissions. *Buildings*, 13(1), 157.

Hunt, Christopher G.; Frihart, Charles R.; Dunky, Manfred; Rohumaa, Anti (2018). Understanding wood bonds–going beyond what meets the eye: a critical review. *Reviews of Adhesion and Adhesives*. 6(4): 369-440.

Lennart, C. (2005) FORENSIC SCIENCES | Fingerprint Techniques, *Encyclopedia of Analytical Science* (Second Edition), Elsevier, Editor(s): Paul Worsfold, Alan Townshend, Colin Poole, 414-423.

Liu, M., Thirumalai , K. G., Venkatesh, R. Yiqiang, W. and Hui, W. (2017). Characterization of the crystalline

https://doi.org/10.52202/080684-0004

regions of cured urea formaldehyde resin. RSC ADVANCES 7. 49536-49541.

Lovdata (2025). LOV-1929-05-24-3 - Lov om delvis forbud mot å bruke blyhvitt m.v. ved malerarbeide – Lovdata.no, accessed 2025-03-21.

NLF (Norske Limtreprodusenters Forening) (2015) Limtreboka, ISBN 978-82-7129-324-6

NS-EN 1995-1-1 (2004) Eurocode 5: Design of timber structures Part 1-1: General Common rules and rules for building

NS-EN 14080 (2013) Timber structures — Glued laminated timber and glued solid timber — Requirements. *Standards Norway.*

NS-EN 14081-1 (2016) Timber structures - Strength graded structural timber with rectangular cross section - Part 1: General requirements. *Standards Norway.*

NS-EN 17418 (2021) Two-component epoxy and polyurethane adhesives for on-site repair of cracked timber structures - Testing, requirements and repair strength verification. *Standards Norway.*

Poljansek and Krajnc (2005) Characterization of phenol-formaldehyde prepolymer resins by in line FT-IR spectroscopy. Acta Chimica Slovenica 52(3):238-244.

Ptiček Siročić, Anita. (2017). Characterization of Casein Fractions – Comparison of Commercial Casein and Casein Extracted from Cow's Milk. *Chemical and Biochemical Engineering Quarterly Journal*. 30. 501-509.

Raknes, E. (1997) Durability of structural wood adhesives after 30 years ageing. *Holz als Roh- und Werkstoff.* 55: 83-90.

Singh, A. P., Causin, V., Nuryawan, A. and Park, B.-D. (2014). Morphological, chemical and crystalline features of urea–formaldehyde resin cured in contact with wood. *European Polymer Journal,* 56: 185–193.

Treteknisk (1999) Kunnskap for fremtiden – Norsk Treteknisk Institutt 1949 – 1999. Anniversary book.

Yahmi, A., Nouri, M., Tahlaiti, M., Khelidj, A., Raffin, C. and Place, N. (2023). Mind-Mapping Assessment of Reuse Potential of Glulam: An Experimental Study. *Buildings*, 13(12), 2929.

Özparpucu, M., Windeisen-Holzhauser, E., Wegener, G. and Richter, K. (2022). A new analytical approach to investigate the influence of wood extracts on the curing properties of phenol-resorcinol–formaldehyde (PRF) adhesives. *Wood Science and Technology*, 56(2), 349-365.

ESTABLISHING THE REAL VALUE CYCLES FOR TIMBER STRUCTURES – FINDINGS FROM CASE PROJECTS

Kristine Nore[1] and Lone Ross[2*]

[1]Omtre AS, Hønefoss, Norway, kristine.nore@omtre.no

[2]Norwegian Institute of Bioeconomy Research, Ås, Norway

*Corresponding Author: lone.ross@nibio.no

ABSTRACT

Background and aim. The four-year project SirkTRE aims to advance circularity in the timber industry, achieving key milestones in standardization, reuse, and market expansion through cooperation, innovation and targeted research. Historically, circular timber construction was common, and modern efforts, like Norway's SirkTRE, work towards reintroducing wood reuse and recycling. SirkTRE is targeting an 8% reduction in CO_2 emissions aligned with Norway's commitments under the Paris Climate Agreement by repurposing 50% of Norway's annual 800,000 tons of counted wood waste by 2030.

Methods and sata. Regulatory frameworks, including the Waste Framework Directive, Construction Products Regulation, and EU Taxonomy, drive towards circularity. Norwegian building regulations (TEK17) also promote climate accounting and reuse mapping. Overcoming market and logistical barriers is crucial. Coordinated efforts across regions, regulations, and industry standards will determine the success of circular timber construction. A new Norwegian standard (NS 3691) facilitates quality assurance for reclaimed wood. In contrast, practical projects—such as circular prefabricated housing, SirkBO, post-consumer wood from demolition and barns, and modular timber skeleton buildings—demonstrate scalable reuse models.

Findings. Studies confirm consumer readiness for recycled wood and highlight significant climate benefits from reuse over incineration. Research on digital product passports (DPPs) emphasizes their role in material tracking and lifecycle management.

Theoretical / Practical / Societal implications. SirkTRE has shown the viability of large-scale timber reuse, yet regulatory, logistical, and technological barriers remain. Future efforts must focus on policy adaptation, industry incentives, and scaling innovative reuse solutions.

KEYWORDS: circular economy, design for disassembly, innovation, post-consumer wood, regulations.

1 INTRODUCTION

1.1 BACKGROUND

Timber is an important contributor to the strategy of decarbonization in the EU. The building sector has major challenges, i.e. climate adaptation of buildings, vast maintenance needs, limited resources, and increasing costs. On the other hand, the building sector has major possibilities and potential in achieving climate, nature, and resource goals in our transition to a more circular economy.

Timber buildings, fit for reuse, were common a century ago in the Nordics; the components could most often be reused, i.e. log buildings. Thus, the non-circular, linear use of building materials has been in the period from industrialization up until our times. Our resources are limited, and our society of increased waste production must end.

In Norway, around 12 million m^3 of certified timber is harvested yearly (SSB 1, 2024). If our society is to succeed in reducing its greenhouse gas emissions, it is expected that the need for bioresources will be significantly greater. Recycled wood can be a crucial input factor through new reuse and material recycling solutions.

The Norwegian Green platform project SirkTRE aims to enhance longevity and reuse in timber construction.

https://doi.org/10.52202/080684-0005

SirkTRE targeting an 8% reduction in Norwegian CO_2 emissions aligned with Norway's commitments under the Paris Climate Agreement (United Nations, 2015). Figure 1 shows Norway's annually collected wood waste, averaging 800.000 tons. Launched in 2021, SirkTRE seeks to repurpose half the wood waste into building products by 2030.

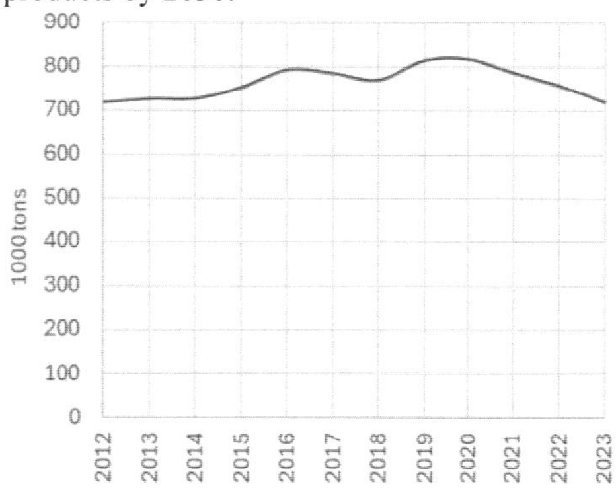

Figure 1 Collected wood waste in Norway from 2012 to 2023 (SSB B, 2024)

1.2 DESCRIPTION OF METHODS

The ambition is to prevent 1 million tons of CO_2 emissions by avoiding incineration of wood waste, while simultaneously sequestering nearly the same amount of in buildings through the reuse and recycling into wooden components.

SirkTRE includes a larger research project, CircWOOD, and five business-driven projects: INN – an innovation center, HELTRE – reuse of solid wood-based solutions, RESSURS – reduced resource use, REAL - realization of circular wood projects and TEK - new technology & new digital production.

SirkTRE ends in June 2025, and the research project CircWOOD ends in June 2026. The funding bodies are the Research Council of Norway, Innovation Norway, and the Industrial Development Corporation of Norway (SIVA), together with 23 business partners that contribute with in-kind hours or cash.

The research project CircWOOD is investigating aspects of wood use in the Norwegian economy, with particular emphasis on the reuse of wood in construction projects, and recycled wood as raw material in today's wood industry. Research results, especially related to resource access and material flows, are linked to the facilitation of the circular flow of goods, handling, environmental impact, design, and production of wood in, and towards, relevant markets in Norway and abroad. The project analyses the sustainability and environmental footprint of the wood-value chain based on strategies and new technologies that contribute to circularity. Furthermore, CircWOOD identifies and investigates ways in which processes can be simplified by using methods for digital collection, analysis, and sharing of data and at the same time addresses the underlying political frameworks and the economic impact cascading of wood. Scientific competence building is one of the main outcomes of the completion of 4 PhD candidates covering material flow and quality of wood resources, policy and governance, socio-economic impacts, and digital tracking and tracing of wood.

1.3 WHY IS THE USE OF WOOD NOT CIRCULAR?

Despite wood's potential as a renewable and sustainable material, its value chains remain among the largest sources of bioenergy incineration (Ellen MacArthur Foundation, 2021). The principles of circularity—extending lifespan, enabling cascade use, and facilitating reuse—have not yet been fully integrated into the wood sector. Several barriers that hinder the transition to circular wood use are collected in SirkTRE:

- Logistical challenges: The distribution and storage of reclaimed building materials pose significant logistical difficulties, require efficient collection systems and storage solutions to prevent material degradation (European Commission, 2020).
- Underdeveloped market: The market for reused building materials remains immature, with limited demand, supply-chain inefficiencies, and a lack of established business models (Gorgolewski, 2017).
- Technological and processing limitations: Current technology and handling systems for returning, sorting, and reprocessing used materials into high-quality, reusable building components are insufficient or underdeveloped (Ellen MacArthur Foundation, 2021).
- Quality assurance and regulatory concerns: There is uncertainty regarding the quality and structural integrity of reused materials, requiring extensive documentation and testing to meet building regulations and standards (Anastasiades, 2021).
- Design-phase constraints: The potential for reuse is often determined early in a product's life cycle, particularly through material selection and construction techniques that may hinder disassembly and repurposing (European Commission, 2020). Delayed time of involvement and thus, not time to influence and plan circular built, neither design, nor use of reclaimed materials (Deloitte, 2022).
- Economic barriers: Virgin raw materials are often more cost-competitive than secondary materials, reducing financial incentives for reuse and making circular alternatives less attractive in the market (OECD, 2024).
- Traditional LCA does not always fully account for circular construction, multiple use, or increased robustness for prolonged lifetime, but extended LCA methodologies and circular

economy-integrated LCA approaches can include these aspects (van Stijn, 2021).

These barriers underscore the need for systemic changes in policy, technology, and market structures – challenges that SirkTRE addresses.

1.4 AIM

As SirkTRE wraps up, the transition to a circular use of wood is complex, however, the relatively low-hanging fruit of the reuse of waste wood is analyzed and more clearly defined.
This paper sums up the upcoming legislation, technology, and needs to ensure that waste wood becomes a valuable source for the timber industry in the future. Furthermore, it also gives insight into innovations and new solutions that have been produced within SirkTRE together with some of the scientific findings that can have a direct impact on the transition to a more circular use of wood.

2 EMERGING FRAMEWORK

Adapting to a more circular approach requires targeted developments and concerted efforts to overcome the barriers outlined in Chapter 1.3. Regulatory frameworks play a crucial role in facilitating this transition.
First, the Paris Agreement (United Nations, 2015) establishes the overarching global ambitions for sustainable development, aiming to ensure that human activities remain within planetary boundaries.
Second, the European building sector is subject to evolving EU regulations designed to promote circularity. Through the European Economic Area (EEA) Agreement, these regulations also apply to Norway, requiring national adaptation to align with European sustainability goals.
Third, each EEA member state integrates EU regulations into its national legal and policy frameworks. In Norway, regulatory developments increasingly reflect principles of the circular economy, including within the building sector.
Beyond regulatory adaptation, the establishment and implementation of standardized guidelines are essential. Industry standards must be developed to provide clear directives, ensuring that the construction sector effectively interprets and complies with new regulatory frameworks, thereby achieving the intended sustainability objectives.

2.1 EU REGULATIONS

Since the EU's launch of the Green Deal in 2019, massive developments in EU regulations have been notified and implemented. The reasoning behind the Green Deal is growth and competition, self-sufficiency, and climate and environment (European Green Deal, 2020).
Below is an excerpt of already ratified regulations relevant to improving the circular activities concerning the transformation of waste wood into a source for the timber industry.
New regulations on the circular economy are notified. An Action Plan for Circular Construction and a Circular Economy Act are upcoming (CEAP, 2020).

2.1.1 Waste Framework Directive

This Waste Framework Directive (WFD, 2019) establishes a waste hierarchy where reuse and recycling are given priority over recovery and disposal. EU countries are obliged to facilitate the separate collection, sorting, and material recovery of waste, including wood.
The EU Waste Framework Directive set a 70% material recovery target for construction and demolition waste by 2020. As of 2023, Norway had achieved 46% material recovery, falling short of this goal (SSB B, 2024).

2.1.2 Construction Products Regulation

The Construction Products Regulations (CPR) set requirements for construction materials, including reused and recycled wood materials (CPR, 2024). This revised version of the CPR strengthens the requirements for circular economy, digital tracking, and market watch and includes more products and a wider scope compared to the former CPR (CPR, 2011).

2.1.3 EU Taxonomy and Sustainable Finance

EU Taxonomy and Sustainable Finance supports the transition to a circular economy, including design for disassembly and reuse of wood materials (EU Taxonomy, 2020). The reuse of wood in construction qualifies as a sustainable investment.

2.1.4 Energy Performance of Buildings Directive and the revised Energy Efficiency Directive

The EU aims for all new buildings to be zero-emission by 2030 and for all existing homes to achieve zero-emission status by 2050 (EPBD, 2024). These energy directives mandate the calculation of life cycle carbon emissions for buildings. Incorporating reused and recycled wood materials can significantly reduce a building's carbon footprint, offering a competitive advantage to developers.

2.1.5 Deforestation-free products regulation

The EU Timber Regulation (EUDR, 2023) applies to imports of wood and wood products, requiring proof that they have not contributed to deforestation. This regulation may incentivize the increased use of recycled and reused wood as an alternative to virgin timber, thereby reducing compliance challenges.

2.1.6 The Ecodesign for Sustainable Products Regulation

The EU has established a framework to enhance the sustainability of products on the European market (ESPR, 2024). The ESPR outlines requirements for the durability, reparability, and reuse of wood materials in sectors such as furniture, construction, and others.

2.1.7 Carbon Capture and Storage

The Carbon Removal Certification Framework (CRCF, 2024) sets rules for carbon sequestration in buildings and the reuse of wood. It has been established to certify carbon sequestration, including bio-based carbon sequestration in

https://doi.org/10.52202/080684-0005

construction. Reusing wood in buildings can contribute to long-term carbon sequestration, qualifying for carbon credits. As such, the CRCF aims to create incentives for the conservation and reuse of wood, encouraging alternatives to incineration or landfilling.

2.2 NORWEGIAN REGULATIONS VALID AND EMERGING

Norwegian building regulations (TEK17, 2023), latest adaptations, valid from July 1st, 2023, are towards improving the built environment's climate footprint, and circular construction. These changes include:

1. Climate accounts for materials in building projects
2. Reuse mapping of buildings prior to demounting
3. Construction work shall be designed and prepared for later dismantling when this can be carried out within a practical and economically justifiable framework.

For these regulations to have an impact, all these changes need supplementary legal requirements and standards:

1. Boundary lines are allowed for climate gas emissions.
2. Mapped used materials, fit for reuse, should be made available in an open information flow. The digital passport development of CPR provides the basis for this development.
3. Design for disassembly must become regulatory. Thus, further standardization in this field is required to define the future design of the different materials and building categories.

2.3 CIRCULAR CONSTRUCTION CEN/TC 350/SC 1

The CEN/TC 350/SC 1 refers to the standardization committee under the European Committee for Standardization (CEN), specifically focused on sustainability in construction, with a subcommittee dedicated to circular construction practices.

The standardization committee will specify circular principles, guidelines, and requirements to support more sustainable practices. The entire life cycle of construction works is covered, from design and construction to dismantling and end-of-life, including both new and existing buildings and structures.

2.4 IMPACT ON THE CIRCULAR BUILDING SECTOR

The construction sector is increasingly incentivized to incorporate reused wood, driven by evolving regulations and market dynamics. The European Union (EU) has implemented several measures to promote the use of recycled and reused materials, including wood, in construction projects.

EU Regulations and Incentives:

- **Cascading Use of Biomass:** The revised Renewable Energy Directive (RED III), effective from November 2023, introduces principles and rules for Member States to prioritize the use of woody biomass in material applications before burning wood for energy. This approach encourages the use of wood in construction and other material applications, potentially reducing the availability of wood for energy purposes and thereby increasing its value in the construction sector (ECOS, 2024).

National Initiatives:

- **Germany:** The Investment and Support Bank of Hamburg (IFBHH) offers financial support for the use of wood in construction. For residential buildings, the bank provides 30 cents EUR per kg of wood, amounting to approximately 5,000-6,000 EUR for a 140m² apartment. This initiative serves as a strong incentive for builders to incorporate wood into their projects. (Interregeurope, 2021).

Implications for Norway:

While specific national incentives in Norway are not detailed in the provided sources, the EU's overarching policies and frameworks are likely to influence Norwegian construction practices. The emphasis on carbon storage in buildings and the promotion of reused materials, including wood, aligns with Norway's sustainability goals. As EU regulations often set precedents, similar incentives and requirements may be adopted or adapted within Norway's regulatory framework, encouraging the use of reused wood in construction projects.

In summary, both EU-level regulations and national initiatives are creating a conducive environment for the increased use of reused wood in the construction sector. These developments are expected to play a significant role in promoting sustainable building practices in Norway and across the EU.

3 RESULTS FROM SIRKTRE

The SirkTRE project aimed to make the timber industry more circular over four years. Though its scope was broad and several key actions have advanced the circular timber sector, including: the development of standards for evaluating reclaimed wood, reducing the carbon footprint of reuse, and potentially expanding the timber industry's market share through larger sourcing volumes. The different initiatives test the way of the transition to a fully circular, green shift in the timber and construction industries. Some examples from the SirkTRE are highlighted.

3.1 STANDARDIZATION – REUSE OF WOOD

As part of SirkTRE, the new Norwegian Standard series for recycled wood aims to simplify the evaluation and quality assurance of used wood and wood-based materials for reuse in new construction products. NS 3691, 'Evaluation of Reclaimed Wood,' (NS 3691, 2025) defines reclaimed wood as material sourced from dismantled structures, packaging, or offcuts, including surplus materials from construction activities. It does not include surplus materials or by-products from sawmills or forestry operations. A standard method for evaluating reclaimed wood can streamline quality assurance and potentially reduce the costs of reused wood.

The standard series consists of three parts:

- NS 3691-1 Evaluation of reclaimed wood – Part 1: Terminology and general rules
- NS 3691-2 Evaluation of reclaimed wood – Part 2: Impurity
- NS 3691-3 Evaluation of reclaimed wood – Part 3: Visual strength sorting

The goal is to support the transition to a more circular wood products industry and reduce valuable waste. In addition to the three existing standards, two more are being developed: one for composite timber components, such as nail-plate frames or glulam, and another for the competencies required to evaluate reclaimed wood.

The Norwegian mirror committee is active in the CEN and ISO development of circular standards.

3.2 NEW SOLUTIONS

The SirkTRE project is driving innovation to make the timber industry more circular, focusing on sustainability and waste reduction. Through various initiatives and activities, SirkTRE is developing new technologies, products and practices, as well as building necessary scientific knowledge, to support the reuse and recycling of wood materials. Each project and activity within SirkTRE has its unique focus and ambition, and short films highlighting their work are available to explore. For more details, visit www.sirktre.no. Below are some selected outcomes presented.

3.2.1 Circular prefabricated house

SirkTRE partner Haugen/Zohar Architects has developed a circular housing series, offering innovative solutions for sustainable housing. The SirkBO concept, see Figure 2, focuses on designing recyclable and reusable components to enter new cycles, reducing environmental impacts. The use of modular and standardized elements simplifies construction and shortens assembly time significantly. (*retracted for peer-review*).

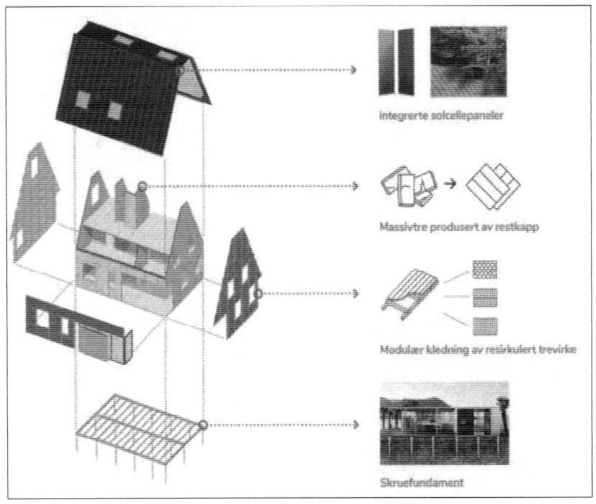

Figure 2. Circular building components of the prefabricated house. Credits: Haugen/Zohar Arkitekter.

3.2.2 Cattle barn built in post-consumer wood

The first SirkTRE case project, a barn in Noresund, Norway, was the first and largest modern agricultural building built in solid wood with a large proportion of recycled wood, see Figure 3. The recycled wood is exposed in the walls on the inside of the barn. The screwed solid timber wall elements consist of 75% recycled wood of 48x98 mm, edge-set, and screwed together in vertical lengths. The initial desire was 100% recycled wood, but to ensure structural safety, every fourth lamella is continuous and made of virgin wood.

Figure 3. Barn raised with 75% reused lamellas in screwed solid timber wall elements. Photo: Authors

3.2.3 Sirkulær Ressurssentral – Circular Resource Hub

Located in a 4,500 m^2 tent at Økern, Oslo, Sirkulær Ressurssentral has created one of Europe's largest recycling hubs. As part of SirkTRE, Sirkulær Ressurssentral processes reclaimed wood from building projects, reintegrating it as a raw material in the wood industry or into new building projects. Reuse requires value creation in an upcoming market. Regions like Oslo show prosperous solutions for such initiatives.

3.2.4 Flexible building in timber

Rammeverk is a new, modular building model, shown in Figure 4, of an apartment building, a so-called "open building", which allows for resident participation in the planning and gives both the developer and residents great

https://doi.org/10.52202/080684-0005

flexibility throughout the life of the building. This timber frame house is an apartment building that can be ordered flat-packed from a glulam factory.

Figure 4. Frame house for flexible housing. Photo rendering: Fragment

3.2.5 Aanesland Treindustri – production facility in timber

Aanesland Treindustri's new steel-free production building in Lillesand is 1,900 m² and was completed in 2022 (Figure 5). The facility was designed by the timber architects at Helen and Hard, Stavanger, Norway. The load-bearing systems for the building were supplied by Aanesland's glulam-production, Sørlaminering, and self-produced oak dowels. Thus, the use of steel was reduced to a minimum. In the SirkTRE -project of Aanesland has been to further develop moment-stiff timber frame connections with oak or beech dowel connections. The use of this technology is available for design in the software for timber engineering TimberTech (2024). https://en.timbertech.eu/.

Figure 5. Timber frame from Aanesland Treindustri production facility during assembly, before installing oak dowels, built in 2022. Photo: Aanesland Treindustri.

3.2.6 Circular interior wall system

Today, interior office sectioning walls are often changed every seventh year in Oslo, Norway. Thus, a reusable interior wall system using timber, suitable for use in both rehabilitation and new building projects, is developed by the architect's office Grape. This wall system offers the flexibility to be disassembled, moved within the same building, or reused in other contexts. The design aims to achieve strict sound insulation, targeting a reduction of approximately 48-49 dB. A wall with an open lock system is shown in Figure 6.

Figure 6. Circular wall system with reused wood. Photo rendering: Grape Architects

3.2.7 Reblåkk

Reblåkk is a patented building block system based on CLT off-cuts (Figure 7). With the increasing CLT production capacity, the cut-off lacks refined use. Reblåkk is optimized for mass production, where a compact production can be attached to an existing CLT production. The development of Reblåkk is currently a scale-up, defining its potential for re-use and in a circular business strategy. Reblåkk fulfils the Safe and Sustainable by Design framework (SSbD) (European Commission, 2022)

Figure 7. Mounting Reblåkk. Photo: Authors

3.3 CONSUMER ATTITUDE

The construction sector has a significant environmental impact, which can be reduced by increasing the use of recycled materials. This has been the backdrop for a study that explored Norwegian consumers' willingness to use recycled wood in houses and cabins, applying the Diffusion of Innovations theory (Khatri et al, 2025). Surveys of 913 homeowners revealed strong consumer readiness, with Relative Advantage and Compatibility as key adoption drivers. Perceived Risk had minimal impact, while Green Values influenced adoption indirectly. These insights can help businesses design appealing recycled products and inform policies that promote circular practices in construction.

3.4 CLIMATIC IMPACT

Reusing waste wood can be a better climate solution than incinerating it for energy recovery, and Hansen et al. (2023) showed that a scenario involving reusing waste wood without processing and distributing it through a reuse center near new construction activities had the highest avoided greenhouse gas emissions. Another reuse scenario, where waste wood is sorted at waste facilities and requires cutting and additional machinery, also results in significantly greater avoided emissions than the reference scenario. A sensitivity analysis, based on reuse through a dedicated station, indicated that increasing the reuse rate from 20% to 80% could lead to approximately 50% higher avoided emissions.

3.5 DIGITAL PRODUCT PASSPORTS

Lyse (2024) explored the role of digital product passports (DPPs) in promoting the reuse, recycling, and recovery of building materials, thereby reducing the need for new raw material extraction. DPPs can help consumers and manufacturers make sustainable choices by providing essential product information and minimizing waste. The study focuses on how DPPs can effectively track and manage building materials at the end of their life cycle to support a circular economy in the construction industry. The research identified a lack of existing studies on how DPPs should be structured, what information they should contain, and how they should be validated. A literature review and a survey of 51 respondents from different sectors (reuse, recycling, repurposing, research, and education) were conducted to fill this gap. Findings highlight five key information categories for DPPs: 1) Product information, 2) Manufacturer details, 3) Installation and assembly data, 4) Maintenance guidelines, and 5) End-of-life information. Additionally, Lyse (2024) proposed a framework for DPPs, outlining necessary validation procedures. Several information models were developed to support the implementation of DPPs in the construction sector. This research provides insights into how digital tracking systems can improve resource efficiency and circularity in the industry. XX had from the start ambitions to show the practical effect of the reuse of large volumes of timber. Lack of investment due to weak construction activity has, however, led to low demand.

4 CONCLUDING REMARKS

The circular timber building sector is within reach. The results show a multitude of examples of how circularity can develop. SirkTRE recommends further development of standardization on all levels, from products and materials to material sourcing and building operations. The analysis indicates substantial potential for carbon savings through SirkTRE initiatives. Initial calculations supporting the 8% reduction target are currently critically reviewed and compared with updated data from ongoing activities, suggesting a promising trajectory toward achieving these climate goals and emphasizing the effectiveness of circular strategies in timber construction. The results enhance theoretical frameworks on circular economies while providing practical insights for policymakers and industry stakeholders. It underscores the importance of innovative resource management in mitigating climate change, fostering sustainable practices in the construction sector, and promoting a circular economy in Norway and beyond.

https://doi.org/10.52202/080684-0005

4.1 INCREASING INTEREST

SirkTRE has drawn inspiration from previous projects and has also served as a catalyst and initiator for numerous consortiums, leading to funded projects. A selection is listed below.

4.1.1 Basajaun, Horizon 2020, 2020-24

Basajaun focused on demonstrating how sustainable wood construction can contribute to a circular bioeconomy. It developed innovative wooden building systems using locally sourced materials, optimizing the entire value chain from forest management to construction. The project showcased two full-scale demonstrators in France and Spain, proving the feasibility of high-performance, low-carbon timber buildings. By integrating digital tools and sustainable practices, Basajaun aimed to enhance resource efficiency and promote the use of wood in European construction.

4.1.2 InnoTLT, Bioeconomy in the North, 2023-2026

InnoTLT aims to advance cross-laminated timber (CLT) into Tailored Laminated Timber (TLT). The project focuses on developing innovative TLT panels for walls and floors, enhancing structural performance, stiffness, and load-bearing capacity. InnoTLT seeks to revolutionize the timber industry by promoting a circular economy and sustainable construction practices.

4.1.3 DRASTIC, Horizon Europe, 2023-2027

DRASTIC aims to reduce whole life cycle GHG emissions in new construction and deep-energy retrofits by demonstrating affordable, innovative circular solutions across five geographical zones and multiple building layers. The project will develop and apply a multi-cyclic performance assessment framework, integrating multi-cycle LCA, multi-cycle LCC, circularity, and sufficiency for construction and building-related products. It will also demonstrate the feasibility of promising, cost-effective technologies, processes, and business models to accelerate market adoption. This will contribute to more sustainable buildings with lower life-cycle carbon emissions, enhanced performance, and reduced costs.

4.1.4 Woodcircles, Horizon Europe, 2023-2027

Transforming sustainability in the construction industry, Woodcircles pioneers' circular solutions for sustainable wood construction. By reducing Europe's reliance on non-renewable resources, the initiative lowers greenhouse gas emissions and minimizes waste. Key innovations include the eco-friendly and efficient construction of an 'urban sawmill' and the use of digital twins. Through integrated, circular, and digitally supported solutions, Woodcircles advances waste reduction and carbon capture in buildings, ushering in a new era of eco-conscious construction that benefits both the environment and the economy.

4.1.5 CIRCULess, Horizon Europe, 2024-2027

CIRCULess aims for a circular solution for construction and manufacturing waste. The process industry needs to embrace a circular economy and restorative feedback loops to optimize resource use and reduce supply costs. The construction sector is expected to benefit significantly from these models. However, reclaimed materials must meet the same standards as new materials. The goal is to promote circularity in the construction and manufacturing industries by reducing construction, demolition, and manufacturing waste, focusing on mineral and timber-based material streams. The project will develop new circular products and processing techniques to improve the quality and performance of secondary materials.

4.1.6 RAW project, EIC Pathfinder, 2024-2027

The full name of the project is 'Computation For A New Age Of Resource Aware Architecture: Waste-Sourced And Fast-Growing Bio-Based Materials'. The project aims for a paradigm-shifting new digital infrastructure that combines non-destructive material sensing technologies with adaptive design and fabrication. This will allow the building industry for the first time to assess and use natural materials in the variability with which they are grown or have been reclaimed, minimizing the current energy consumption and wasteful practices of material homogenization. With a focus on reclaimed timber, biopolymers from agricultural waste, and composites from fast-growing hemp fibers the project will help to reduce CO_2 emissions, support the circular economy, and create new aesthetic possibilities for architecture.

4.1.7 WoodStock, Horizon Europe, 2024-2028

The WoodStock project aims to promote climate-smart, circular, and zero-waste utilization of underutilized wood from forests and existing buildings in the construction sector, supporting the New European Bauhaus initiative. The project focuses on quantifying and mapping wood resources, including underutilized streams, using Harvested Wood Products accounting, dynamic Material Flow Analysis and LCA to assess wood utilization potential, climate mitigation impacts, and resource availability. Through six Living Labs across different European regions, WoodStock develops zero-waste and circular building designs, leveraging digital twins and co-creation activities to enhance human health and well-being. The project also establishes the European Wood Construction Observatory, an AI-powered hub for best practices and innovative solutions, ensuring long-term impact beyond its duration. Empowering climate-smart, circular, and zero-waste use of underutilized wood from the forest and building stock in the construction sector.

4.2 TRANSITION TO A CIRCULAR ECONOMY IN THE EU AND THE NORDIC COUNTRIES

EU's Circular Economy Action Plan (CEAP) (European Commission, 2020) challenges the member states to test out incentives to promote a circular economy.

In the Nordics, Sweden responded to this challenge in 2024, with two incentives for the built environment (Söderholm et al., 2024).

1) Incentives for reuse, collection, and material recycling. Thus, increasing the producer's

responsibility for building products. Proposals that lead to increased collection and recycling should be analyzed further.
2) Incentives for economically motivated renovations and efficient use of outer building elements. Several regulations and regulations should be reviewed based on how they provide incentives for socially and economically motivated renovations and efficient use of space.

Norway has a panel of government-appointed experts who are writing a report planned for May 2025 that will cover incentives to increase circularity, proposed new and simplified regulations, tax and different support schemes. This will be the start of a follow-up mission of holistic implementation of the circular economy in Norway. SirkTRE and CircWOOD are represented in both the expert group and the reference group.

ACKNOWLEDGEMENTS

The authors inform that AI tool Chat GPT 4.0 has been used during the preparation of this paper for literature search and language improvement.

REFERENCES

Anastasiades, K. et al. (2021) Standardisation: An essential enabler for the circular reuse of construction components? A trajectory for a cleaner European construction industry, Journal of Cleaner Production, (298) https://doi.org/10.1016/j.jclepro.2021.126864.

CEAP, 2020. https://eur-lex.europa.eu/legal-content/EN/TXT/?qid=1583933814386&uri=COM:2020:98:FIN

CRCF (2024) Carbon Removal Certification Framework (EU/2024/3012 https://eur-lex.europa.eu/legal-content/EN/TXT/?uri=OJ:L_202403012

CPR (2024) EU 2024/3110. https://eur-lex.europa.eu/eli/reg/2024/3110/oj/eng

Deloitte (2020) Circular economy in the construction, civil engineering and real estate industries. Sirkulær økonomi i bygg-, anlegg- og eiendomsnæringen Directorate for Public Administration and Financial Management (DFØ) In Norwegian. www.anskaffelser.no/sites/default/files/2023-01/Sirkular-okonomi-i-BAE-naringen.pdf

ECOS 2024. An emerging framework for timber buildings policy and standards briefing, www.ecostandard.org/wp-content/uploads/2024/09/2024-09-09_ECOS_Framework-timber-buildings.pdf

Ellen MacArthur Foundation (2021). Circular Economy in Construction: Rethinking Materials and Waste.

EPBD (2024) energy performance of buildings EU/2024/1275; EU/2023/1791 https://eur-lex.europa.eu/legal content/EN/TXT/?uri=OJ:L_202401275&pk_keyword=Energy&pk_content=Directive

ESPR (2024) EU 2024/1781 Ecodesign requirements for sustainable products. https://eur-lex.europa.eu/legal-content/EN/TXT/?uri=CELEX%3A32024R1781&qid=1719580391746

EUDR (2023) Deforestation and forest degradation and repealing https://eur-lex.europa.eu/eli/reg/2023/1115/oj/eng

European Commission (2020). Circular Economy Action Plan: For a Cleaner and More Competitive Europe.

European Commission (2022). Directorate-General for Research and Innovation, *Safe and sustainable by design chemicals and materials – A European assessment framework*, Publications Office of the European Union, https://data.europa.eu/doi/10.2777/86120

European Green Deal (2020) Communication from the commission to the European parliament, the european council, the council, the European economic and social committee and the committee of the regions the European green deal https://commission.europa.eu/strategy-and-policy/priorities-2019-2024/european-green-deal_en

EU Taxonomy (2020) https://eur-lex.europa.eu/legal-content/EN/TXT/?uri=CELEX:32020R0852

Gorgolewski, M. (2017). Resource Salvation: The Architecture of Reuse. Wiley.

Hanssen, M. R, Thorkildsen, T. E, Nyrud, A. Q. (2023). Life Cycle Analysis of Reuse of Waste Wood. In Norwegian: Livsløpsanalyse av ombruk av returtre. *MINA fagrapport 91.* Ås: Norges miljø- og biovitenskapelige universitet, Fakultet for miljøvitenskap og naturforvaltning, 35 s. https://static02.nmbu.no/mina/publikasjoner/mina_fagrapport/pdf/mif91.pdf

Interregeurope.eu (2021) Wood as a sustainable construction material. https://www.interregeurope.eu/find-policy-solutions/stories/wood-as-a-sustainable-construction-material?

Khatri, S., Sjølie, H. K., & Nyrud, A. Q. (2025). Consumer drivers for intended adoption of recycled wood as construction material. *Cleaner and Circular Bioeconomy*, 10. https://doi.org/10.1016/j.clcb.2024.100129.

Lyse, S. M. (2024). Digital Product Passports for Efficient Tracking and Monitoring of Building Materials. In Norwegian: Digitale produktpass for effektiv sporing og følging av byggematerialer. Master's thesis. Norwegian University of Science and Technology (NTNU).

NS 3691 (2025). Evaluation of return wood. Evaluering av retur tre. In Norwegian. Norsk Standard

SSB A, (2024). Avfallsregnskapet. https://www.ssb.no/natur-og-miljo/avfall/statistikk/avfallsregnskapet

SSB B (2024). Skogavvirkning-for-salg https://www.ssb.no/jord-skog-jakt-og-fiskeri/skogbruk/statistikk/skogavvirkning-for-salg

 https://doi.org/10.52202/080684-0005

Söderholm P. et al. (2025). Om ekonomiska styrmedel för en mer cirkulär ekonomi Betänkande av Kommittén om ekonomiska styrmedel för att främja omställningen till en cirkulär ekonomi. In Swedish. SOU 2024:67, Stockholm, Sweden.

TEK 17 (2023). Regulations on technical requirements for construction works. In Norwgian. Norwegian building authority. ttps://www.dibk.no/regelverk/byggteknisk-forskrift-tek17

TimberTech (2024) Software for timber structure design. https://en.timbertech.eu/.

United Nations. (2015). Paris Agreement. United Nations.

van Stijn, A., et al. (2021) A Circular Economy Life Cycle Assessment (CE-LCA) model for building components, Resources, Conservation and Recycling, 174, https://doi.org/10.1016/j.resconrec.2021.105683.

WFD (2019). Waste Framework Directive https://eur-lex.europa.eu/eli/dec_impl/2019/1004/oj#:~:text=Commission%20Implementing%20Decision%20(EU)%202019,2384%20(notified%20under%20document%20C(

A FRAMEWORK FOR CIRCULAR BUILDING RENOVATION: INTEGRATING LCA, C2C, 10R, AND STAKEHOLDER ENGAGEMENT

Bojana Petrovic[1]*, Georgios Pardalis[2], and Migena Sula[3]

[1] Norwegian Institute for Sustainability Research (NORSUS), Norway

[2] International Institute for Industrial Environmental Economics (IIIEE), Lund University, Sweden

[3]Department of Built Environment and Energy Technology, Linnaeus University, Sweden

*Corresponding Author: bojana@norsus.no

ABSTRACT

Background and aim. Circular economy (CE) practices in the built environment require integrating strategies such as life cycle assessment (LCA), cradle-to-cradle (C2C) principles, stakeholder collaboration, and the 10R framework to enhance resource efficiency and minimize environmental impacts across the entire building. However, existing research lacks a comprehensive framework that systematically combines these elements while demonstrating their practical application and addressing stakeholder alignment in real-world scenarios. The aim of this study is to propose a novel framework that integrates LCA, C2C principles, the 10R framework, and stakeholder engagement to advance CE practices in building renovation.

Methods and Data. By applying a mixed-methods approach, this study combines qualitative and quantitative analyses to evaluate CE strategies. The qualitative analysis involves material suitability for reuse, refurbishment, or storage, and explores stakeholder roles within the 10R and C2C frameworks. The quantitative analysis, based on LCA, measures GHG emissions comparing two scenarios using new and reused materials, highlighting potential carbon savings. A case study of a single-family building renovation from Sweden illustrates the practical application of these strategies and emphasizes the importance of stakeholder collaboration in overcoming barriers.

Findings. The findings underscore the importance of strategic material selection and the transformative role of material reuse in achieving long-term carbon savings and minimizing GHG emissions. Incorporating reused materials into building renovation practices can lead to a substantial 94% reduction in GHG emissions compared to using newly produced materials.

Theoretical / Practical / Societal implications. The study demonstrates how circular economy strategies can drive a low-emission building sector, offering practical insights and replicable method for real building projects.

KEYWORDS: Circular economy, Cradle-to-cradle, Life cycle assessment, Reuse, Stakeholder engagement

1 INTRODUCTION AND RESEARCH CONTEXT

The built environment plays a crucial role in the global push toward sustainability, as it is responsible for a significant portion of resource consumption, waste generation, and greenhouse gas (GHG) emissions (Joensuu et al., 2020). The European Union (EU) has introduced various programs and initiatives to encourage stakeholders to transition from a linear to a circular economy (CE), acknowledging the building sector as the largest waste producer and a significant consumer of resources (Giorgi et al., 2022). Both the European Commission and EU member states actively support the adoption of circular strategies, with a goal of full implementation by 2050 (Al-Obaidy, Courard, & Attia, 2022). As such, the transition from a linear economy, characterized by a "take, make, dispose" model, to a CE has become essential for mitigating environmental impacts in this sector (Illankoon & Vithanage, 2023). CE principles aim to optimize the use of resources by designing systems that minimize waste and allow for the

https://doi.org/10.52202/080684-0006

continuous reuse and recycling of materials. This transition is especially urgent in the building sector, which accounts for nearly 42% of final energy use and approximately 36% of EU-wide GHG emissions (Fabbri et al., 2023). While the potential of CE principles to transform the built environment is increasingly recognized, their practical implementation remains limited, especially to addressing the entire lifecycle of buildings (AlJaber et al., 2023).

To illustrate the practical application of CE strategies, this study focuses on a single-family house building as a case study. Single-family houses represent a significant share of the built environment, contributing notably to resource consumption, energy use, and GHG emissions due to their prevalence and specific design and material requirements (Arceo, 2023; Soust-Verdaguer et al., 2016).

Central to advancing circular practices in the built environment is the application of Life Cycle Assessment (LCA), a widely used methodology that evaluates the environmental impacts of buildings throughout their lifecycle—from material extraction and construction to operation and eventual demolition. LCA provides a comprehensive understanding of a building's environmental footprint, helping stakeholders identify opportunities to reduce resource consumption and minimize negative environmental impacts (Yadav et al., 2024). However, traditional LCA approaches have focused primarily on new buildings and often fail to account for CE principles such as material reuse and recycling (Larsen et al., 2022). Furthermore, LCA is typically confined to system boundaries that do not capture long-term environmental benefits, such as the effects of material recovery and reprocessing (Xing et al., 2022). Consequently, there is a need to expand LCA methodologies to fully integrate CE strategies, particularly in assessing the environmental performance of buildings under circular scenarios.

In fact, the LCA of building materials serves as a valuable tool for addressing this issue and can be applied within different system boundaries (Silvestre et al., 2014). First, "Cradle-to-Gate" focuses on the impacts associated with the production process of building materials. Second, "Cradle-to-Grave" encompasses the impacts of production, the transportation, the operational phase, and the disposal. Finally, "Cradle-to-Cradle" extends to include all impacts from production to the end-of-life, involving avoided emissions beyond the system boundary, as captured in the D module (Petrovic et al., 2024).

Minunno et al. (2020) conducted a study comparing the environmental benefits of reusing and recycling building components, revealing that reused components can reduce GHG emissions by up to 88% compared to recycling. While the recycling of materials such as steel, concrete, and plasterboard is well-established and regulated by policies in several countries, reuse practices offer even greater contributions to a CE. It can be noticed that components designed for disassembly can achieve reuse rates of up to 95%, allowing these products to be restored and reintroduced to the market at the end of their previous service life (Galvez-Martos et al., 2018). Recent LCA studies highlight the critical role of building materials in the overall life cycle of buildings. Consequently, the end-of-life (EOL) phase has gained prominence in the building industry, as currently only 20–30% of construction and demolition waste is reused or recycled (Honic et al., 2021).

A promising framework for supporting the CE transition in buildings is Cradle-to-Cradle (C2C) design, which emphasizes the continuous reuse of materials without degradation (Futas et al., 2019). While C2C principles have been applied to individual building materials and products, their integration into the entire building lifecycle—covering design, construction, and material recovery—remains undiscovered (Allam & Nik-Bakht, 2023). C2C principles could significantly enhance the environmental performance of buildings by ensuring that all materials are reclaimed, reused, or recycled at the end of their service life, thus supporting the broader goals of circularity in construction.

Another critical factor in driving the transition to a CE is effective stakeholder engagement (Munaro & Tavares, 2023). The built environment is inherently fragmented, involving multiple stakeholders, including architects, engineers, contractors, suppliers, policymakers, and building owners. Each stakeholder has distinct interests, expertise, and incentives, which can create barriers to collaboration and hinder the implementation of circular practices (Kaewunruen et al., 2024; Lee et al., 2024). While stakeholder engagement is widely recognized as essential for promoting sustainability, research on integrating diverse perspectives into decision-making—particularly in LCA and C2C contexts—is limited (Larsen et al., 2022; Lee et al., 2024).

Despite the growing emphasis on stakeholder engagement, there is a lack of research on structured methodologies that integrate stakeholder input with established CE assessment tools. Specifically, limited studies explore how stakeholder-driven decision-making can be systematically embedded within LCA and C2C frameworks to facilitate circularity in the built environment.

This paper addresses these gaps by proposing a novel framework that integrates LCA, C2C principles, stakeholder engagement, and the 10R framework to advance CE practices in the built environment. Unlike previous studies, which typically focus on either technical assessments or stakeholder perspectives separately, this research bridges the two by embedding stakeholder collaboration directly into the LCA process, ensuring that decision-making aligns with both environmental performance and practical feasibility.

To demonstrate the framework's applicability, this study evaluates GHG emissions in a real-world case building renovation project, comparing scenarios that utilize new versus reused building materials. Using LCA , the study assesses carbon savings and resource efficiency achieved through material reuse across different lifecycle stages. A real-world case study of a single-family house renovation demonstrates the practical application of this integrated approach, presenting how circular strategies—such as refuse, reduce, reuse, refurbish, and recycle—can be implemented in building practices. Furthermore, the study emphasizes the critical role of stakeholder collaboration in overcoming practical barriers and aligning diverse interests, offering actionable insights toward achieving a sustainable, low-carbon built environment.

2 METHODOLOGICAL APPROACH

This study involves a mixed-methods approach, integrating both qualitative and quantitative analyses to evaluate circular strategies for the built environment. The qualitative aspect focuses on assessing the suitability of materials for reuse, refurbishment, or long-term storage within circular scenarios. It also considers the roles of key stakeholders and broader frameworks such as the 10R and C2C principles to provide contextual insights. The quantitative aspect, driven by LCA, offers measurable data on environmental impacts such as GHG emissions and material efficiency. Combining these approaches ensures a comprehensive understanding of the potential environmental, practical, and strategic benefits of transitioning to circular practices in the building sector. This integrated methodology enables a holistic evaluation of the case study, emphasizing both the technical feasibility and the broader implications of circular strategies. This study adopts a multi-faceted approach to evaluate the environmental benefits of integrating circular economy (CE) principles into building renovation processes. A combination of quantitative and qualitative methods is used to analyze greenhouse gas (GHG) emissions and stakeholder roles across the building lifecycle. The methodology includes LCA, application of the 10R framework, and stakeholder mapping to comprehensively assess circular renovation strategies.

2.1 CASE STUDY OVERVIEW

The methodology is demonstrated through a case study of a single-family building with a gross floor area (GFA) of 182 m^2, constructed in the 1970s, located in the city of Växjö, Sweden. The house is of generic classification and has been selected as a representative example of common residential typologies of that era. The building (Figure 1) is currently inhabited and features bearing and non-bearing elements such as partitions, finishes, and cladding, which present significant opportunities for reuse, refurbishment, or recycling.

Figure 1: Exterior view of the case study building.

Comprehensive data, including the building's blueprints in Figure 2 and a draft material quantity in Table 1, provides a foundation for evaluating its material composition and identifying circular strategies.

2.2 SCENARIO DEFINITION

A cradle-to-cradle (C2C) system boundary is used to quantify GHG emissions across all stages of the building lifecycle. Two renovation scenarios are defined for comparative analysis:

- Scenario 1 (Conventional): Utilizes all newly produced materials except concrete.
- Scenario 2 (Circular): Assumes the building is composed of all reused materials, aligning with CE principles.

This approach allows for a comparison of the environmental impacts of new versus reused materials in the building following the C2C approach. Further, Scenario 1 explores how the GHG emissions from newly produced building materials vary between different lifecycle stages. In Scenario 2, the study evaluates GHG emissions from reused materials across different lifecycle stages. This scenario incorporates CE principles and the 10R framework to explore strategies that minimize waste and extend material lifecycles. The environmental impacts of these circular strategies are analyzed using LCA, comparing them to conventional renovation practices in Scenario 1 to highlight the potential benefits in reducing GHG emissions. Both scenarios are developed to align with C2C design principles, emphasizing the circularity of materials without degradation.

https://doi.org/10.52202/080684-0006

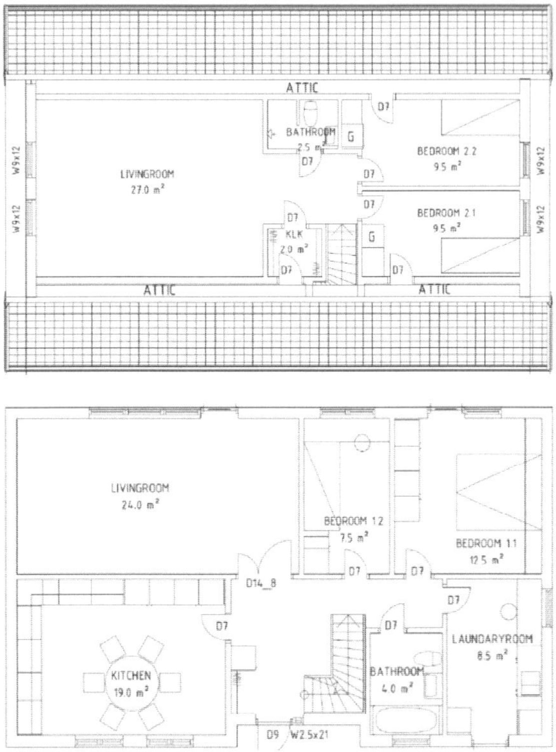

Figure 2: Drawings including first floor and ground floor.

The analysis is conducted using One Click LCA software following the C2C system boundary, assess GHG emissions across all life cycle stages, including production (A1-A3), transport (A4), construction waste (A5), material replacement and refurbishment (B4-B5), end-of-life processes (C2-C4), and avoided emissions (D).

The lifespan chosen for assessment is 50 years. Only windows (40 years), asphalt layer (30 years) and PE layer (20 years) were assumed to be replaced during the building's lifetime. Further, the construction waste percentage and the end-of-life waste treatment for each material is based on Swedish market practices, including energy recovery scenario for all materials that were incinerated (Table 1).

2.3 MATERIAL ANALYSIS

A detailed material inventory is created by analyzing the draft material quantities shown in Table 1. This inventory identifies non-bearing elements by type—such as brick, wood, gypsum, concrete and metal—and evaluates their condition and suitability during the end-of-life processes. Each material is assessed against the 10R framework, prioritizing strategies such as reuse, refurbishment, and recycling to maximize circularity. Factors such as the ease of disassembly, material durability, and the potential for repurposing are considered to determine their viability within the two scenarios. This material analysis serves as a critical input for both the environmental impact evaluation and the proposed circular strategies.

Table 1: Input materials and waste processing.

Building material	Volume (m^3)	Mass (kg)	Construction Waste %	End of life process
Brick	12,36	24712	5,0	Crushed to aggregate
Particle board	1,77	1219	16,7	Incineration
Concrete (foundation)	11,10	26633	0	Crushed to aggregate
Windows	0,04	93	Not available	Not available
Gypsum	13,44	10753	12,5	Recycling
Metal stairs	0,04	287	7,5	Recycling
Glass wool insulation	78,12	10156	8,0	Landfilling
PE Layer	0,28	252	10,0	Landfilling
Roof ceramic tiles	3,73	7464	5,0	Crushed to aggregate
Asphalt layer	0,93	970	Not available	Not available
Timber	22,11	12379	17,9	Incineration
Wood Board	24,92	16694	17,9	Incineration

2.4 INTEGRATION OF LCA, C2C, AND THE 10R FRAMEWORK

LCA method is used to quantify the environmental impacts of the two scenarios, with system boundaries extending from cradle-to-grave to cradle-to-cradle. The analysis emphasizes GHG emissions, resource efficiency, and waste generation, providing a quantitative basis for comparing circular renovation strategies with traditional practices. C2C principles guide the design of these strategies, ensuring that materials are reused or recycled in a manner that avoids degradation and reduces reliance on virgin resources. The 10R framework further informs the analysis by mapping circular strategies—such as refusal, reduction, reuse, refurbishment, and recycling—across each stage of the building lifecycle. Together, these frameworks enable a comprehensive evaluation of CE potential within the case study.

2.5 STAKEHOLDER CONSIDERATION IN SCENARIOS

Although this study does not engage stakeholders directly, it identifies key participants and their roles are essential for implementing circular strategies. Key stakeholders are identified and their roles are analyzed across lifecycle phases: production, transport, construction, use, and end-of-life. Stakeholders include architects, designers, contractors, deconstruction specialists, homeowners, material banks, and policymakers. Their contributions are evaluated based on their influence on material choices, resource management, and emissions control. Collaborative strategies are proposed to align stakeholder actions with CE objectives.

In the renovation scenario, architects and designers are responsible for incorporating reused or refurbished materials into the renovation plan while maintaining the building's functional and aesthetic quality (Passoni et al., 2021). Contractors play a critical role in disassembling and preparing materials for reuse (Dams et al., 2021), while homeowners influence the adoption of circular practices as the primary decision-makers (Kaewunruen et al., 2024). In the future renovation scenario, deconstruction specialists ensure the careful recovery of materials to preserve their quality, and material banks facilitate long-term storage and tracking of reusable components (Oliveira et al., 2024). Policymakers and regulators are also highlighted as pivotal in creating standards and incentives to support material reuse (Nußholz et al., 2019). These considerations provide a framework for understanding the collaborative nature of circular practices in the building sector, even without direct engagement.

2.6 EVALUATION METRICS

The evaluation of the two scenarios is based on a combination of quantitative and qualitative analysis, derived from a structured analysis of material use, environmental impact, and circularity potential. The quantitative analysis evaluates GHG emissions and explores their reduction potential by comparing two renovation scenarios. Using the LCA method, the study provides a comprehensive, data-driven evaluation of positive (released) and negative (avoided) impacts.
Qualitative metrics, informed by literature and established frameworks such as the 10R and C2C principles, assess the practicality and circularity potential of the proposed strategies. This includes evaluating factors such as the feasibility of material recovery, storage, and reuse based on typical practices in the building sector, as well as alignment with industry standards and policy trends. By integrating these perspectives, the analysis provides a comprehensive evaluation of the scenarios, highlighting their technical, environmental, and strategic implications without relying on direct stakeholder input.

3 FINDINGS

3.1 ANALYSIS OF SELECTED SCENARIOS

The results presented in Figure 3 compare GHG emissions across LCA stages using new materials (Scenario 1) versus reused materials (Scenario 2). The production process (A1-A3) significantly contributes to emissions in Scenario 1, while Scenario 2 demonstrates a significant reduction in this stage due to the use of reused materials and used cut-off method. The transport emissions and construction waste emissions are zero as the Scenario 2 assumes that all materials from the existing building are reused. Thus, the total impact is substantially lower in Scenario 2, representing 94% reduction and highlighting the environmental benefits of material reuse. Additionally, the D module (avoided emissions) offsets emissions in Scenario 1 (new materials) by accounting for the avoided impacts of future material recycling, reusing and using as energy substitution in district heating after incineration process. However, in Scenario 2 (reused materials), the D module does not account for additional avoided emissions, as the credits are already allocated during for primarily used materials. This underscores the long-term carbon savings potential of material reuse compared to the reliance on newly produced materials.

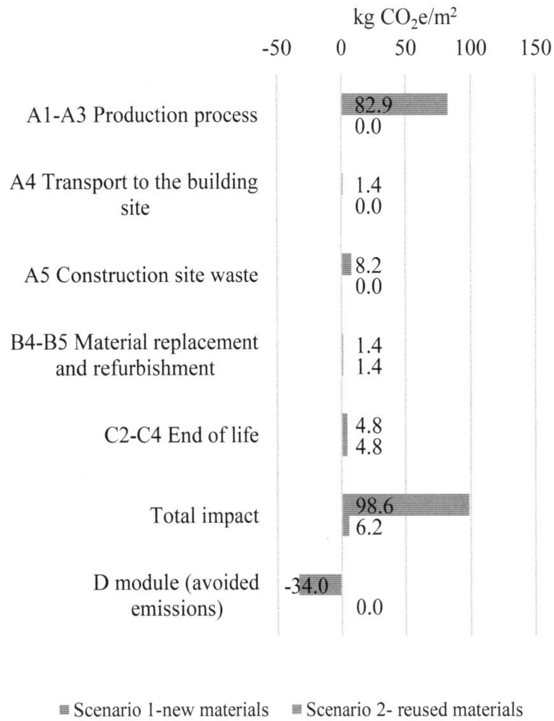

Figure 3: Comparison of GHG emissions between new and reused materials across different LCA stages.

Figure 4 shows GHG emissions across LCA stages for new materials in Scenario 1. The production stage (A1-A3) is the largest contributor, especially for brick, gypsum, and roof ceramic tiles. The only material that is not changed during renovation is concrete installed in the foundation. Transport (A4), construction waste (A5), replacement (B4-B5) and end-of-life (C2-C4) stages have relatively minor emissions. The D module indicates avoided emissions for materials such as timber, wood board and metals, highlighting potential future benefits through recycling or energy recovery. This emphasizes the importance of material choice in minimizing GHG impacts.

https://doi.org/10.52202/080684-0006

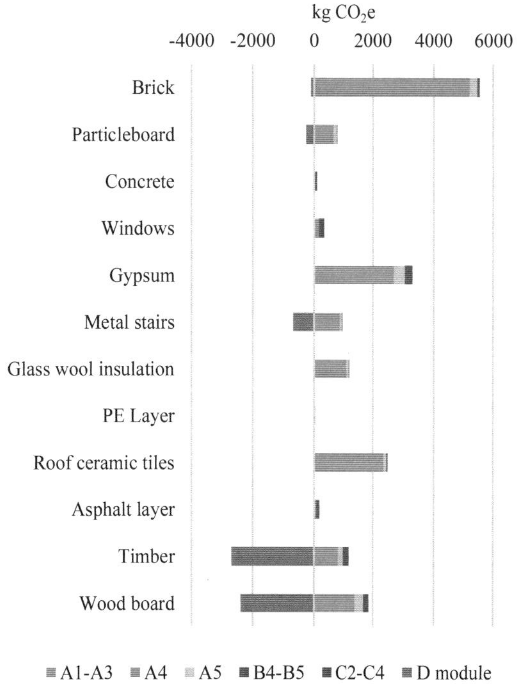

Figure 4: Scenario 1 - GHG emissions across LCA stages for new materials.

Figure 5 presents GHG emissions across LCA stages for all reused materials in Scenario 2. The production stage (A1-A3) has zero emission contribution as the cut-off method is used. Further, the transport emissions (A4) and construction waste emissions (A5) remain zero as the reused materials are inserted from the existing building. The highest emissions are noticed at the end of life (C2-C4) for gypsum, timber, and wood board during the waste processing stage, followed by the replacement of materials (B4-B5). The D module remains zero as the benefits during recycling/reusing/energy recovery processes are counted in the newly produced products.

3.2 INTEGRATION OF THE 10R FRAMEWORK AND STAKEHOLDER CONSIDERATION IN SCENARIOS

To assess the environmental impacts across the building lifecycle in our different scenarios, we have analysed the application of circular strategies, framed within the 10R principles. The primary objective was to identify and map theoretically how different stakeholders contribute to achieving the environmental benefits of material reuse, as opposed to new material use, while integrating the principles of Refuse, Rethink, Reduce, Reuse, Repair, Refurbish, Remanufacture, Repurpose, Recycle, and Recover. Each stakeholder's role is aligned with specific stages of the building lifecycle, ensuring that the implementation of these strategies is both practical and effective.

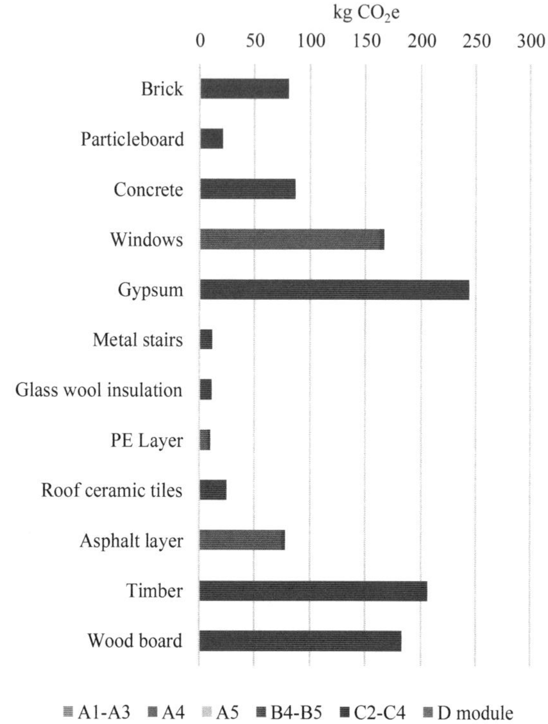

Figure 5: Scenario 2 - GHG emissions across LCA stages for reused materials.

The identification of relevant stakeholders was informed by insights gained from the literature review and an understanding of common practices in the building industry. The focus was on stakeholders typically involved in the design, construction, operation, and demolition of a single-family house. These include architects, designers, contractors, deconstruction specialists, homeowners, material banks, and policymakers. The selection of these stakeholders was based on their established influence over material choices, resource management, waste reduction, and emissions control at each lifecycle stage, as identified in previous studies and industry reports. By considering their roles, it was possible to link specific circular strategies to each stakeholder's influence on the project's environmental outcomes.

3.2.1 Production Phase

Including the principles of Reduce, Reuse, and Repair, the industry can significantly lower energy consumption, minimize waste, and cut GHG emissions, fostering a more sustainable and efficient construction process. Table 2 summarizes these principles, compares emissions across two scenarios, and highlights key stakeholders in sustainable material production. It emphasizes the significant impact of material choices on emissions and underscores the role of stakeholders in promoting practices that reduce the environmental footprint of the building industry.

https://doi.org/10.52202/080684-0006

Table 2: Production phase strategies and impacts.

Element	Description
Principles	**Reduce**: Minimize the need for new materials by choosing alternatives such as reused materials. **Reuse**: Use materials that are still in good condition, recovered from demolition or deconstruction. **Repair**: Restore damaged or deteriorated materials instead of replacing them, extending the life of existing materials.
Scenario 1	Relies on new materials, leading to high emissions due to energy-intensive manufacturing processes and raw material extraction.
Emissions Impact (Scenario 1)	Significant emissions from material production (A1-A3) and raw material extraction, contributing to a large environmental footprint.
Scenario 2	Utilizes reused materials, reducing the need for new production and lowering emissions.
Emissions Impact (Scenario 2)	Major reduction in emissions from material production.
Stakeholders	**Architects and designers**: Incorporate reused and refurbished materials into designs. **Contractors**: Reuse materials locally to minimize transport emissions and contribute to disassembling buildings for reuse. **Deconstruction specialists**: Recover reusable materials from existing structures. **Policymakers**: Support circular production processes and incentivize the use of reused materials to reduce emissions.

3.2.2 Transport Phase

By adopting the principle of Reduce, the industry can lower transportation emissions and enhance sustainability. Table 3 compares the emissions of sourcing materials locally versus long-distance transportation. It also identifies key stakeholders involved in reducing transportation emissions, emphasizing the optimization of logistics and the promotion of local sourcing to minimize the overall carbon footprint of construction projects.

Table 3: Transport phase strategies and impacts.

Element	Description
Principle	**Reduce**: Minimize transportation emissions by sourcing materials locally and optimizing logistics, reducing the need for long-distance transportation.
Impact of Reduce	Reduces emissions associated with material transport, lowers the carbon footprint of construction projects, and minimizes inefficiencies in logistics.
Scenario 1 and 2	Sourcing materials locally, reducing the need for long-distance transport and improving logistics to lower emissions.
Emissions Impact (Scenario 1 and 2)	Substantial reduction in transportation emissions through localized sourcing and improved logistics efficiency.
Stakeholders	**Contractors**: Optimize logistics and coordinate local sourcing to minimize transportation emissions. **Logistics managers**: Ensure efficient delivery and minimize fuel consumption. **Policymakers**: Introduce policies that incentivize local sourcing and transportation efficiency to reduce emissions.
	Architects & Designers: Source local reused products/materials to incorporate in their projects to minimize transportation needs.

3.2.3 Construction Phase

By applying the principles of Reuse, Repair, and Refurbish, emissions and resource consumption can be significantly minimized. Table 4 identifies key stakeholders involved in implementing these principles to reduce waste and resource use. Focusing on reusing materials, repairing components, and refurbishing of materials allows the building industry to lower its impacts and contribute to a CE.

Table 4: Construction phase strategies and impacts-integration.

Element	Description
Principles	**Reuse**: Saves materials from previous projects to reduce demand for new resources and prevent waste. **Repair**: Restores or fix building components to extend their lifespan. **Refurbish**: Updates outdated materials to avoid full replacement.
Impact of Principles	**Reuse**: Reduces emissions from material production by avoiding new manufacturing. **Repair**: Saves resources and minimizes waste by extending material lifespans. **Refurbish**: Promotes resource efficiency and reduces the need for new material production.
Scenario 1	Relies on new materials and potential waste during construction phase
Emissions Impact (Scenario 1)	High emissions from material production (A1-A3) and significant contributions from waste generation during construction (A5).
Scenario 2	Integrates reused materials, repairs, and refurbishments to reduce the need for new production, thereby lowering emissions to zero.
Emissions Impact (Scenario 2)	Zero emissions from production phase due to cut-off method, and no waste in (A5) as the all materials are sourced from existing building.
Stakeholders	**Contractors**: Execute repairs and integrate salvaged materials. **Deconstruction specialists**: Salvage and prepare materials for reuse. **Material banks**: Store and track reusable materials.

3.2.4 Use Phase

Using the principles of Reuse, Repair, Refurbish, and Remanufacture, the building industry can significantly reduce the need for new resources, lower emissions, and minimize waste. These practices ensure that materials are maintained, repurposed, and extended in their lifecycle, contributing to sustainability and the promotion of a CE. Table 5 summarizes these principles, outlines their impacts on emissions, compares different scenarios in terms of sustainability, and identifies the stakeholders involved in maintaining and extending the lifecycle of materials.

https://doi.org/10.52202/080684-0006

Table 5: Use phase strategies and impacts.

Element	Description
Principles	**Reuse**: Ongoing reuse of materials to extend their lifecycle and reduce the need for new resources. **Repair**: Regular repair and maintenance to preserve the functionality of materials. **Refurbish**: Upgrading or improving existing materials to meet modern standards with minimum replacement. **Remanufacture**: Processing old materials into new components to reduce waste and demand for raw materials.
Impact of Principles	**Reuse**: Reduces the need for new material production, lowers emissions and conserves resources by extending the lifecycle of building materials. **Repair**: Minimizes resource consumption and waste generation by repairing rather than replacing. **Refurbish**: Extends the life of existing systems materials, reducing the demand for new materials and if necessary, replacing them with reused options. **Remanufacture**: Supports a CE by converting old materials into usable new products, reducing raw material extraction.
Scenario 1 and 2	Relies on new materials for repairs and replacements, leading to emissions from material production and resource extraction.
Emissions Impact (Scenario 1 and 2)	Emissions due to the production of new replaced materials (B4-B5). Uncertain which materials will be replaced-depending on occupant preferences and the nature of building materials.
Stakeholders	**Homeowners**: Make decisions regarding the replacement, repair, and refurbishment of materials to extend their lifecycle. **Contractors**: Offer consultancy on which materials can be repaired, refurbished, or replaced. They provide technical expertise on how to maintain or upgrade materials to meet modern standards while minimizing waste and emissions. Additionally, they execute repairs and refurbishments to extend the lifespan of materials, contributing to sustainability and efficient resource use. **Deconstruction Specialists:** Provide consultancy on which materials can be salvaged for reuse or remanufacture. They evaluate building elements for potential repurposing, contributing to the circular economy by minimizing waste during demolition and renovation projects. Additionally, they facilitate material recovery at the end of the building's life, ensuring materials are available for reuse or remanufacturing.

3.2.5 End-of-Life Phase

By applying the principles of Repurpose, Recycle, and Recover, construction projects can effectively reduce waste, minimize the demand for new resources, and contribute to a CE. Table 6 summarizes these key principles, their impacts on emissions, and the stakeholders involved in managing materials at the end of their lifecycle.

Table 6: End-of-life strategies and impacts.

Element	Description
Principles	**Repurpose**: Find new uses for building (adaptive reuse) components that are no longer serving their original function, reducing waste. **Recycle**: Process materials into new products to reduce the need for virgin materials. **Recover**: Extract energy or materials from waste to be used in other industries, reducing overall environmental impact.
Impact of Principles	Repurposing extends the life of materials, reducing disposal waste. Recycling reduces the demand for raw materials and conserves resources. Recovery minimizes the environmental footprint by extracting useful energy or materials from waste.
Scenario 1	Focuses on recycling (C2-C4) and energy recovery processes (D module). Wooden materials are firstly incinerated, then used as energy recovery in district heating. While metals are recycled.
Emissions Impact (Scenario 1)	Recycling and energy recovery in Scenario 1 show potential benefits for long-term carbon savings, although it doesn't account for the carbon savings of material reuse.
Scenario 2	Emphasizes the reuse of materials, with some waste processing required for materials such as gypsum, timber, and wood boards.
Emissions Impact (Scenario 2)	Reuse in Scenario 2 reduces the need for new production, contributing to carbon savings.
Stakeholders	**Deconstruction specialists**: Disassemble buildings, recover materials for repurposing, recycling, or energy recovery. **Contractors**: Manage disposal, recycling, or recovery at the end of the lifecycle. **Material banks**: Store and repurpose materials for future use. **Policymakers**: Set standards and offer incentives to support adaptive reuse, recycling, and energy recovery. **Architects and Designers:** Design to adapt existing spaces to new functions and repurpose materials in existing buildings for new design projects.

4 DISCUSSION AND CONCLUSIONS

This study critically evaluated the environmental impacts of materials in building renovation by comparing two scenarios: new materials (Scenario 1) and reused materials (Scenario 2) in a building. The results illustrate the environmental benefits of material reuse, with Scenario 2 demonstrating a significant reduction in GHG emissions, primarily due to the cut-off method applied to reused materials. By incorporating reused materials, Scenario 2 achieved an approximate 94% reduction in GHG emissions compared to Scenario 1, which relies on newly produced materials. This reduction was most evident in the production phase (A1-A3), where new materials, particularly resource-intensive ones contribute substantially to the overall carbon footprint. In contrast, the emissions in Scenario 2 during this phase were zero, as the cut-off method is used.

While the production phase reveals the most significant differences between the two scenarios, other lifecycle stages—A4, A5, B4-B5, and C2-C4 show relatively low GHG emissions. These stages are influenced by various factors, such as transport distances, the choice of waste management practices, and the extent to which materials are replaced or refurbished.

When deciding to use reused materials instead of newly produced ones, several uncertainties arise over the building's lifecycle. For example, the replacement rate is influenced by the nature of materials used (new or reused) during the use phase, as well as by the occupant preferences. Furthermore, transport distances for new materials (Scenario 1) are based on distances by using generic Nordic data. While the transport distance for reused materials is not applied as all materials are derived from the existing building. The construction waste was only calculated for newly produced materials, while for reused materials it remains zero due to utilized materials from the existing building. Thus, prioritizing Scenario 2 over Scenario 1, presents a great potential to prolong the service life of building materials.

The avoided emissions in the D module, accounting for the future recycling, reuse, or energy recovery of materials, further highlight the long-term carbon savings potential of material reuse. For Scenario 1 (new materials), the D module captures the avoided emissions from future material recycling and energy recovery, but these benefits do not exist in Scenario 2, where material reuse has already been credited for the new products.

These findings highlight the importance of strategic material selection of reducing the environmental impact in renovation projects. The choice of materials, whether new or reused, directly influences the GHG emissions associated with a building's lifecycle.

To achieve the full environmental benefits of material reuse, collaborative efforts among stakeholders are essential. The integration of CE principles, such as reuse, repair, and refurbishment, into building renovation requires the active involvement of a diverse group of stakeholders, each with a unique role in influencing material choices and construction practices. Architects and designers, for instance, play a crucial role in specifying and integrating reused materials into building designs, ensuring that the potential environmental benefits of these materials are fully realized. Their decisions during the design phase have a direct impact on the feasibility and effectiveness of material reuse strategies in the construction phase. Furthermore, contractors and deconstruction specialists are vital in sourcing, disassembling, and salvaging materials for reuse, ensuring that valuable materials are not wasted but rather reincorporated into the building cycle. These professionals must also address the technical challenges associated with material reuse, such as ensuring that reused materials meet the required performance standards for safety and durability.

In addition to these industry stakeholders, policymakers play a key role in shaping the broader framework within which material reuse can increase. The development of supportive regulations and financial incentives is essential for encouraging the adoption of reused materials and CE practices across the building sector. Policymakers can foster an environment where the use of reused materials is not only encouraged but made economically viable through incentives such as tax credits, subsidies, or grants. Furthermore, the establishment of regulations that require or incentivize the reuse of materials, as well as the recycling and recovery of materials at the end of their life, will drive the industry towards more sustainable and circular practices. This includes setting standards for material recovery, creating certification systems for reused materials, and promoting the development of material banks to store and track reusable resources.

In conclusion, this study confirms that material reuse is a powerful strategy for reducing the carbon footprint of building renovation projects and advancing the principles of the CE. The integration of reused materials into construction practices offers substantial GHG reductions, particularly in the production phase. However, the potential of material reuse can only be fully realized through collaborative action across all sectors of the building industry, supported by strong policy frameworks that incentivize sustainable practices. By aligning the efforts of architects, designers, contractors, deconstruction specialists, and policymakers, the building industry can significantly contribute to reducing GHG emissions and promoting a more sustainable built environment.

REFERENCES

Allam, A. S., & Nik-Bakht, M. (2023). From demolition to deconstruction of the built environment: A synthesis of the literature. Journal of Building Engineering, 64, 105679. https://doi.org/10.1016/j.jobe.2022.105679

AlJaber, A., Martinez-Vazquez, P., & Baniotopoulos, C. (2023). Barriers and enablers to the adoption of circular economy concept in the building sector: a systematic literature review. Buildings, 13(11), 2778. https://doi.org/10.3390/buildings13112778

Arceo, A. (2023). Uncertainty in Quantification of Material Use and Embodied Greenhouse Gas Emissions of Single-Family Dwellings (Doctoral dissertation, University of Toronto (Canada)).

Dams, B., Maskell, D., Shea, A., Allen, S., Driesser, M., Kretschmann, T., ... & Emmitt, S. (2021). A circular construction evaluation framework to promote designing for disassembly and adaptability. Journal of Cleaner Production, 316, 128122. https://doi.org/10.1016/j.jclepro.2021.128122

Fabbri, M.; Kockat, J.; Jankovic, I.; Sibileau, H. (2023). How to stay warm and save energy: Insulation opportunities in European homes. Buildings Performance Institute Europe (BPIE): Brussels, Belgium.

Futas, N., Rajput, K., & Schiano-Phan, R. (2019). Cradle to Cradle and Whole-Life Carbon assessment–Barriers and opportunities towards a circular economic building sector. In IOP Conference Series: Earth and Environmental Science (Vol. 225, No. 1, p. 012036). IOP Publishing. doi:10.1088/1755-1315/225/1/012036

Gálvez-Martos, J. L., Styles, D., Schoenberger, H., & Zeschmar-Lahl, B. (2018). Construction and demolition waste best management practice in Europe.

https://doi.org/10.52202/080684-0006

Resources, conservation and recycling, 136, 166-178. https://doi.org/10.1016/j.resconrec.2018.04.016

Giorgi, S., Lavagna, M., Wang, K., Osmani, M., Liu, G., & Campioli, A. (2022). Drivers and barriers towards circular economy in the building sector: Stakeholder interviews and analysis of five European countries policies and practices. Journal of cleaner production, 336, 130395. https://doi.org/10.1016/j.jclepro.2022.130395

Honic, M., Kovacic, I., Aschenbrenner, P., & Ragossnig, A. (2021). Material Passports for the end-of-life stage of buildings: Challenges and potentials. Journal of Cleaner Production, 319, 128702. https://doi.org/10.1016/j.jclepro.2021.128702

Illankoon, C., & Vithanage, S. C. (2023). Closing the loop in the construction industry: A systematic literature review on the development of circular economy. *Journal of Building Engineering*, 107362. https://doi.org/10.1016/j.jobe.2023.107362

Joensuu, T., Edelman, H., & Saari, A. (2020). Circular economy practices in the built environment. Journal of cleaner production, 276, 124215. https://doi.org/10.1016/j.jclepro.2020.124215

Kaewunruen, S., Teuffel, P., Donmez Cavdar, A., Valta, O., Tambovceva, T., & Bajare, D. (2024). Comparisons of stakeholders' influences, inter-relationships, and obstacles for circular economy implementation on existing building sectors. Scientific Reports, 14(1), 11046. https://doi.org/10.1038/s41598-024-61863-0

Larsen, V. G., Tollin, N., Sattrup, P. A., Birkved, M., & Holmboe, T. (2022). What are the challenges in assessing circular economy for the built environment? A literature review on integrating LCA, LCC and S-LCA in life cycle sustainability assessment, LCSA. Journal of Building Engineering, 50, 104203. https://doi.org/10.1016/j.jobe.2022.104203

Lee, P. H., Han, Q., & de Vries, B. (2024). Advancing a Sustainable Built Environment: A Comprehensive Review of Stakeholder Promotion Strategies and Dual Forces. Journal of Building Engineering, 110223. https://doi.org/10.1016/j.jobe.2024.110223

Minunno, R., O'Grady, T., Morrison, G. M., & Gruner, R. L. (2020). Exploring environmental benefits of reuse and recycle practices: A circular economy case study of a modular building. Resources, Conservation and Recycling, 160, 104855. https://doi.org/10.1016/j.resconrec.2020.104855

Munaro, M. R., & Tavares, S. F. (2023). A review on barriers, drivers, and stakeholders towards the circular economy: The construction sector perspective. Cleaner and Responsible Consumption, 8, 100107. https://doi.org/10.1016/j.clrc.2023.100107

Nußholz, J. L., Rasmussen, F. N., & Milios, L. (2019). Circular building materials: Carbon saving potential and the role of business model innovation and public policy. Resources, Conservation and Recycling, 141, 308-316. https://doi.org/10.1016/j.resconrec.2018.10.036

Oliveira, J. D., Schreiber, D., & Jahno, V. D. (2024). Circular Economy and Buildings as Material Banks in Mitigation of Environmental Impacts from Construction and Demolition Waste. Sustainability, 16(12), 5022. https://doi.org/10.3390/su16125022

Passoni, C., Marini, A., Belleri, A., & Menna, C. (2021). Redefining the concept of sustainable renovation of buildings: State of the art and an LCT-based design framework. Sustainable Cities and Society, 64, 102519. https://doi.org/10.1016/j.scs.2020.102519

Petrović, B., Eriksson, O., Zhang, X., & Wallhagen, M. (2024). Carbon Assessment of a Wooden Single-Family Building—Focusing on Re-Used Building Products. Buildings, 14(3), 800. https://doi.org/10.3390/buildings14030800

Silvestre, J. D., De Brito, J., & Pinheiro, M. D. (2014). Environmental impacts and benefits of the end-of-life of building materials–calculation rules, results and contribution to a "cradle to cradle" life cycle. Journal of cleaner production, 66, 37-45. https://doi.org/10.1016/j.jclepro.2013.10.028

Soust-Verdaguer, B., Llatas, C., & García-Martínez, A. (2016). Simplification in life cycle assessment of single-family houses: A review of recent developments. Building and Environment, 103, 215-227. https://doi.org/10.1016/j.buildenv.2016.04.014

Xing, W., Tam, V. W., Le, K. N., Hao, J. L., & Wang, J. (2022). Life cycle assessment of recycled aggregate concrete on its environmental impacts: A critical review. Construction and Building Materials, 317, 125950.

Yadav, J., Singh, V. P., & Kumar, A. (2024). Life Cycle Assessment of sustainable building materials. Sustainable Technologies for Energy Efficient Buildings, 64. https://doi.org/10.1016/j.conbuildmat.2021.125950

LUND UNIVERSITY | LTH FACULTY OF ENGINEERING

STRATEGIC DECISION-MAKING IN UNCERTAINTY: INTEGRATING FORWARD-LOOKING SCENARIO PLANNING AND MULTI-CRITERIA ANALYSIS FOR ADAPTIVE REUSE PROJECTS

Brian van Laar[1*], Angela Greco[1], Hilde Remøy[1], and Vincent Gruis[1]

[1]Department of Management in the Built Environment, Faculty of Architecture and the Built Environment, Delft University of Technology, The Netherlands

[*]Corresponding Author: b.r.vanlaar@tudelft.nl

ABSTRACT

Background and aim. Adaptive reuse enhances circularity by repurposing buildings, reducing carbon emissions, and preserving heritage. However, decision-making is complex due to stakeholder conflicts, regulations, and uncertainties. This study introduces an integrated framework combining Cross-Impact Balance (CIB) analysis, the Analytic Hierarchy Process (AHP), and Fuzzy-TOPSIS to support structured, participatory decision-making.

Methods and Data. A mixed-method approach integrates CIB for scenario development, AHP for stakeholder-driven prioritization, and Fuzzy-TOPSIS for ranking reuse scenarios. A hypothetical case study demonstrates the framework's applicability.

Findings. The integration of CIB, AHP, and Fuzzy-TOPSIS provides a structured decision-making approach that enhances scenario coherence, aligns decisions with stakeholder priorities, and improves scenario ranking robustness. The framework enables systematic exploration of adaptive reuse scenarios, ensuring alignment with stakeholder objectives.

Theoretical / Practical / Societal implications. Theoretically, this study advances scenario-based decision-making by integrating scenario development and decision-making approaches, addressing gaps in adaptive reuse decision frameworks. Practically, it provides policymakers, urban planners, and developers with a structured tool to navigate complex decision-making in adaptive reuse projects. Societally, it supports sustainable and inclusive urban development by fostering consistent, long-term strategies that balance environmental, economic, and social considerations.

KEYWORDS: Adaptive Reuse, Circularity, Cross-Impact Balance (CIB) Analysis, Multi-criteria Decision-Making, Scenario Planning,

1 INTRODUCTION

The adaptive reuse of buildings has become a cornerstone strategy for promoting circularity in the built environment (Foster, 2020). By repurposing existing structures, adaptive reuse significantly reduces CO_2 emissions, curbs the extraction of virgin materials, and conserves valuable resources (Shahi et al., 2020). This approach directly supports global sustainability goals and addresses critical urban challenges, including resource scarcity and environmental degradation (Conejos, 2013). However, despite its promise, adaptive reuse decision-making processes remain complex and uncertain (Yung & Chan, 2012). These projects often involve a diverse set of stakeholders with conflicting interests and must navigate a range of regulatory, economic, and technical constraints (Wilkinson, 2014). Consequently, the strategies chosen for adaptive reuse are often limited to short-term perspectives and a narrow set of options, hindering their potential to achieve long-term sustainability and circularity (Greco et al., 2024; Vardopoulos et al., 2021). To address the intricacies and uncertainties of adaptive reuse decision-making, a range of tools and methodologies has been developed (Nedeljkovic et al., 2023). Among these, multi-criteria decision-making (MCDM) models have gained considerable traction for evaluating adaptive reuse projects (Nadkarni & Puthuvayi, 2020). These models provide a structured framework for assessing and comparing alternatives by incorporating multiple criteria (Love et al., 2023). However, existing decision-making approaches tend to adopt either overly generalized frameworks; focused solely on functional reuse, or overly specific ones, which prioritize granular design considerations (van Laar et al., 2024). Both approaches often overlook the broader, long-

 https://doi.org/10.52202/080684-0007

term objectives required to achieve true sustainability and circularity. Furthermore, most frameworks rely either on quantitative methods like cost-benefit analyses (Sanchez et al., 2019), and lifecycle assessments (Foster, 2020), or on generic qualitative approaches to evaluate the feasibility of proposed interventions (Wilkinson, 2014). While these methods offer valuable insights into resource efficiency and financial viability, they often fail to account for nuanced, context-specific factors or integrate forward-looking scenario planning essential for addressing the dynamic nature of urban development.

Scenarios are particularly valuable for adaptive reuse decision-making because they offer comprehensive, future-oriented perspectives. They enable decision-makers to explore how various reuse strategies might perform under different environmental, social, and economic conditions (Weimer-Jehle, 2023). This foresight helps ensure that decisions are robust, flexible, and aligned with long-term sustainability and community goals (Bottero et al., 2022). Normative scenarios, which outline pathways to achieve specific objectives (van Notten et al., 2003), are especially relevant for adaptive reuse. They help stakeholders collaboratively develop a broad range of desirable futures, ensuring that decisions reflect shared values and strategic priorities. Despite their potential, scenario-based methods are underutilized in adaptive reuse (van Laar et al., 2024), often resulting in decisions that fail to anticipate future challenges or opportunities.

There is a pressing need for decision-making frameworks that are both future-oriented and capable of addressing the inherent uncertainty and complexity of adaptive reuse projects. Such frameworks must enable the development of nuanced, context-specific scenarios that incorporate normative objectives, reflect stakeholder priorities, and facilitate the ranking of alternatives based on quantitative and qualitative criteria. To address these gaps, this study introduces an integrated decision-making framework that combines Cross-Impact Balance (CIB) analysis with the Analytic Hierarchy Process (AHP) and Fuzzy-TOPSIS methods.

This research highlights the strength of combining these methodologies into a cohesive, stepwise framework, demonstrating how they can guide adaptive reuse decision-making in a structured yet flexible manner. Using a hypothetical adaptive reuse project, the study showcases how this approach facilitates scenario development, interdependency analysis, and the evaluation of alternatives under uncertainty. The main finding illustrates how these tools can be integrated into a systematic process that supports stakeholders in collaboratively designing and prioritizing adaptive reuse scenarios. This framework offers a practical pathway for addressing the complexity of adaptive reuse while aligning decisions with long-term sustainability and social responsibility goals.

2 BACKGROUND LITERATURE

Scenario development and Multi-Criteria Decision Making (MCDM) analysis are two complementary methodologies extensively used in decision-making processes involving complex systems, such as adaptive reuse. Scenario development enables the exploration of possible futures by considering various uncertainties (Weimer-Jehle, 2023), while MCDM provides a structured framework for evaluating and ranking alternatives against multiple criteria (Saaty, 1990). The integration of these methodologies has gained significant attention, for its potential to improve decision-making outcomes by combining qualitative and quantitative insights (Stewart et al., 2013).

2.1 SCENARIO DEVELOPMENT

Scenario development is a structured approach for envisioning possible future states of a system under uncertainty. Scenarios, described as: coherent, consistent, and plausible descriptions of potential futures, are categorized as exploratory, predictive, or normative (van Notten et al., 2003). Exploratory scenarios examine possible futures based on varying assumptions, aiding in visualizing outcomes. Predictive scenarios forecast likely futures based on current trends, while normative scenarios prescribe pathways to achieve specific goals (van Notten et al., 2003). The normative approach is particularly valuable for adaptive reuse decision-making, where alignment with sustainability goals and community values is essential (Gassner & Steinmüller, 2018). Scenario development methods can be categorized into quantitative, qualitative, and mixed-method approaches, each suited to different needs. Quantitative methods rely on mathematical modeling for precision but often limit stakeholder involvement and are less effective over long-term projections, as they tend to extrapolate trends and may give a false sense of certainty (Amer et al., 2013). In contrast, qualitative methods, like Intuitive Logics (IL), excel in addressing complex issues through nuanced, context-specific insights. However, they can oversimplify systems by focusing on a limited number of uncertainties, potentially overlooking critical factors (Rowe et al., 2017).

Mixed-method approaches effectively combine the strengths of both, integrating data-driven analysis with stakeholder input to foster comprehensive discussions about future possibilities (Symstad et al., 2017). An example is Cross-Impact Balance (CIB) analysis, a semi-quantitative method that uses systems theory to model integrative and holistic scenarios (Weimer-Jehle, 2006). By employing formal logic to structure quantitative and qualitative inputs, CIB generates internally consistent narrative scenarios based on interactions among drivers of change, making it particularly suitable for complex socio-technical systems (Weimer-Jehle, 2023).

2.2 MULTI-CRITERIA DECISION-MAKING

Multi-Criteria Decision-Making (MCDM) methods, such as AHP (Analytic Hierarchy Process), Fuzzy TOPSIS, PROMETHEE, and VIKOR, are widely used for evaluating and ranking alternatives across multiple conflicting criteria (Sahoo & Goswami, 2023). AHP excels in hierarchically structuring complex problems, prioritizing criteria through pairwise comparisons, and aggregating stakeholder preferences into a unified priority structure, fostering consensus while respecting diverse perspectives (Saaty, 1990). Fuzzy TOPSIS, which extends the classical TOPSIS method, effectively manages vagueness and subjectivity by using fuzzy set theory to rank alternatives based on their closeness to ideal and negative ideal solutions (Chen, 2000).

Combining AHP and Fuzzy TOPSIS enhances decision-making by integrating AHP's hierarchical structuring and consistency checks with Fuzzy TOPSIS's capacity for handling uncertainty (Efe, 2016). This hybrid approach is particularly valuable for complex, uncertain environments, as it provides a structured yet flexible evaluation framework (Mathew et al., 2020). Such integrations have been applied successfully in fields like supply chain management (Patil & Kant, 2014), and urban planning (Dang et al., 2019), demonstrating their versatility and effectiveness. While the combination of AHP and Fuzzy TOPSIS effectively ranks uncertain alternatives based on stakeholder preferences, it often relies on externally provided options, highlighting the need for an integrated approach that develops and ranks scenarios concurrently.

2.3 INTEGRATION OF SCENARIO DEVELOPMENT AND MULTI-CRITERIA DECISION ANALYSIS

The integration of scenario development and MCDM addresses the limitations of each methodology when applied independently. Scenario development often lacks a structured mechanism to prioritize options within each scenario, while MCDM can be overly deterministic without considering the broader context of future uncertainties (Sahoo & Goswami, 2023). By combining these methods, decision-makers can evaluate the robustness of alternatives across different scenarios, incorporate qualitative and quantitative dimensions of uncertainty, and enhance stakeholder engagement by providing a more holistic view of decision impacts (Sahoo & Goswami, 2023).

Numerous frameworks integrate scenario development with multi-criteria decision-making (MCDM), typically following one of two approaches. The scenario-driven MCDM approach develops scenarios first and applies MCDM to rank alternatives within each scenario (Bottero et al., 2022). In contrast, the MCDM-driven approach uses MCDM criteria to shape scenarios, aligning them with decision priorities (Della Spina, 2020). These frameworks have been applied in various fields, including urban planning, supply chain management, and engineering .

However, these studies often face limitations, such as relying on a limited number of scenarios that fail to capture the full range of possibilities. Many frameworks lack consistency calculations, reducing the coherence and realism of the scenarios (Weimer-Jehle, 2006). Additionally, there is an overemphasis on predictive scenarios and mathematical models, prioritizing quantitative precision over qualitative insights and stakeholder perspectives (Weimer-Jehle, 2023). These shortcomings diminish the robustness and practical applicability of the scenarios in addressing complex challenges.

The Cross-Impact Balance (CIB) method overcomes these challenges by generating numerous consistent and plausible scenarios through a combination of qualitative and quantitative inputs (Weimer-Jehle, 2023). This makes it particularly effective for exploring complex systems. However, CIB has not been fully integrated with MCDM methods like AHP or Fuzzy TOPSIS, which excel at prioritizing and ranking alternatives. Combining these approaches offers significant potential, enabling the systematic creation, evaluation, and prioritization of scenarios within a unified framework. In a participatory setting, this integration enhances stakeholder engagement by involving them in the entire process, from scenario development to ranking, ensuring scenarios are aligned with diverse preferences and easing the adoption of the chosen scenario through consensus and trust in the outcomes.

3 STEPWISE APPROACH FOR COMBINING CROSS-IMPACT BALANCE ANALYSIS, AHP AND THE FUZZY TOPSIS METHODS

This section outlines a structured, multi-step framework tailored for decision-making in normative, uncertain, and complex contexts such as adaptive reuse projects (Figure 1). By integrating Cross-Impact Balance (CIB) analysis, the Analytic Hierarchy Process (AHP), and Fuzzy-TOPSIS methodologies, this approach effectively addresses the uncertainties inherent in adaptive reuse. It enables stakeholders to collaboratively assess, develop, and prioritize reuse scenarios, demonstrating its application through a hypothetical example of an adaptive reuse project.

3.1 STEP 1: DEFINE THE AIM AND OBJECTIVES

The first step establishes the foundation for the decision-making process by ensuring a clear understanding of the project's scope and goals. To create normative scenarios; future pathways that are achievable (van Notten et al., 2003), this step focuses on defining objectives that will guide subsequent scenario development. Stakeholders collaborate to articulate the overarching goal and themes, identify desired objectives, and determine the criteria

https://doi.org/10.52202/080684-0007

necessary to evaluate progress toward these objectives. To balance adequacy and completeness in the scenario analysis, it is recommended to include 9–15 objectives for the development of descriptors and variants in Step 2, in line with the methodological guidelines of Weimer-Jehle, (2023). By addressing these critical elements, this step provides a structured and goal-oriented process fostering clarity, alignment, and a shared vision among all stakeholders.

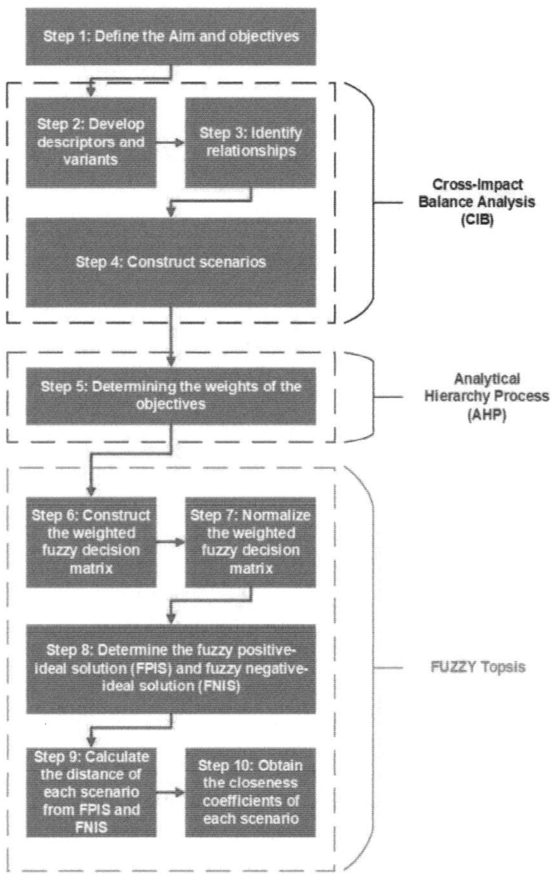

Figure 1: Stepwise approach for combining cross-impact balance analysis (CIB), AHP and the Fuzzy TOPSIS methods

3.2 STEP 2: DEVELOP DESCRIPTORS AND VARIANTS

The CIB method uses systems theory and formal logic to create internally consistent scenarios based on interacting drivers of change, integrating both qualitative and quantitative inputs (Weimer-Jehle, 2006). A key step in this process is identifying descriptors; 'critical factors defining the system' and their associated variants, which represent specific states these factors can assume (Weimer-Jehle, 2023). Descriptors should be developed at a high aggregation level (Weimer-Jehle, 2023), with each descriptor representing one objective, that can be supported by related criteria and / or a narrative that explains the descriptor's role and significance within the system. Variants then enable systematic exploration of scenarios by capturing the range of possible outcomes for each descriptor. For example, in adaptive reuse projects, "Environmental impact" could be a descriptor for the objective: 'Reducing environmental impact of the building', with variants such as "Low," "Medium," and "High." Stakeholders are encouraged to assign descriptive names and narratives to variants for clarity and effective communication, keeping 2–4 variants per descriptor as recommended by (Weimer-Jehle, 2023). The CIB analysis requirements of completeness (descriptor variants must cover all possible futures), mutual exclusivity (each development aligns with only one variant), and absence of overlap (variants of different descriptors must address distinct topics) should also be taken into account when developing variants (Weimer-Jehle, 2023).

Although the CIB methodology supports variants with various characteristics (ordinal, nominal, or ratio) this paper focuses on descriptors with ordinal measurement scales. For instance, "user demand" as a descriptor might include ordinal variants like "Low," "Medium," and "High," reflecting their ranked importance. This approach simplifies the system, making it possible to translate qualitative ordinal variants into linguistic variables essential for integration with the Fuzzy TOPSIS method. Using ordinal descriptors ensures consistency in both the CIB analysis and fuzzy TOPSIS methods, enabling structured evaluation of interactions and their influence on adaptive reuse scenario outcomes.

3.3 STEP 3: IDENTIFY RELATIONSHIPS BETWEEN DESCRIPTORS AND VARIANTS

Identifying the interrelationships between descriptor variants is critical in Cross-Impact Balance (CIB) analysis, as it ensures the logical coherence and plausibility of the scenarios generated (Weimer-Jehle, 2023). These interrelationships capture how one variant influences or is influenced by another, reflecting the underlying dynamics of the system. Without this step, the analysis risks inconsistencies or contradictions, undermining the reliability of the scenarios (Weimer-Jehle, 2023).

To identify these relationships, the scale recommended by (Weimer-Jehle, 2006) provides a structured and systematic approach. This scale uses a range from -3 to +3 to denote the influence of one variant on another: +3 indicates a strong positive impact, 0 signifies no impact, and -3 represents a strong negative impact. These values are assigned within a cross-impact matrix, ensuring all potential interactions are considered (Table 1). This elicitation of data can be conducted in a participatory group setting with stakeholders, fostering collaboration and shared understanding. Alternatively, other methods such as expert surveys (Weimer-Jehle et al., 2012), Delphi techniques (Tori et al., 2023), or literature reviews (Weimer-Jehle, 2023), can be employed to gather the required input systematically. By following this method, the CIB process produces scenarios that are not only internally consistent but also reflective of the real-world, project-specific dynamics among the factors studied (Weimer-Jehle, 2023).

Table 1: Example of a cross-impact balance judgement section

	Political and Community support		
Environmental Impact	High	Medium	Low
Low	3	2	-2
Medium	2	1	-1
High	-3	-1	2

-3 Strongly hindering   0 Neutral   Strongly promoting +3

3.4 STEP 4: CONSTRUCT SCENARIOS

In Cross-Impact Balance (CIB) analysis, constructing scenarios involves generating combinations of descriptor variants and assessing their internal consistency. The consistency of each scenario is determined using the impact sum, which quantifies the cumulative influence of all variants in a scenario on one another (Weimer-Jehle, 2006). This sum, derived from the cross-impact matrix, indicates whether the combination of variants aligns with the specified interdependencies among descriptors. Without considering interdependencies, any combination of descriptor variants could form a scenario. While the CIB methodology tolerates marginal inconsistencies due to the qualitative nature of input data (Weimer-Jehle, 2023), high inconsistency values suggest contradictions, whereas low values indicate internally consistent and plausible scenarios. To determine the acceptable inconsistency threshold the following Equation (1) can be used (Weimer-Jehle, 2023), in which IC_s is the acceptable inconsistency value and n is the number of descriptors:

$$IC_s \approx \frac{1}{2}\sqrt{n-1} \quad (1)$$

The calculation process can be facilitated using the ScenarioWizard software[1], which automates the assessment of consistency across all possible combinations of descriptor variants. The software produces a scenario tableau as an outcome of this calculation. The tableau displays all consistent scenarios, highlighting the selected variants for each descriptor, and serves as input for the decision analysis in Steps 6–10. This structured representation enables researchers and stakeholders to identify and analyse the most plausible scenarios, ensuring that the results are both rigorous and actionable. By employing this method, CIB analysis supports the systematic exploration of potential futures and aids decision-making processes based on robust, internally consistent scenarios.

3.5 STEP 5: DETERMINING THE WEIGHTS OF THE OBJECTIVES

To pick the most appropriate scenario for a project, it is important that the preferences of the stakeholders are reflected in the outcomes of the decision model. The Analytical Hierarchy Process (AHP) is a robust method for multicriteria decision-making that ensures decisions align with stakeholder priorities through a structured stepwise approach. The process begins with pairwise comparisons, where stakeholders evaluate the relative importance of the objectives from step 2, using Saaty's 9-point Likert scale, ranging from equal importance (1) to extreme superiority (9) (Saaty, 1990). These comparisons populate a matrix that reflects the relative weights of each objective, following Equation (2).

$$M = \begin{bmatrix} a_{11} & \cdots & a_{1n} \\ \vdots & \ddots & \vdots \\ a_{n1} & \cdots & ann \end{bmatrix} \left\{ \frac{1/9 \leq a_{i,j} \leq 9\, if\ i \sim j}{a_{i,j} = 1,\ if\ i\ =\ j} \right\} \quad (2)$$

Once the matrices are completed, weights are calculated by normalizing the values within each column to reflect the relative importance of the objective. This involves summing the values in each column $\sum a_{kj}$ for $\boldsymbol{n}$, dividing each objective a_{kj} by the total of its column, and then averaging the normalized scores for each row. The weight for each objective, is computed using Equation (3):

$$w_k = \frac{1}{n}\sum_{j=1}^{n}\frac{a_{kj}}{\sum_{i=1}^{n} a_{ij}}\ for\ k\ =\ 1,2,3,\ldots,n \quad (3)$$

where n is the number of objectives. This structured normalization process aggregates the scores to derive the final weights, ensuring a systematic approach that integrates both qualitative judgments and quantitative analysis into the decision-making framework.

The AHP then employs the Consistency Ratio (CR) to assess the coherence of decision-makers' judgments. The CR is determined by comparing the Consistency Index (CI) to the Random Index (RI), which represents the average consistency expected by chance for matrices of a given size (Saaty, 1990). If the CR exceeds a commonly accepted threshold, typically 0.10, it signals that the judgments are not adequately consistent and may need to be revised or reevaluated to ensure reliability.

For instance, if a stakeholder considers 'Environmental Impact' more important than: 'Cost', and 'Cost' more important than: 'Social Impact', it is logically expected that 'Environmental Impact' would also be prioritized over 'Social Impact'. The Consistency Ratio (CR) quantifies the coherence of such pairwise comparisons. A CR below 0.10 indicates a satisfactory level of consistency in the judgments, while a CR exceeding 0.10 suggests inconsistencies that require revision. This evaluation should be performed independently for each matrix and stakeholder to ensure precision and reliability

[1] https://www.cross-impact.org/english/CIB_e_ScW.htm

https://doi.org/10.52202/080684-0007

in the decision-making process. The CR is calculated using Equation (4):

$$CR = \frac{CI}{RI} \tag{4}$$

The Consistency Index (CI) is a key metric in the AHP used to measure the logical coherence of judgments in pairwise comparison matrices, while the Random Index (RI) represents the average CI derived from 500 reciprocal matrices populated with values from Saaty's fundamental 1–9 scale (Saaty, 1990). The RI varies based on the number of criteria in a matrix, as outlined in Table (2). The CI is calculated using Equation (5):

$$CI = \frac{\lambda_{max} - n}{n - 1} \tag{5}$$

where λ_{max} is the maximum eigenvalue of the comparison matrix, and n is the number of objectives. To compute the eigenvalue for a pairwise comparison matrix in AHP, multiply the pairwise comparison matrix M by the priority vector w using Equation (6):

$$M * w = \lambda max * w \tag{6}$$

Here, w represents the normalized priority weights of the criteria. For each row i in the resulting matrix $A * w$ computed by using Equation (7):

$$\lambda_i = \frac{(A * w)_i}{w_i} \tag{7}$$

Where $(A * w)_i$ is the *i-th* element of the resulting vector, and w_i is the *i-th* element of the priority vector. The maximum eigenvalue of the comparison matrix is then calculated by taking the average value of all λ_i, Where n is the number of objectives using Equation (8):

$$\lambda_{max} = \frac{\sum_{i=1}^{n} \lambda_i}{n} \tag{8}$$

The consistency check is essential to ensure that judgments are logically consistent, as inconsistencies can compromise the validity of the decision-making process, leading to unreliable outcomes. This process reinforces robust decision-making by encouraging stakeholders to critically evaluate their judgments, ensuring coherence and reliability throughout the analysis.

Table 2: Random Index (RI) for different numbers of objectives (Saaty, 1990)

Number of criteria	Random Index (RI)
2	0
3	0.58
4	0.90
5	1.12

3.6 STEP 6: CONSTRUCT THE WEIGHTED FUZZY DECISION MATRIX

Following the elicitation of decision-makers' preferences through the AHP method, the subsequent step involves conducting decision analysis utilizing the Fuzzy TOPSIS method. The Fuzzy TOPSIS method is an extension of the traditional Technique for Order of Preference by Similarity to Ideal Solution (TOPSIS) that incorporates fuzzy set theory to handle uncertainty and vagueness in decision-making (Chen, 2000). This approach is particularly useful when preferences are expressed in qualitative terms, such as linguistic variables, which are subjective and imprecise by nature, such as with scenarios in the CIB analysis. Fuzzy sets enable the representation of linguistic variables such as "Low," "Medium," and "High" as fuzzy numbers. Among the different forms of fuzzy sets, triangular fuzzy numbers (TFNs) are most commonly used due to their simplicity and computational efficiency (Chen, 2000). A triangular fuzzy number is represented as $\widetilde{A}_{ij} = \left(l_{ij}, m_{ij}, u_{ij}\right)$ where l is the lower bound, m is the most likely value, and u is the upper bound, forming a triangular membership function. A fuzzy number $\widetilde{A}$ on R is defined as a triangular fuzzy number (TFN) it its membership function $\mu_{\widetilde{A}}(x) : R \rightarrow [0,1]$ is expressed as follows in Equation (9):

$$\mu_{\tilde{A}}(x) = \begin{cases} \frac{x - l}{m - l}, & l \leq x \leq m \\ \frac{u - x}{u - m}, & m \leq x \leq u \\ 0, & Otherwise \end{cases} \tag{9}$$

To convert linguistic variables into fuzzy numbers, a predefined fuzzy scale should be developed which assigns specific TFNs to each linguistic term based on expert judgment or domain knowledge. This allows qualitative assessments to be transformed into quantitative data that can be processed within the Fuzzy TOPSIS framework, enabling a more nuanced and flexible evaluation of scenarios under uncertainty. A fuzzy scale is employed to transform the qualitative ordinal variants from the consistent scenarios in Step 4 into fuzzy numbers, which are subsequently used to construct the decision matrix.

Following the fuzzification process, construct the fuzzy pairwise decision matrix by first calculating the relative importance of each objective (w_j) following step 5, using Equation (10):

$$\tilde{V} = \tilde{A}_{ij} \times w_j = \left(l_{ij} * w_j * m_{ij} * w_j * u_{ij} * w_j\right) \tag{10}$$

The overall weighted fuzzy decision matrix can then be constructed using Equation (11), Where: m is the number of scenarios, and n is the number of objectives.

$$\tilde{V} = \begin{bmatrix} (l_{11} * w_1, m_{11} * w_1, u_{11} * w_1) & \cdots & (l_{1n} * w_n, m_{1n} * w_n, u_{1n} * w_n) \\ \vdots & \ddots & \vdots \\ (l_{m1} * w_1, m_{m1} * w_1, u_{m1} * w_1) & \cdots & (l_{mn} * w_n, m_{mn} * w_n, u_{mn} * w_n) \end{bmatrix} \tag{11}$$

3.7 STEP 7: NORMALIZE THE WEIGHTED FUZZY DECISION MATRIX

To normalize the weighted fuzzy decision matrix $\widetilde{V}$, each objective $\widetilde{V}_{ij} = (l_{ij}, m_{ij}, u_{ij})$ is normalized based on the type of objective (benefit or cost).
For benefit objectives (higher values are preferred) Equation (12) can be used:

$$\tilde{R}_{ij} = (\frac{l_{ij}}{u_j}, \frac{m_{ij}}{u_j}, \frac{u_{ij}}{u_j}) \quad (12)$$

For the cost criteria (lower values are preferred) Equation (13) is used:

$$\tilde{R}_{ij} = (\frac{l_j^-}{u_{ij}}, \frac{m_j^-}{m_{ij}}, \frac{u_j^-}{l_{ij}}) \quad (13)$$

Where u_j^+ is the maximum upper bound for the $j-th$ objective, and l_j^- is het minimum lower bound for the $j-th$ objective.

3.8 STEP 8: DETERMINE THE FUZZY POSITIVE-IDEAL SOLUTION (FPIS) AND FUZZY NEGATIVE-IDEAL SOLUTION (FNIS)

In the Fuzzy TOPSIS method, the FPIS (Fuzzy Positive Ideal Solution) represents the optimal fuzzy values for each objective, while the FNIS (Fuzzy Negative Ideal Solution) reflects the least desirable outcomes. These are determined by identifying the best and worst fuzzy scores across all scenarios for each objective. Scenarios are ranked based on their proximity to the FPIS and distance from the FNIS, with the closest scenario to the FPIS and farthest from the FNIS considered the best choice.
If x_{ij} represents the fuzzy evaluation of the i-th alternative with respect to the j-th objective, the FPIS for each criterion can be represented as:

$$A_j^* = \begin{cases} max_i x_{ij} \text{ if the objective is beneficial} \\ min_i x_{ij} \text{ if the objective is non} - \text{beneficial} \end{cases} \quad (14)$$

Conversely, If x_{ij} represents the fuzzy evaluation of the i-th alternative with respect to the j-th objective, the FNIS for each objective can be represented as:

$$A_j^- = \begin{cases} min_i x_{ij} \text{ if the objective is beneficial} \\ max_i x_{ij} \text{ if the objective is non} - \text{beneficial} \end{cases} \quad (15)$$

3.9 STEP 9: CALCULATE THE DISTANCE OF EACH SCENARIO FROM FPIS AND FNIS

The distances from each scenario to the FPIS (d_i^*) and FNIS (d_i^-) are calculated using the fuzzy distance measure: Euclidian distance using Equation (16).
The distance d between two fuzzy numbers $\widetilde{A}_1 = (l_1, m_1, u_1)$ and $\widetilde{A}_2 = (l_2, m_2, u_2)$ is:

$$d(\tilde{A}_1, \tilde{A}_2) = \sqrt{\frac{1}{3}} = [(l_1 - l_2)^2 + (m_1 - m_2)^2 + (u_1 - u_2)^2] \quad (16)$$

To calculate the distances from FPIS and FNIS to each scenario the following Equations (17&18) can be used:

Distance from FPIS:

$$(d_i^+) : d_i^+ \sqrt{\sum_{j=1}^{n} (\tilde{R}_{ij} - \tilde{A}_j^+)^2} \quad (17)$$

Distance from FNIS

$$(d_i^-) : d_i^- \sqrt{\sum_{j=1}^{n} (\tilde{R}_{ij} - \tilde{A}_j^-)^2} \quad (18)$$

Here, *n* is the number of objectives, x_{ij} is the fuzzy score of the *i-th* scenario on the *j-th* objective, and A_j^+ is the score of the FPIS on the *j-th* objective, and A_j^- is the score of the FNIS on the *j-th* objective. Using these distances, each scenario's relative closeness to the ideal solution is calculated, which is used to rank the scenarios. The alternative with the shortest distance to the FPIS and the longest distance from the FNIS is considered the optimal choice.

3.10 STEP 10: OBTAIN THE CLOSENESS COEFFICIENTS OF EACH SCENARIO

In the Fuzzy TOPSIS method, the closeness indicator is a metric for ranking scenarios by measuring their proximity to the Fuzzy Positive Ideal Solution (FPIS) and their distance from the Fuzzy Negative Ideal Solution (FNIS). This ranking provides decision makers with a clear understanding of which scenario best aligns with their preferences and objectives. By summarizing each scenario's performance across all objectives, the closeness indicator supports informed, consensus-driven decisions, highlighting not only the best options but also how closely each one approaches the ideal conditions. The closeness indicator is calculated by using the following Equation (19):

$$CC_i = \frac{d_i^-}{d_i^+ + d_i^-} \quad (19)$$

Where $\boldsymbol{d_i^+}$ is the distance of the *i-th* alternative from the FPIS, and $\boldsymbol{d_i^-}$ is its distance from the FNIS. The closeness indicator, $\boldsymbol{CC_i}$, ranges from 0 to 1, where a value closer to 1 indicates that the scenario is closer to the FPIS and farther from the FNIS, making it a more preferable option.

4 HYPOTHETICAL EXAMPLE

The application of the newly introduced mixed-method approach is demonstrated using a hypothetical example of an adaptive reuse project.

https://doi.org/10.52202/080684-0007

4.1 STEP 1: DEFINE THE AIM AND OBJECTIVES

For the hypothetical example we have developed the following aim, objectives and criteria. For the selection of the objectives and criteria we have drawn inspiration from van Laar et al., (2024), who conducted an extensive literature review on criteria and objectives in the decision-making process of adaptive reuse. For practical reasons we have limited the number of objectives to five.

Table 3: The project aim and objectives of the hypothetical example

Project Aim	The aim of this project is to adaptively reuse an existing building to meet functional, environmental, and social needs while preserving its historical, significance.
Objectives	O1) To increase social impact O2) To reduce environmental impact O3) To reduce cost O4) To improve the physical quality and durability of the building O5) To preserve the historic and cultural value of the building

4.2 STEP 2: DEVELOP DESCRIPTORS AND VARIANTS

Based on the objectives chosen, comprehensive descriptors and variants were developed that included names, description, objective and criteria (Appendix A). For all descriptors, 3 ordinal variants were drawn up: a strong variant in which the objective within the descriptor is definitely reached, a medium variant in which the objective is partially reached, and a weak variant in which the objective is not reached.

The same linguistic variables were chosen for each objective to simplify the FUZZY translation in Step 6.

An example for the descriptor Social impact is provided in Table (4).

Table 4: The descriptor: "Social Impact" and its corresponding variants

Descriptor: Social Impact		**Variants**
Objective	To increase social Impact	**A1: Social Heaven (strong variant)** The adaptive reuse project enhances social impact by addressing socio-economic factors like house prices, gentrification, and perceived safety while boosting neighbourhood liveability. It fosters social cohesion by serving as a community hub and improves surrounding public spaces.
Criteria	Social cohesion Public spaces Liveability Socio-economic conditions	**A2: Socially Acceptable (medium variant)** The adaptive reuse project avoids negative socio-economic impacts, with some focus on enhancing public spaces and liveability. While not central to fostering community, it offers spaces for social interaction. **A3: Socially Limited (weak variant)** The adaptive reuse project negatively impacts socio-economic conditions, potentially raising housing prices and driving gentrification. It fails to improve liveability, public spaces, or social cohesion.

4.3 STEP 3: IDENTIFY RELATIONSHIPS BETWEEN DESCRIPTORS AND VARIANTS

We have mapped the interactions between all descriptor-variant combinations using the scale from Weimer-Jehle, (2006). This resulted in the following Cross-impact balance matrix (Table 5):

Table 5: The completed CIB matrix for the hypothetical example

CIB Matrix	**A) Social Impact**			**B) Environmental Impact**			**C) Cost**			**D) Physical quality**			**E) Historic and Cultural value**		
	A1	A2	A3	B1	B2	B3	C1	C2	C3	D1	D2	D3	E1	E2	E3
A) Social Impact:															
A1) Social heaven				3	2	-2	1	0	0	0	0	0	3	2	-2
A2) Socially acceptable				2	1	-1	0	0	0	0	0	0	2	1	-1
A3) Socially limited				-2	-1	2	-1	0	1	0	0	0	-2	0	2
B) Environmental Impact:															
B1) Sustainability heaven	2	1	0				-3	-1	3	0	0	0	0	-1	0
B2) Environmentally friendly	0	0	0				-1	0	1	0	0	0	1	1	-1
B3) Environmentally unfriendly	-2	-1	1				-1	1	-1	0	0	0	1	0	0
C) Cost:															
C1) Cost Efficient	1	1	-1	-1	2	1				3	2	1	-2	-1	2
C2) Moderately costly	0	0	0	-1	2	1				1	0	-1	1	1	0
C3) Very costly	-2	-1	1	2	-2	2				2	1	-2	-3	-1	3
D) Physical quality:															
D1) Strong and Durable	1	0	0	3	2	0	2	-1	-2				3	2	-2
D2) Sufficiently durable	0	0	0	2	1	0	1	1	-1				2	1	0
D3) Poor building quality	0	0	1	-2	0	1	-2	1	2				-3	-1	2
E) Historic/Cultural value															
E1) Preserving History	2	1	-1	-1	0	0	-2	-1	2	3	2	-2			
E2) Attention to history	1	0	0	0	0	0	-1	0	1	2	1	-1			
E3) Ignoring history	-2	-1	2	1	0	0	1	0	-1	-1	0	2			
Impact Sum	6	3	-2	4	6	-1	-2	-3	3	6	4	-1	4	2	-2

https://doi.org/10.52202/080684-0007

4.4 STEP 4: CONSTRUCT SCENARIOS

The consistency analysis was performed using the ScenarioWizard software, with a consistency value of 1 following Equation (1). This resulted in 4 consistent scenarios that are included for decision analysis (Figure 2). Each scenario consists of a consistent combination of variants that is characterised by strong (green), medium (yellow), or weak (red) in relation to the objective of the descriptor.

Scenario No. 1	Scenario No. 2	Scenario No. 3	Scenario No. 4
Social Impact: Social heaven	Social Impact: Socially limited		
Environmental Impact: Sustainability heaven	Environmental Impact: Environmentally unfriendly	Environmental Impact: Sustainability heaven	Environmental Impact: Environmentally unfriendly
Cost: Very costly	Cost: Moderately costly	Cost: Very costly	Cost: Moderately costly
Physical quality: Strong and Durable			Physical quality: Poor building quality
Historic and Cultural value: Preserving History	Historic and Cultural value: Attention to history	Historic and Cultural value: Ignoring history	

Figure 2: The scenario tableau for the hypothetical example

4.5 STEP 5: DETERMINING THE WEIGHTS OF THE OBJECTIVES

We determined the weights of the objectives through the AHP methodology by using the Saaty's 9-point Likert scale (Saaty, 1990). The relative importance of the objectives is displayed in the pairwise comparison matrix:

$$M = \begin{matrix} \\ O_1 \\ O_2 \\ O_3 \\ O_4 \\ O_5 \end{matrix} \begin{bmatrix} O_1 & O_2 & O_3 & O_4 & O_5 \\ 1 & (3) & (5) & (7) & (9) \\ (0.33) & 1 & (3) & (5) & (7) \\ (0.2) & (0.33) & 1 & (3) & (5) \\ (0.14) & (0.2) & (0.33) & 1 & (3) \\ (0.11) & (0.14) & (0.2) & (0.33) & 1 \end{bmatrix} \quad (20)$$

The pairwise comparison matrix was normalized by dividing each entry by the sum of its column using Equation (3), which results in the normalized pairwise comparison matrix: Equation (21).

$$M_n = \begin{matrix} \\ O_1 \\ O_2 \\ O_3 \\ O_4 \\ O_5 \end{matrix} \begin{bmatrix} O_1 & O_2 & O_3 & O_4 & O_5 \\ (0.520) & (0.545) & (0.476) & (0.437) & (0.400) \\ (0.173) & (0.182) & (0.286) & (0.312) & (0.311) \\ (0.104) & (0.061) & (0.095) & (0.188) & (0.222) \\ (0.073) & (0.036) & (0.032) & (0.062) & (0.133) \\ (0.057) & (0.027) & (0.019) & (0.041) & (0.089) \end{bmatrix} \quad (21)$$

The relative weight w_k of each objective was calculated by averaging the normalized values across each row. Table (6). presents the final weights. The objective "To reduce cost" is the most important, while the objective "To increase social impact" is the least important.

Table 6: The weights for each objective following AHP

Objective (O_n)	Weight (w_k)
To reduce cost (O_3)	0.476
To preserve the historic & cultural value of the building (O_5)	0.253
To improve the physical quality/ durability of the building (O_4)	0.134
To reduce environmental impact (O_2)	0.067
To increase social impact (O_1)	0.045

To ensure the judgments were consistent, the largest eigenvalue was computed using Equation (22) along with the Consistency Index (CI); Equation (23) and Consistency Ratio (CR); Equation (24):

Largest Eigenvalue:

$$\lambda_{max} = \frac{\lambda_1 + \lambda_2 + \lambda_3 + \lambda_4 + \lambda_5}{5} =$$

$$\lambda_{max} = \frac{5.83 + 5.78 + 5.51 + 5.42 + 4.07}{5} = 5.333 \quad (22)$$

Consistency Index (CI):

$$CI = \frac{\lambda_{max} - n}{n - 1} = \frac{5.333 - 5}{5 - 1} = 0.083 \quad (23)$$

The Random Index (RI) value with 5 objectives is: 1.12 (Table 2).

Consistency Ratio (CR)

$$CR = \frac{CI}{RI} = \frac{0.083}{1.12} = 0.074 \quad (24)$$

The Consistency Ratio (CR) = 0.074 is below the threshold of 0.1, indicating that the pairwise comparison matrix is acceptably consistent.

4.6 STEP 6: CONSTRUCT THE WEIGHTED FUZZY DECISION MATRIX

The decision matrix D with linguistic variables is constructed based on the outcome of the consistency analysis from Step 4 (Figure 2), with S_n being the scenarios; Equation (25).

$$D = \begin{matrix} \\ O_1 \\ O_2 \\ O_3 \\ O_4 \\ O_5 \end{matrix} \begin{bmatrix} S_1 & S_2 & S_3 & S_4 \\ Strong & Weak & Weak & Weak \\ Strong & Weak & Strong & Weak \\ Weak & Medium & Weak & Medium \\ Strong & Strong & Strong & Weak \\ Strong & Medium & Weak & Weak \end{bmatrix} \quad (25)$$

To transform the linguistic decision matrix into a fuzzy matrix, the following conversion scale is used that incorporates triangular fuzzy numbers (Table 7).

Table 7: The linguistic variable conversion table

Linguistic variable	Corresponding triangular fuzzy numbers (l, m, u)
Weak	(1,3,5)
Medium	(3,5,7)
Strong	(5,7,9)

After conversion the following fuzzy decision matrix D_f was constructed using Equation (26):

https://doi.org/10.52202/080684-0007

$$D_f = \begin{matrix} O_1 \\ O_2 \\ O_3 \\ O_4 \\ O_5 \end{matrix} \begin{bmatrix} S_1 & S_2 & S_3 & S_4 \\ (5,7,9) & (1,3,5) & (1,3,5) & (1,3,5) \\ (5,7,9) & (1,3,5) & (5,7,9) & (1,3,5) \\ (1,3,5) & (3,5,7) & (1,3,5) & (3,5,7) \\ (5,7,9) & (5,7,9) & (5,7,9) & (1,3,5) \\ (5,7,9) & (3,5,7) & (1,3,5) & (1,3,5) \end{bmatrix} \quad (26)$$

To arrive at the weighted fuzzy decision matrix the relative weights of the objectives w_k were multiplied with the triangular fuzzy numbers (Table 8):

4.7 STEP 7: NORMALIZE THE WEIGHTED FUZZY DECISION MATRIX

Using Equation (12) we can then normalize the weighted fuzzy decision matrix (Table 9).

Table 8: The weighted fuzzy decision matrix for the hypothetical example

	Scenario 1 (S_1)	Scenario 2 (S_2)	Scenario 3 (S_3)	Scenario 4 (S_4)
Social Impact (O_1)	(0.225, 0.315, 0.405)	(0.045, 0.135, 0.225)	(0.045, 0.135, 0.225)	(0.045, 0.135, 0.225)
Environmental impact (O_2)	(0.335, 0.469, 0.603)	(0.067, 0.201, 0.335)	(0.335, 0.469, 0.603)	(0.067, 0.201, 0.335)
Cost (O_3)	(0.476, 1.428, 2.380)	(1.428, 2.380, 3.332)	(0.476, 1.428, 2.380)	(1.428, 2.380, 3.332)
Physical quality (O_4)	(0.670, 0.938, 1.206)	(0.670, 0.938, 1.206)	(0.670, 0.938, 1.206)	(0.134, 0.402, 0.670)
Historic/ cultural value (O_5)	(1.265, 1.771, 2.277)	(0.759, 1.265, 1.771)	(0.253, 0.759, 1.265)	(0.253, 0.759, 1.265)

Table 9: The normalized weighted fuzzy decision matrix for the hypothetical example

	Scenario 1 (S_1)	Scenario 2 (S_2)	Scenario 3 (S_3)	Scenario 4 (S_4)
Social Impact (O_1)	(0.095, 0.132, 0.170)	(0.014, 0.041, 0.068)	(0.019, 0.057, 0.095)	(0.014, 0.041, 0.068)
Environmental impact (O_2)	(0.141, 0.197, 0.253)	(0.020, 0.060, 0.101)	(0.141, 0.197, 0.253)	(0.020, 0.060, 0.101)
Cost (O_3)	(0.200, 0.600, 1.000)	(0.429, 0.714, 1.000)	(0.200, 0.600, 1.000)	(0.429, 0.714, 1.000)
Physical quality (O_4)	(0.282, 0.394, 0.507)	(0.201, 0.282, 0.362)	(0.282, 0.394, 0.507)	(0.040, 0.121, 0.201)
Historic/ cultural value (O_5)	(0.532, 0.744, 0.957)	(0.228, 0.380, 0.532)	(0.106, 0.319, 0.532)	(0.076, 0.228, 0.380)

4.8 STEP 8: DETERMINE THE FUZZY POSITIVE-IDEAL SOLUTION (FPIS) AND FUZZY NEGATIVE-IDEAL SOLUTION (FNIS)

Using the normalized weighted fuzzy decision matrix from Step 7 (Table 9), the FPIS and FNIS were calculated for each objective using Equation 14 and 15 resulting in Table (10).

Table 10: The FPIS and FNIS values for each objective in the hypothetical example

Objectives (O_n)	FPIS	FNIS
Social Impact (O_1)	(0.095, 0.132, 0.170)	(0.014, 0.041, 0.068)
Environmental impact (O_2)	(0.141, 0.197, 0.253)	(0.020, 0.060, 0.101)
Cost (O_3)	(0.429, 0.714, 1.000)	(0.200, 0.600, 1.000)
Physical quality (O_4)	(0.282, 0.394, 0.507)	(0.040, 0.121, 0.201)
Historic/ cultural value (O_5)	(0.532, 0.744, 0.957)	(0.076, 0.228, 0.380)

4.9 STEP 9: CALCULATE THE DISTANCE OF EACH SCENARIO FROM FPIS AND FNIS

Using the Euclidean distance each scenario from the FPIS and FNIS were computed using Equation (16). Distances were calculated for each scenario based on the FPIS $D^+$and FNIS D^- (Table 11).

Table 11: The distance from each scenario to the FPIS and FNIS

Scenario (S_n)	D^+ (FPIS Distance)	D^- (*FNIS Distance*)
Scenario 1 (S_1)	0.543	0.802
Scenario 2 (S_2)	0.786	0.617
Scenario 3 (S_3)	0.643	0.732
Scenario 4 (S_4)	0.849	0.503

4.10 STEP 10: OBTAIN THE CLOSENESS COEFFICIENTS OF EACH SCENARIO

Once the distances from FPIS and FNIS are determined, the Closeness Coefficients can be obtained using Equation (27). An example calculation for Scenario 2 is given:

$$CC_i = \frac{d_i^-}{d_i^+ + d_i^-} = \frac{0.617}{0.786 + 0.617} = 0.440 \quad (27)$$

This results in the following scenario ranking, with scenario 1 ultimately ranking on top (Table 12).

Table 12: The Closeness Coefficient for each scenario

Scenario (S_n)	Closeness Coëfficiënt CC_n	Rank
Scenario 1 (S_1)	0.596	1
Scenario 2 (S_2)	0.440	3
Scenario 3 (S_3)	0.532	2
Scenario 4 (S_4)	0.372	4

5 CONCLUSION AND REMARKS

This study has introduced an integrated decision-making framework that combines Cross-Impact Balance (CIB) analysis, the Analytic Hierarchy Process (AHP), and Fuzzy-TOPSIS to enhance scenario development and multi-criteria evaluation in adaptive reuse projects. By incorporating scenario-based methodologies within a structured decision-making process, this approach enables stakeholders to systematically explore future-oriented reuse options while addressing uncertainty, complexity, and competing priorities. The framework was demonstrated through a hypothetical adaptive reuse project, illustrating how these methods interact to generate, assess, and rank consistent scenarios.

The findings highlight the benefits of integrating different methodologies to strengthen decision-making. CIB analysis ensures scenario consistency, reducing the likelihood of incoherent or contradictory planning outcomes. AHP provides a structured means to weight stakeholder preferences, ensuring that diverse perspectives are reflected in the evaluation process. Meanwhile, Fuzzy-TOPSIS offers a robust ranking mechanism that accounts for uncertainty, allowing decision-makers to prioritize alternatives more effectively. The integration of these methods enhances future-oriented decision-making by ensuring that adaptive reuse strategies consider long-term sustainability, economic feasibility, and social impact rather than being constrained by immediate limitations. Additionally, the approach fosters stakeholder engagement and transparency by actively involving participants in defining objectives, developing descriptors, and evaluating scenarios, leading to a more inclusive and aligned decision-making process. The structured methodology also enhances practical applicability, making it adaptable for real-world projects where trade-offs must be assessed, and priorities established.

Despite its advantages, certain limitations should be acknowledged. The methodology relies significantly on subjective inputs, particularly in scenario development and the conversion of linguistic variables in the Fuzzy-TOPSIS method. Its effectiveness depends on the ability of stakeholders and experts to define meaningful descriptors and variants, assess interactions accurately, and translate qualitative insights into quantitative measures. Any inconsistencies or biases in these subjective judgments could influence the final rankings. Moreover, for the methodology to function effectively, it is crucial to ensure active stakeholder participation at multiple stages, including defining objectives, developing scenario descriptors, weighting criteria, and ranking scenarios. Without sufficient engagement, the approach risks overlooking critical real-world considerations and diminishing the legitimacy of its outcomes. Future research should explore participatory mechanisms to strengthen stakeholder involvement and ensure a balanced representation of perspectives.

5.1 FUTURE RESEARCH DIRECTIONS

To further validate the proposed approach, real-world case studies should be conducted to test its practical applicability. Future research could also focus on:

- Improving the linguistic variable conversion process by developing standardized fuzzy scales that minimize subjectivity.
- Automating parts of the methodology to reduce the complexity of data input and improve usability.
- Exploring hybrid decision-support tools that integrate participatory scenario development with computational methods to enhance consistency and scalability.

The proposed framework demonstrates the potential of integrating scenario planning and multi-criteria decision-making, yet its full impact can only be realized through real-world applications. As the built environment continues to evolve, future efforts should focus on refining participatory methods and optimizing decision-support tools to promote practical applicability, ensuring that adaptive reuse strategies are data-driven, inclusive, and aligned with long-term sustainability goals.

REFERENCES

Amer, M., Daim, T. U., & Jetter, A. (2013). A review of scenario planning. *Futures*, *46*, 23–40. https://doi.org/10.1016/j.futures.2012.10.003

Bottero, M., Datola, G., Fazzari, D., & Ingaramo, R. (2022). Re-Thinking Detroit: A Multicriteria-Based Approach for Adaptive Reuse for the Corktown District. *Sustainability*, *14*(14), Article 14. https://doi.org/10.3390/su14148343

Chen, C.-T. (2000). Extensions of the TOPSIS for group decision-making under fuzzy environment. *Fuzzy Sets and Systems*, *114*(1), 1–9. https://doi.org/10.1016/S0165-0114(97)00377-1

Conejos, S. (2013). Optimisation of future building adaptive reuse design criteria for urban sustainability. *Journal of Design Research*, *11*(3), 225–242.

Dang, V. T., Wang, J., & Dang, W. V.-T. (2019). An Integrated Fuzzy AHP and Fuzzy TOPSIS Approach to Assess Sustainable Urban Development in an Emerging Economy. *International Journal of Environmental Research and Public Health*, *16*(16), Article 16. https://doi.org/10.3390/ijerph16162902

Della Spina, L. (2020). Adaptive Sustainable Reuse for Cultural Heritage: A Multiple Criteria Decision Aiding Approach Supporting Urban Development Processes. *Sustainability*, *12*(4), Article 4. https://doi.org/10.3390/su12041363

Efe, B. (2016). An integrated fuzzy multi criteria group decision making approach for ERP system selection. *Applied Soft Computing*, *38*, 106–117. https://doi.org/10.1016/j.asoc.2015.09.037

Foster, G. (2020). Circular economy strategies for adaptive reuse of cultural heritage buildings to reduce environmental impacts. *Resources, Conservation and*

 https://doi.org/10.52202/080684-0007

Recycling, *152*, 104507. https://doi.org/10.1016/j.resconrec.2019.104507
Gassner, R., & Steinmüller, K. (2018). *Scenarios that tell a Story. Normative Narrative Scenarios – An Efficient Tool for Participative Innovation-Oriented Foresight*. 37–48. https://doi.org/10.1007/978-3-658-25074-4_3
Greco, A., van Laar, B., Remøy, H., & Gruis, V. (2024). Accelerating circularity systemically: Three directions for impactful research. *Npj Urban Sustainability*, *4*(1), 1–3. https://doi.org/10.1038/s42949-024-00183-8
Love, P. E. D., Ika, L. A., & Pinto, J. K. (2023). Fast-and-frugal heuristics for decision-making in uncertain and complex settings in construction. *Developments in the Built Environment*, *14*, 100129. https://doi.org/10.1016/j.dibe.2023.100129
Mathew, M., Chakrabortty, R. K., & Ryan, M. J. (2020). A novel approach integrating AHP and TOPSIS under spherical fuzzy sets for advanced manufacturing system selection. *Engineering Applications of Artificial Intelligence*, *96*, 103988. https://doi.org/10.1016/j.engappai.2020.103988
Nadkarni, R. R., & Puthuvayi, B. (2020). A comprehensive literature review of Multi-Criteria Decision Making methods in heritage buildings. *Journal of Building Engineering*, *32*. Scopus. https://doi.org/10.1016/j.jobe.2020.101814
Nedeljkovic, D., Jurenic, T., & Djokic, L. (2023). Comparative analysis of multi-criteria models for decision-making in the process of building adaptation. *Heliyon*, *9*(6), e16620. https://doi.org/10.1016/j.heliyon.2023.e16620
Patil, S. K., & Kant, R. (2014). A fuzzy AHP-TOPSIS framework for ranking the solutions of Knowledge Management adoption in Supply Chain to overcome its barriers. *Expert Systems with Applications*, *41*(2), 679–693. https://doi.org/10.1016/j.eswa.2013.07.093
Rowe, E., Wright, G., & Derbyshire, J. (2017). Enhancing horizon scanning by utilizing pre-developed scenarios: Analysis of current practice and specification of a process improvement to aid the identification of important 'weak signals'. *Technological Forecasting and Social Change*, *125*, 224–235. https://doi.org/10.1016/j.techfore.2017.08.001
Saaty, T. L. (1990). How to make a decision: The analytic hierarchy process. *European Journal of Operational Research*, *48*(1), 9–26. https://doi.org/10.1016/0377-2217(90)90057-I
Sahoo, S. K., & Goswami, S. S. (2023). A comprehensive review of multiple criteria decision-making (MCDM) Methods: Advancements, applications, and future directions. *Decision Making Advances*, *1*(1), 25–48.
Sanchez, B., Esnaashary Esfahani, M., & Haas, C. (2019). A methodology to analyze the net environmental impacts and building's cost performance of an adaptive reuse project: A case study of the Waterloo County Courthouse renovations. *Environment Systems and Decisions*, *39*(4), 419–438. https://doi.org/10.1007/s10669-019-09734-2
Shahi, S., Esnaashary Esfahani, M., Bachmann, C., Haas, C. (2020). A definition framework for building adaptation projects. *Sustainable Cities and Society*, *63*. https://doi.org/10.1016/j.scs.2020.102345
Stewart, T. J., French, S., & Rios, J. (2013). Integrating multicriteria decision analysis and scenario planning—Review and extension. *Omega*, *41*(4), 679–688. https://doi.org/10.1016/j.omega.2012.09.003
Symstad, A. J., Fisichelli, N. A., Miller, B. W., Rowland, E., & Schuurman, G. W. (2017). Multiple methods for multiple futures: Integrating qualitative scenario planning and quantitative simulation modeling for natural resource decision making. *Climate Risk Management*, *17*, 78–91. https://doi.org/10.1016/j.crm.2017.07.002
Tori, S., te Boveldt, G., Keseru, I., & Macharis, C. (2023). Brussels mobility in 2050: Participatory mixed-method scenario building to explore alternative futures. *Transportation Research Interdisciplinary Perspectives*, *22*, 100919. https://doi.org/10.1016/j.trip.2023.100919
van Laar, B., Greco, A., Remøy, H., & Gruis, V. (2024). What matters when? – An integrative literature review on decision criteria in different stages of the adaptive reuse process. *Developments in the Built Environment*, 100439. https://doi.org/10.1016/j.dibe.2024.100439
van Notten, P. W. F., Rotmans, J., van Asselt, M. B. A., & Rothman, D. S. (2003). An updated scenario typology. *Futures*, *35*(5), 423–443. https://doi.org/10.1016/S0016-3287(02)00090-3
Vardopoulos, I., Tsilika, E., Sarantakou, E., Zorpas, A. A., Salvati, L., & Tsartas, P. (2021). An Integrated SWOT-PESTLE-AHP Model Assessing Sustainability in Adaptive Reuse Projects. *Applied Sciences*, *11*(15), Article 15. https://doi.org/10.3390/app11157134
Weimer-Jehle, W. (2006). Cross-impact balances: A system-theoretical approach to cross-impact analysis. *Technological Forecasting and Social Change*, *73*(4), 334–361. https://doi.org/10.1016/j.techfore.2005.06.005
Weimer-Jehle, W. (2023). *Cross-Impact Balances (CIB) for Scenario Analysis: Fundamentals and Implementation*. Springer Nature Switzerland. https://doi.org/10.1007/978-3-031-27230-1
Weimer-Jehle, W., Deuschle, J., & Rehaag, R. (2012). Familial and societal causes of juvenile obesity—A qualitative model on obesity development and prevention in socially disadvantaged children and adolescents. *Journal of Public Health*, *20*(2), 111–124. https://doi.org/10.1007/s10389-011-0473-8
Wilkinson, S. (2014). The preliminary assessment of adaptation potential in existing office buildings. *International Journal of Strategic Property Management*, *18*, 77–87. https://doi.org/10.3846/1648715X.2013.853705
Yung, E. H. K., & Chan, E. H. W. (2012). Implementation challenges to the adaptive reuse of heritage buildings: Towards the goals of sustainable, low carbon cities. *Habitat International*, *36*(3), 352–361. https://doi.org/10.1016/j.habitatint.2011.11.001

LIFE CYCLE ASSESSMENT OF STRUCTURAL MATERIAL REUSE IN ON-SITE PRESERVATION OF A CONCRETE STRUCTURE WITH TIMBER ADDITIONS

Jonn Are Myhren[1,3*], Signe Elgedin[1], Claudia Österlund[1], Amjad Almusaed[2], André Augusto[3,4], Åsa Wahlström[5], Jingchun Shen[1,3], and Alaa Khadra[1,3]

[1]Construction Technology, School of Information and Engineering, Dalarna University, Sweden,

[2]Jönköping University, Sweden

[3]Sustainable Energy Research Centre, Dalarna University, Sweden

[4]Energy Engineering, School of Information and Engineering, Dalarna University, Sweden

[5]Lund University, Sweden

*Corresponding Author: jam@du.se.

ABSTRAC

Background and aim: The construction industry contributes approximately 19% of global greenhouse gas (GHG) emissions and accounts for one-third of worldwide energy consumption, underscoring its pivotal role in addressing climate change. This study evaluates the environmental impact of preserving an existing concrete structure versus constructing a new one with cross-laminated timber (CLT) or virgin concrete.

Methods and data: The effectiveness of environmental comparison in mitigating carbon emissions and reducing resource consumption is investigated through a comparative lifecycle analysis of reuse and replacement scenarios. Utilizing the Life Cycle Assessment (LCA) framework, three scenarios were analysed: (1) preserving existing concrete floors on-site and adding two cross-laminated timber (CLT) extensions, (2) demolishing the existing concrete structure to construct an entirely new five story building using CLT, and (3) demolishing and constructing a new five story structure with cast-in-place virgin concrete. The analysis comprehensively quantifies the Global Warming Potential (GWP) across the production, operational, and end-of-life stages.

Findings: Results demonstrate that reusing existing concrete floors reduces approximately 40 kg CO_2e/m^2 gross floor area compared to a new timber construction and 121 kg CO_2e/m^2 tons compared to new concrete construction.

Theoretical/practical/societal implications: The results highlight the environmental benefits of implementing circular economy principles into construction practices.

Keywords: Life Cycle Assessment, concrete reuse, cross-laminated timber, circular economy, carbon emissions.

1 INTRODUCTION

The construction industry is a cornerstone of economic and infrastructure development. However, it significantly contributes to global greenhouse gas (GHG) emissions and resource consumption, accounting for approximately one-third of global energy use (Kumar & Zhang, 2024). Boverket (2023) reports that the construction and real estate industries account for 21% of Sweden's annual CO_2 emissions, highlighting their critical role in achieving national climate objectives. Addressing these challenges through sustainable material usage can reduce emissions and resource depletion, particularly amid growing infrastructure demands (Akan, Dhavale & Sarkis, 2017). Yet, balancing economic growth with environmental sustainability remains complex, especially when transitioning to low-carbon economies. As the largest consumer of natural resources, the Architecture, Engineering, and Construction (AEC) sector plays a critical role in ecological sustainability. Climate change, a pressing 21st-century challenge, underscores the urgency of action, with Sustainable Development Goal (SDG) 13 emphasizing climate mitigation (Magazzino et al., 2022). Transitioning to a circular economy that optimizes resource use, minimizes waste, and reduces environmental impacts across material lifecycles offers a potential pathway. However, technological, institutional, market, and cultural barriers inder this shift (Grafström & Aasma, 2021). Moving from a linear

https://doi.org/10.52202/080684-0008

"take, make, dispose" model to a circular framework based on recycling and reuse is imperative (Elisha, 2020).

Initiatives to reduce environmental impact assessment of building materials have been recently introduced in EU (EU 2024). The EU Directive on energy performance of buildings (recast) emphasised measures to reduce the whole life-cycle greenhouse gas (GHG) emissions of buildings including material production, construction, operation, renovation and end of life stages. In Sweden, the climate declaration of buildings regulation (Boverket 2020) requires assessment of the carbon footprint of new buildings.

This study examines the environmental implications of adopting circular economy principles in the construction sector, focusing on the structural materials of concrete, steel, and cross-laminated timber (CLT). Specifically, the research addresses the optimization and reuse solution for the "Lumi" project, a five-story office building of 21 000 m^2 gross floor area in Uppsala, Sweden. Three stories of an old building are reused and CLT is used to construct two additional stories. Due to the structural limitations of the pre-existing foundation, constructing a large and heavy structure, such as one utilizing cast-in-place concrete, was deemed unfeasible. Assessing the environmental impact of structural systems, including cast-in-place concrete and CLT, is pivotal for advancing sustainable construction practices and addressing climate-related challenges.

The study focuses on three distinct scenarios to determine the option with the lowest environmental impact:

- **Scenario 1:** Retain three concrete floors and add two cross-laminated timber (CLT) floors.
- **Scenario 2:** Demolish concrete floors and construct a new five-story CLT building.
- **Scenario 3:** Demolish concrete floors and construct a new five-story concrete building.

Figure 1 illustrates the building and floors. Scenarios 2 and 3 necessitate the complete demolition of the concrete structure and the construction of an entirely new building.

The study examines greenhouse gas emissions, quantified in terms of carbon dioxide equivalents (CO-eq), to evaluate each scenario's environmental performance.

2 LITERATURE REVIEW

In the European Union, more than 20% of residential buildings were constructed before 1945 and are now approaching the end of their expected lifespans. This situation necessitates either the renovation or demolition of these structures (Czarnecki & Rudner, 2023). When decommissioning old buildings, approximately 70% of the waste from high-rise buildings has the potential to be reused or recycled (Umar, Shafiq, & Ahmad, 2020). Adopting a circular economy approach in construction and demolition waste management can offer both environmental and economic advantages. However, the sustainability of such efforts depends on site-specific factors including the type of material, building components, transportation distances, and the economic and political context (Ghisellini et al., 2017).

A recent report by the Swedish board of housing lists the typical climatic impact of different parts of buildings by focusing on phases A1 to A5 of the building life cycle (Malmqvist, T 2023). The findings reveal that the foundation and load bearing structure are the most significant contributors to the climate impact, often accounting for more than half of the emissions. This underscores the importance in efforts to reduce emissions from structural material. Conversely, energy usage and structural completeness contribute less to the impact of climate change, indicating opportunities for targeted improvements.

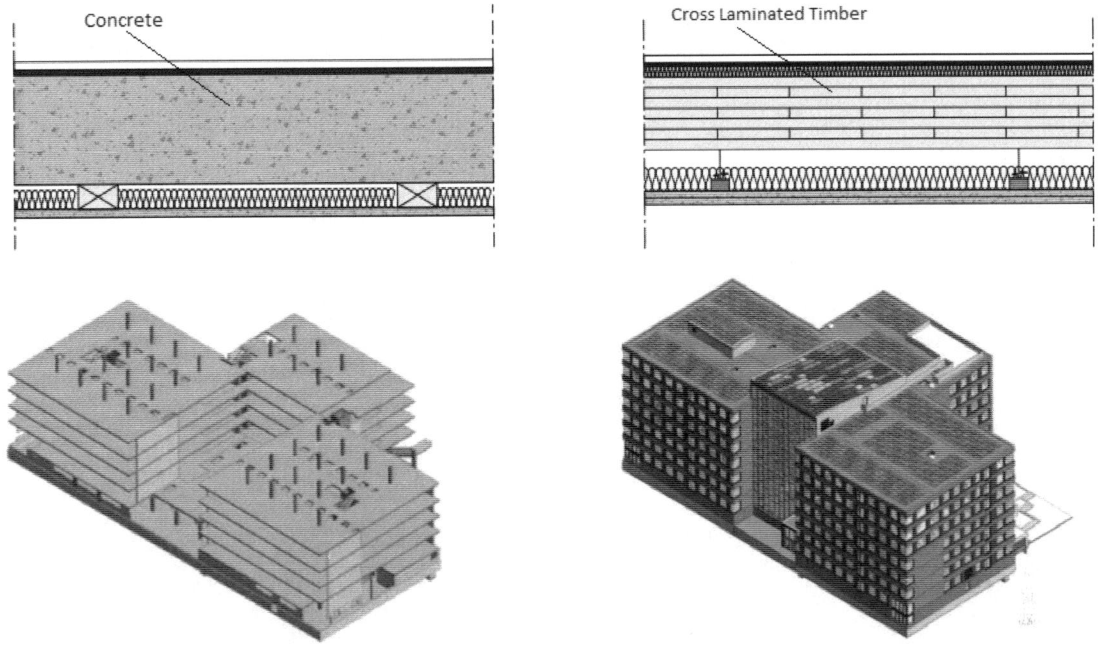

Figure 1: Above: Illustration of concrete and CLT floors Below: Illustrations of three retained concrete floors and the completed building with two additional cross-laminated timber (CLT) floors (scenario 1).

These results emphasize the critical need for life cycle assessment and material optimization to foster sustainable construction practices and achieve long-term ecological benefits.

Concrete is the most common material in building foundations and load bearing structures. The environmental consequences of concrete production are profound, primarily due to the carbon-intensive nature of cement manufacturing, a critical concrete component. Cement production, predominantly driven by the calcination of limestone, is responsible for approximately 8% of global CO_2 emissions, presenting a formidable challenge for climate change mitigation (Amran et al., 2022). Accelerating decarbonization and implementing improved solutions are imperative for achieving net-zero emissions, particularly in addressing the structural and foundational demands of the construction industry (Amran et al., 2022). As an example, design for deconstruction (DfD) offers environmental benefits 1.8 to 2.8 times greater than those of recycled aggregate concrete (RAC) (Xia et al., 2020). Life cycle assessment (LCA) models are essential in establishing sustainable cement standards. Terán-Cuadrado et al. (2024) underscore the significance of supplementary cementitious materials (SCMs), functional units, and supply chain dynamics in enhancing the sustainability of blended cement. Concrete production involves energy-intensive processes, including raw material extraction and high-temperature kiln operations, exacerbating its environmental footprint (Boakye et al., 2024). Furthermore, the environmental impact of concrete extends to its usage and disposal phases. Although concrete is highly durable, demolishing concrete structures generates considerable waste, much of which is downcycled or landfilled. While carbonation during its lifecycle absorbs a portion of CO_2, this compensates for only a fraction of the emissions generated during production (Alhawat et al., 2022).

Timber is distinguished by its renewable nature and carbon-sequestering properties. It acts as a carbon sink during its growth phase, capturing atmospheric CO. Life cycle assessment studies consistently indicate that timber exhibits a lower Global Warming Potential (GWP) than concrete, particularly during the production and construction phases. Its lightweight nature further contributes to reduced transportation emissions and enhanced construction efficiency. Duan et al. (2022) report that, despite the higher embodied energy of mass timber, it achieves 43% lower greenhouse gas (GHG) emissions than reinforced concrete (RC) (Duan et al., 2022). However, the ecological benefits of timber are contingent upon sustainable forestry practices. Unsustainable logging can result in deforestation, biodiversity loss, and carbon release, significantly undermining timber's advantages. Deforestation accounts for approximately 15% of global GHG emissions, contributing substantially to climate change (Kumar et al., 2022). Innovations such as cross-laminated timber (CLT) enhance timber's potential for construction while retaining its environmental benefits. Younis and Dodoo (2022) highlight the advantages of CLT, including a low carbon footprint, high strength-to-weight ratio, and ease of installation (Younis & Dodoo, 2022). A smart combination of CLT and the preserved existing concrete structure was used in the Lumi case. A

https://doi.org/10.52202/080684-0008

heavier structure than CLT would not have allowed the reuse of the three floors from the decommissioned structure on-site. Reusing concrete offers an avenue for reducing emissions associated with raw material extraction, cement production, and waste disposal.

3 METHODOLOGY

This study systematically quantifies the Global Warming Potential (GWP) of building constructions over the lifecycle, which includes raw material extraction, manufacturing and end-of-life disposal, using Life Cycle Assessment (LCA). Expressed in carbon dioxide equivalents (CO_2e), GWP standardizes the radiative forcing effects of various greenhouse gases into a single measure, thus enabling a scientifically robust evaluation of climate impact. Given its significant role in sustainability assessments, this study prioritizes GWP as the main environmental impact category when comparing structural alternatives. The analysis adheres to the EN 15978 standard (CEN (2011)) which defined the lifecycle phases as material manufacturing (A1–A3), construction processes (A4–A5), operational use (B1–B7), and end-of-life considerations (C1–C4). The LCA ensured a robust, systematic, and objective environmental performance evaluation by incorporating all lifecycle stages, as illustrated in Figure 2. This study comprehensively evaluated greenhouse gas emissions associated with four structural material scenarios over a 100-year lifecycle and defines 1 m² of gross floor area (GFA) per residential unit as the functional unit. This clear definition ensures methodological consistency in the Life Cycle Assessment (LCA) and enables comparability between the construction scenarios analyzed.

Operational energy and maintenance were excluded to focus exclusively on material-related emissions in a similar way to the Swedish climate declaration of buildings regulation. But in contract to the regulations, this study only included the load-bearing structure excluding foundation and roof etc.

The analysis was structured across three key lifecycle stages, as defined below:

- Phases A1–A5: Encompassing raw material extraction, production, transportation, and construction activities.
- Phase B: Addressing materials' energy-free durability and longevity during the operational phase.
- Phases C1–C4: Covering end-of-life processes, including demolition, waste management, recycling, and disposal.

3.1 DATA COLLECTION

The data collection within the study incorporates both primary and secondary sources to ensure precision, reliability, and standardization in assessing environmental impacts. The study includes:

- Primary Data: Lumi project structural design and material amounts.
- Secondary data: sources that include emission factors and material attributes from EPDs, literature, and industry sources like OneClick LCA.
- Additional inputs: Transportation distances and construction activities with fuel usage and emission factors.

3.2 CALCULATION OF GWP

Each material's GWP was calculated

- Material Inventory Assessment: Quantify materials utilized.
- Lifecycle Phase Assessment: Assess emissions throughout production (A1-A3), construction (A4-A5), use (B1-7), and end-of-life (C1-C4) stages.
- Verify GWP emission factors from EPDs for emission factor application.
- End-of-Life Considerations: Assess emissions from demolition, recycling, and material reuse credits.
- Results Aggregation: Each material's total phase emissions.

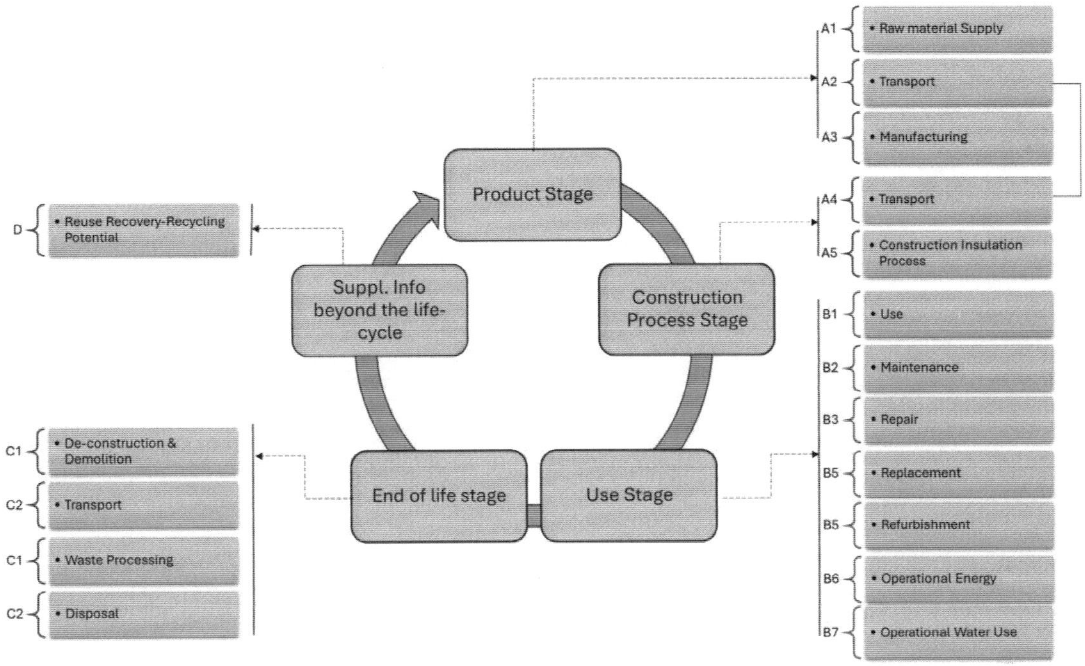

Figure 2: Life Cycle Stages in Construction and Their Environmental Impact Considerations

4 RESULTS & DISCUSSION

An analysis of the Global Warming Potential (GWP) of three structural alternatives revealed significant differences in environmental performance, expressed in kg CO_2e/m^2 gross floor area (Figure 3). Among the structures evaluated, Scenario 1, a preserved concrete structure with two added stories in CLT demonstrated the lowest GWP, calculated at 36 kg CO_2e/m^2, representing a 77% reduction compared to Scenario 3, a building with virgin concrete, which has a GWP of 157 kg CO_2e/m^2. This substantial reduction of 121 kg CO_2e/m^2, or a total of 2,800 tons $CO_2$for the whole building, was attributed to eliminating energy-intensive processes such as raw material extraction and cement manufacturing during life cycle stages A1–A3. Despite its widespread use, versatility, and durability, standard concrete's high environmental impact made it less suitable for sustainable construction practices. By reusing existing materials, the demand for new cement production was almost nullified, thereby mitigating emissions.

A complete five story timber construction (scenario 2), with a GWP of 76 kg CO_2e/m^2, offers a 52% reduction in emissions compared to standard concrete, resulting in an absolute reduction of 81 kg CO_2e/m^2. Even though the GWP assessments of timber do not account for biogenic carbon storage, which would further enhance its ecological benefits, timber provides a more sustainable alternative to standard concrete. However, the carbon savings with timber (Scenario 2) compared to standard concrete (Scenario 3) were less substantial than those achieved by reusing an existing concrete structure (Scenario 1). Moreover, current GWP assessments of timber do not account for biogenic carbon storage, which would further enhance its ecological benefits. While Scenario 2, timber, provides a more sustainable alternative to Scenario 3, standard concrete, its carbon savings were less substantial than those achieved by Scenario 1, reusing an existing concrete structure.

A comparison between preserved concrete structures on-site and new timber structures further highlights the superior environmental performance of reused concrete. With an additional reduction of 40 kg CO_2e/m^2 compared to timber (76 kg CO_2e/m^2 for timber versus 36 kg CO_2e/m^2 for reused concrete), reused concrete

https://doi.org/10.52202/080684-0008

demonstrates a greater capacity to minimize carbon emissions. For the whole building the reduction amounts to 1,800 tons CO_2. These findings underscore the critical role of material reuse in advancing sustainable construction practices and reducing the climate impact of the built environment. Although timber was a viable low-carbon alternative to standard concrete, reused concrete provides the most significant reductions in GWP, aligning more effectively with circular economy principles and sustainable development. In practice, the reuse of structural components not only reduces embodied emissions but also preserves the urban fabric and cultural value of existing architecture—providing social and aesthetic benefits alongside environmental gains.

The product stage represents the dominant share of the life cycle impacts in Scenario 3 with a concrete construction. As we introduce more timber the end-of-life stage becomes a more notable contribution to the building's overall environmental effects.

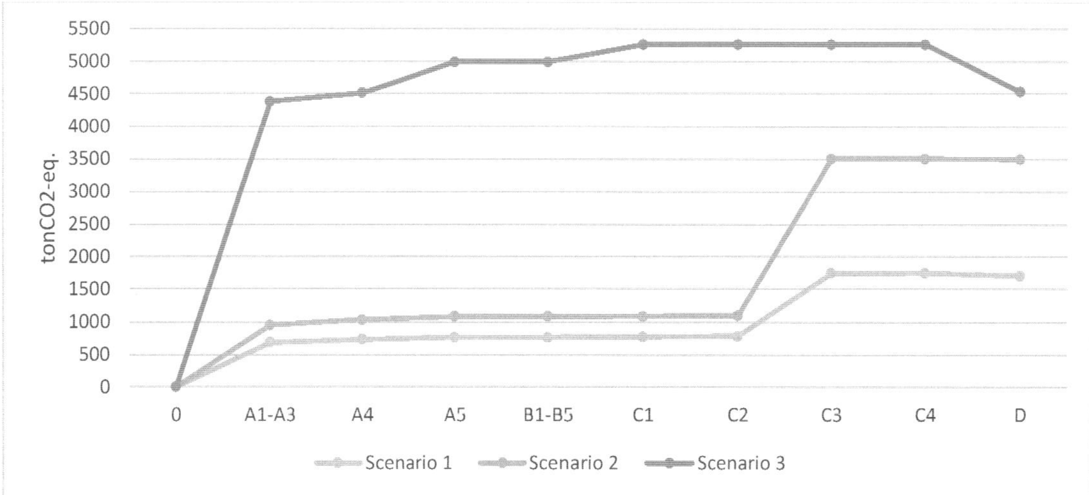

Figure 3: Comparative Analysis of Global Warming Potential (GWP) Across Lifecycle Phases for All Scenarios Over 100 Years.

This study expands beyond concrete to encompass cross-laminated timber (CLT), enhancing Life Cycle Assessment (LCA) by exploring material reuse in rehabilitation and new construction. Unlike other studies, such as De Wolf et al. (2020), which primarily focus on specific material lifecycles, this research integrates reuse techniques across various structural components. Incorporating an estimated building lifetime and correlating $kgCO_2e/m^2$ annually will further refine the assessment by providing a more accurate indication of long-term environmental impact.

4.1 CHALLENGES AND FUTURE RESEARCH

Despite promising outcomes, the study had limitations. Regional differences in material availability and transportation constraints limit the generalizability of the results. Future research could address these gaps by employing multi-impact assessments, examining reused concrete's long-term durability and cost-effectiveness, and incorporating biogenic carbon storage in timber life cycle evaluations. These advancements would aid in establishing evidence-based, sustainable construction methods by enhancing the understanding of material performance. While reusing materials offers significant environmental benefits, various practical and financial constraints impede its widespread acceptance. Reuse is often less economically viable than traditional construction due to high labor costs associated with demolition, sorting, and processing recycled materials, which typically outweigh any potential savings. Additionally, strict testing and certification requirements create further financial and logistical challenges. Other obstacles include transportation, on-site storage, and integrating salvaged components into new projects. Additional constraints on implementation involve limited market demand and the lack of consistent regulations. Future research should explore legislative incentives, streamlined regulatory frameworks, and improvements in modular design to boost cost-effectiveness and scalability in circular construction.

5 CONCLUSIONS

This study demonstrates that circular economy practices in the building sector, such as reusing structural materials, can significantly benefit the environment. The research evaluates three structural scenarios one utilizing an existing concrete structure on-site, one involving new cross-laminated timber (CLT) and the last virgin concrete—to assess their Global Warming Potential (GWP). Reusing the existing concrete structure and adding two CLT stories saves around 40 kg CO_2e/m^2 compared to a new CLT construction and 121 kg CO_2e/m^2 compared to a new concrete construction, effectively reducing greenhouse gas emissions. This emphasizes the environmental advantages of extending the lifecycle of existing materials while minimizing resource extraction and processing. Lifecycle emissions are consistently lower for scenarios involving reused concrete and new CLT than traditional concrete buildings, with reused concrete emerging as the most sustainable option. These findings highlight the critical role of material reuse in improving construction sustainability and reducing environmental impacts while acknowledging the practical challenges.

In conclusion, the synergy between structural reuse and renewable materials offers a robust pathway for reducing embodied carbon, particularly in the renovation and densification of existing urban areas. With strategic planning, the construction industry can shift from linear consumption models toward circular systems that prioritize longevity, adaptability, and climate resilience. Future work should continue to refine these assessments, scale demonstration projects, and embed circularity in mainstream architectural and engineering practices.

This study provides valuable insights for developing policies and strategies that align with global climate mitigation targets. Emphasizing resource efficiency and realistic approaches to lifetime emission reduction contributes to advancing the transition to low-carbon economies and promoting sustainable building practices.

ACKNOWLEDGEMENTS

The authors thank the Swedish Energy Agency for financial contributions to this study (grant P2023-01473).

REFERENCES

Kumar, D. & Zhang, C. (2024). Carbon emission reduction in construction industry: qualitative insights on procurement, policies and artificial intelligence. Built Environment Project and Asset Management

Boverket. (2023). Miljöindikatorer för bygg- och fastighetssektorn. Retrieved from https://www.boverket.se

Akan, M.Ö.A., Dhavale, D.G. and Sarkis, J., 2017. Greenhouse gas emissions in the construction industry: An analysis and evaluation of a concrete supply chain. Journal of Cleaner Production, 167, pp.1195-1207

Magazzino, C., Toma, P., Fusco, G., Valente, D. and Petrosillo, I., 2022. Renewable energy consumption, environmental degradation, and economic growth: the greener, the richer? Ecological Indicators, 139, p.108912

Grafström, J. and Aasma, S., 2021. Breaking circular economy barriers. Journal of Cleaner Production, 292, p.126002

Elisha, O.D., (2020). Moving beyond take-make-dispose to take-make-use for a sustainable economy. Int. J. Sci. Res. Educ, 13(3), pp.497-516

European Commission, Directive (EU) 2024/1275 of the European Parliament and of the Council of 24 April 2024 on the energy performance of buildings (recast) (Text with EEA relevance). Official Journal of the European Union, 2024.

Boverket, Regulation on climate declarations for buildings- proposal for a roadmap and limit values. Report number: *2020:28, ISBN print: 978-91-7563-730-3, ISBN pdf: 978-91-7563-731-0.* 2020.

Almeida, L., Silva, A., Veiga, M., Mirão, J., & Vieira, M., 2021. 20th-Century Award-Winning Buildings in Lisbon (Portugal). Study of Plasters, Rendering, and Concrete Materials Aiming Their Sustainable Preservation. Buildings. https://doi.org/10.3390/buildings11080359.

Kapica, B., Targowski, W., & Kulowski, A., 2024. Is the Concept of Zero Waste Possible to Implement in Construction?. Buildings. https://doi.org/10.3390/buildings14020428.

Malmqvist, T., Boverket: Referensvärden för klimatpåverkan vid uppförande av byggnader, Version 3, 2023, KTH Skolan för Arkitektur och Samhällsbyggnad, ISBN: 978-91-8040-754-0

Amran, M., Makul, N., Fediuk, R., Lee, Y. H., Vatin, N. I., Lee, Y. Y., & Mohammed, K. (2022). Global carbon recoverability experiences from the cement industry. Case Studies in Construction Materials, 17, e01439.

Xia, B., Ding, T., & Xiao, J., 2020. Life cycle assessment of concrete structures with reuse and recycling strategies: A novel framework and case study.. Waste management, 105, pp. 268-278 . https://doi.org/10.1016/j.wasman.2020.02.015.

Terán-Cuadrado, G., Tahir, F., Nurdiawati, A., Almarshoud, M. A., & Al-Ghamdi, S. G. (2024). Current and potential materials for the low-carbon cement production: Life cycle assessment perspective. Journal of Building Engineering, 96, 110528

Boakye, K., Simske, S., Bradley, T., Troxell, W., & Goemans, C. (2024). Raw material optimization and co2 sensitivity-predictive analytics in cement manufacturing: a case study at Union Bridge plant, Heidelberg materials, Maryland

Alhawat, M., Ashour, A., Yildirim, G., Aldemir, A. and Sahmaran, M., 2022. Properties of geopolymers sourced from construction and demolition waste: A review. Journal of Building Engineering, 50, p.104104].[Sverdrup, H. U., & Olafsdottir, A. H. (2023). Dynamical Modelling of the Global Cement Production and Supply System, Assessing Climate Impacts of Different Future Scenarios. Water, Air, & Soil Pollution, 234(3), 191

Duan, Z., Huang, Q., & Zhang, Q. (2022). Life cycle assessment of mass timber construction: A review. Building and Environment, 221, 109320

Kumar, R., Kumar, A., & Saikia, P. (2022). Deforestation and forest degradation impact the environment. In Environmental Degradation: Challenges and Strategies for Mitigation (pp. 19-46). Cham: Springer International Publishing

 https://doi.org/10.52202/080684-0008

Younis, A., & Dodoo, A. (2022). Cross-laminated timber for building construction: A life-cycle-assessment overview. Journal of Building Engineering, 52, 104482

CEN. (2011). EN 15978:2011. Sustainability of construction works—Assessment of environmental performance of buildings—Calculation method. Brussels, Belgium: European Committee for Standardization.

Kumar, D., Zhang, C., Holubnyak, E., & Demirkesen, S. (2024). Integrated assessment of levelized hydrogen production costs: Evaluating renewable and fossil pathways with emission costs and tax incentives. International Journal of Hydrogen Energy, 95, 389-401

De Wolf, C., Hoxha, E. and Fivet, C., 2020. Comparison of environmental assessment methods when reusing building components: A case study. Sustainable Cities and Society, 61, p.102322

LIFE CYCLE ASSESSMENT OF DIFFERENT STRUCTURAL FRAMES APPROACH IN SWEDISH ROW HOUSE CONSTRUCTION: RECLAIMED CONCRETE, NEW CONCRETE, AND TIMBER

Jonn Are Myhren[1,3]*, Evelina Andersson[1], Felix Jansson[1], Alaa Khadra[1,3], André Augusto[3,4] Åsa Wahlström[5], Amjad Almusaed[2], and Jingchun Shen[1,3]

[1]Construction Technology, School of Information and Engineering, Dalarna University, Sweden

[2]Jönköping University, Sweden

[3]Sustainable Energy Research Centre, Dalarna University, Sweden

[5]Energy Engineering, School of Information and Engineering, Dalarna University, Sweden

[5]Lund University, Sweden

*Corresponding Author: jam@du.se

ABSTRACT

Background and aim. The Intergovernmental Panel on Climate Change (IPCC) reported in 2019 that the building sector accounts for 21% of global greenhouse gas (GHG) emissions, with 18% originating from producing construction materials such as cement and steel. This highlights the urgent need to address embodied carbon in construction to align with climate goals. This study examines the potential of reusing structural materials, primarily concrete elements, to significantly reduce embodied emissions in the construction sector, which has increasingly focused on embodied carbon alongside operational energy efficiency.

Methods and Data. A lifecycle analysis compared the Global Warming Potential (GWP) of concrete elements reclaimed from an old building, conventional concrete, and timber construction for the structural frame of a row house.

Findings. Reclaimed concrete demonstrated the lowest GWP, achieving a 77% reduction compared to traditional concrete and surpassing timber. These findings indicate that reclaimed concrete elements can rival timber as a sustainable building material.

Theoretical / Practical / Societal implications. Prioritizing sustainable material choices and resource efficiency is crucial for the construction sector to meet increasingly stringent global climate targets. This study emphasizes the importance of reusing structural materials to lower carbon emissions during construction, contributing to a more sustainable built environment.

KEYWORDS: Carbon emissions, Circular economy, Life cycle assessment, Reuse, Structural elements

1 INTRODUCTION

The construction sector significantly influences resource consumption and greenhouse gas emissions (IEA, 2022). Boverket (2023) reports that the construction and real estate industries account for 21% of Sweden's annual CO_2 emissions, highlighting their critical role in achieving national climate objectives. These figures emphasize the urgent need for sustainable strategies to mitigate the environmental impacts of the building industry. The cement sector plays a significant role in global carbon emissions, with energy-intensive

https://doi.org/10.52202/080684-0009

calcination processes constituting about 8% of worldwide CO emissions (Nikolakopoulos et al., 2024). Additionally, the construction sector generates over 40% of global waste, substantially intensifying its environmental footprint (Abubakar et al., 2022). The effects of the construction sector on carbon emissions and waste production present significant challenges to decarbonization goals (Sbahieh et al., 2023). In response, European Union initiatives like the Whole Life Carbon Roadmap and the Recreate project advocate for circular economic approaches, emphasizing the importance of material reuse and reducing reliance on virgin resource extraction in line with broader sustainability objectives (Norouzi & Masoud, 2021; UNEP, 2022a, 2022b). Transitioning to a circular economy is essential for achieving the ambitious targets of the Paris Agreement. Although advances in energy efficiency have lowered operational emissions, the focus has shifted towards embodied carbon, underscoring the vital importance of material choice and construction methods in reducing environmental impacts (Minunno, 2021).

While reusing concrete components from decommissioned structures in new buildings is seldom considered a primary strategy for enhancing sustainability in the construction industry, concrete reuse has a long history with several successful applications demonstrating significant financial and environmental benefits. (Küpfer, Bastien-Masse & Fivet 2023).

Recent several researches highlight the growing emphasis on reusing concrete elements to reduce embodied carbon and advance circular economy principles in construction. Ahmad Al-Najjar and Tove Malmqvist (2025) conducted a Swedish pilot study with reusing concrete elements in new buildings, presenting a significant embodied carbon savings. The study highlights that reusing concrete elements offers greater carbon savings than recycling or using new low-carbon materials. Küpfer et al. (2023) critically reviewed 77 concrete reuse cases from Europe and the USA. They identified that reusing concrete pieces in new structures is not commonly practiced. Building on this, Küpfer et al. (2024) further explored the reuse of saw-cut reinforced concrete (RC) pieces from demolished structures to create new load-bearing floor systems, showcasing technical feasibility through structural testing and life-cycle assessments.

Building on the insights from recent case studies, we now turn our attention to the reference carbon intensity data for Swedish residential buildings, which provides a crucial benchmark for evaluating the environmental impact of construction practices in this region. The total GHG emissions for erecting a traditional concrete structure (lifecycle stage A1-A5) was estimated to be around 350 kg $CO_{2\,e}/m^2$ (tempered floor area) in an LCA study for a six-storey multifamily house by (IVL, 2017). From voluntary building certification system, the up-limits and reference value of the upfront carbon (lifecycle stage A1-A5) of the multifamily building are 260 kg $CO_{2\,e}/m^2$ in Miljöbyggnad 4.0 and 310 kg $CO_{2\,e}/m^2$ in BREEAM (Miljöbyggnad, 2023; BREEAM, 2023)

This is representative of new Swedish energy-efficient multi-family buildings. Single-family houses and row houses with 1-2 stories have around half of that impact. The emissions for this category of houses average 164 kg $CO_{2\,e}/m^2$ and the most significant emissions occur during the A1-A3 phase (Boverket, 2023). The structure, including foundation, structural framework, façade, and roofs, accounts for the majority.

While previous research has thoroughly examined the environmental impact of new construction materials, few studies have systematically evaluated the feasibility of reusing concrete elements as a sustainable construction material alternative to either virgin concrete or timber.

This study examines and compares the Global Warming Potential (GWP) of three structural options for a row house in southern Sweden: locally reclaimed concrete, traditional structure using virgin cast-in-place concrete, and a structural frame utilizing light timber. Repurposing, recovering, and incorporating old concrete elements into new construction projects is often feasible. However, the environmental competitiveness of reclaimed concrete elements from decommissioned buildings as an alternative to timber remains underexplored. By systematically evaluating the climatic impact of the construction phase, the study aims to generate more insights into the environmental performance of structural building materials, emphasizing the importance of material selection in reducing carbon footprints. Through this analysis, the research advances sustainable construction practices, supports the adoption of reclaimed materials in alignment with global climate objectives, and fosters innovation within the industry, providing a robust foundation for informed decision-making in future projects. Given the construction industry's significant contribution to embodied carbon emissions, advancing sustainable building practices and informing policy on low-carbon construction strategies relies on determining whether material reuse offers a viable alternative to conventional and renewable materials.

This work distinguishes itself from other LCA studies of reclaimed concrete by employing a comprehensive methodological approach that includes an empirical evaluation of construction and installation impacts (A5) as well as an extensive sensitivity analysis of transportation emissions (A4). Unlike more conventional studies, this research utilizes real-world case study data to capture all environmental consequences of deconstruction, transportation, and reassembly. The findings provide new insights into emissions reduction and highlight how localized reuse techniques lower embodied carbon, thereby supporting the practical feasibility of reclaimed concrete in circular building designs. This study clarifies reuse techniques, aiding in optimizing low-carbon construction strategies.

The study directly aligns with European policy programs such as the Circular Economy Action Plan and the Whole Life Carbon Roadmap, which focus on reducing embodied carbon and fostering a resource-efficient construction economy through more reuse or recycling from non-hazardous construction and demolition waste (CDW).

2 METHOD

The article employs a Life Cycle Assessment (LCA) framework to quantify and compare the Global Warming Potential (GWP) of three structural alternatives for a row house. The LCA encompasses essential components of the superstructure, including the frame, upper floors, roof, stairs, and external walls. The methodology adheres to European standards, specifically EN 15978:2011 for building-level assessments and all of the product-level datasets in the study follow EN 15804 standard based on CML, ensuring compliance and reliability. The scope of the LCA focuses on life cycle stage A and the results in Section 3 illustrate the life cycle impacts within the GWP impact category, measured in kg CO_{2e} over a specified service life.

2.1 CASE STUDY DESCRIPTION

Built between 1966 and 1969 as part of Sweden's million-program housing initiative, the existing multi-family residential building in Drottninghög, Helsingborg, consists of prefabricated concrete components. The structure supplies structural elements for a new row house project. This study examines the feasibility of reusing these components—including super structure elements in both frame and envelope—within a circular building design to reduce environmental impact. The study forms part of a research project that explores the feasibility of reusing structural concrete elements from donor buildings for a new row house in Helsingborg, Sweden. Figure 1 shows the floor plan of the new apartments with a floor area of 97 m^2 and the wall structures. The walls are all assumed to be designed to have the same U-value.

The analysis employed specific measurements from the donor building, sourced from an inventory of architectural records and structural features. When exact data was lacking, methodological consistency was upheld across instances by utilizing assumed values. These assumptions ensured comparability in LCA among new concrete, reclaimed concrete, and timber construction scenarios.

The study defines 1 m^2 of gross floor area (GFA) per residential unit as the functional unit over a 50-year lifetime. This clear definition ensures methodological consistency in the Life Cycle Assessment (LCA) and enables comparability between the construction scenarios analyzed.

To emphasize the environmental impact of material choices at design phase, only the product phase (A1–A3) and the construction phase (A4-A5) are considered in this study, while Stages B (use phase) and C (end-of-life) were excluded in accordance with Swedish climate declaration method. Although dismantling emissions for reclaimed concrete ensured methodological consistency, the study assumed that used components retain full functionality without additional maintenance.

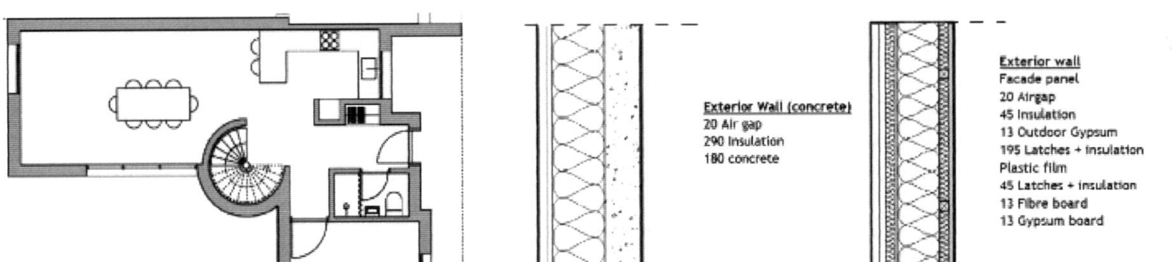

Figure 1. Left: Floor plan (two stories). Middle: Exterior wall in concrete. Left: Exterior wall in light timber

Three structural systems were evaluated:

Case 1: Structural concrete elements from an existing donor building are disassembled, inspected, transported, and reassembled without reprocessing, demonstrating direct reuse and minimizing resource extraction, waste, and embodied energy.

Case 2: A conventional system constructed entirely with virgin cast-in-place concrete is a benchmark for comparing reuse methodologies' performance and environmental impact.

Case 3: A light timber system exemplifying sustainable construction with renewable materials, low embodied carbon, and compatibility with circular construction, providing an additional comparative baseline.

https://doi.org/10.52202/080684-0009

2.2 DATA COLLECTION AND COLLABORATION

Material quantities (A1–A3) were derived from design data, which included building information models, architectural drawings, and structural inventories provided by project representatives, primarily an architecture student from KTH. While the construction team validated logistics and practical aspects, the primary responsibility for material weights and quantities rested with the design contributors. Transport distances (A4) were estimated using standard averages integrated into the One-Click LCA tool, and construction emissions (A5) were based on benchmark data from similar projects. This structured and collaborative approach ensured accuracy and reliability when calculating the environmental impacts of reclaimed materials, adhering to EN 15804 standard based on CML methodology.

2.3 LIFECYCLE ASSESSMENT (LCA)

The LCA study evaluated the environmental impacts of the structural frameworks, highlighting distinct phases (A1–A4) for the reclaimed concrete. The use of reclaimed structural elements refers to components sourced from existing buildings. This ensured that the impacts from disassembly, inspection, transport, and reuse were thoroughly captured for accurate comparison. Calculations were performed using the OneClick-LCA educational edition, generic environmental product declarations (EPDs), and industry-average data, standardized to one square meter (m^2) of gross floor area over a 50-year lifespan. OneClick LCA was chosen as a calculation tool for its holistic functions in terms of comprehensive environmental impact databases and consistent methodology in line with EN 15978. The system boundary included the material production phases (A1–A3), the transport phases (A4), and the construction and installation phases (A5).

Data and information about the properties and quantities of the materials used are provided by the project designers. Since specific product data was unavailable at the early design phase, associated environmental impact data has primarily been obtained from the Swedish National Board of Housing, Building and Planning's climate database. Where generic data from the Swedish National Board of Housing, Building and Planning's climate database is missing, localized generic climate data from One Click LCA has been used. In A4, generic transport data (distance and transport mode) has been used based on typical transport data for each material. Generic transport data has been retrieved from the Swedish National Board of Housing, Building and Planning's Climate Database. In A5, we followed Swedish climate declaration context, encompasses waste management, energy use, and on-site emissions.

Accordingly, the localized generic data from One Click LCA has been used at the construction phase for the inclusion of construction site vehicles, machinery and equipment. Energy impacts in A5 were modeled for fuel, etc. but only for excavation and backfilling, which was calculated using project-scale averages, considering machinery, fuel, and electricity consumption. Construction waste and the management of donor materials were also included, creating a comprehensive framework to evaluate the role of recycled concrete in advancing circular construction practices.

3 RESULTS & DISCUSSION

3.1 DIFFERENT STRUCTURAL MATERIAL'S PERFORMANCE

The comparative GWP results of the three alternative structural structures studied is presented in Figure 2. The results show substantial differences in environmental impact between the three cases, with reclaimed concrete as the most sustainable option.

Reused concrete reduces Global Warming Potential (GWP) by 77% compared to virgin cast-in-place concrete, demonstrating its superior environmental performance. While timber benefits from renewability and carbon sequestration, its emissions remain higher at 75 kg CO_2 e/m^2, whereas reused concrete achieves a significantly lower 36 kg CO_2 e/m^2. This highlights reclaimed concrete as a key low-carbon option in sustainable construction, aligning with previous research (Bertin et al., 2022). By eliminating emissions from cement production and raw material extraction, reused concrete substantially cuts embodied carbon.

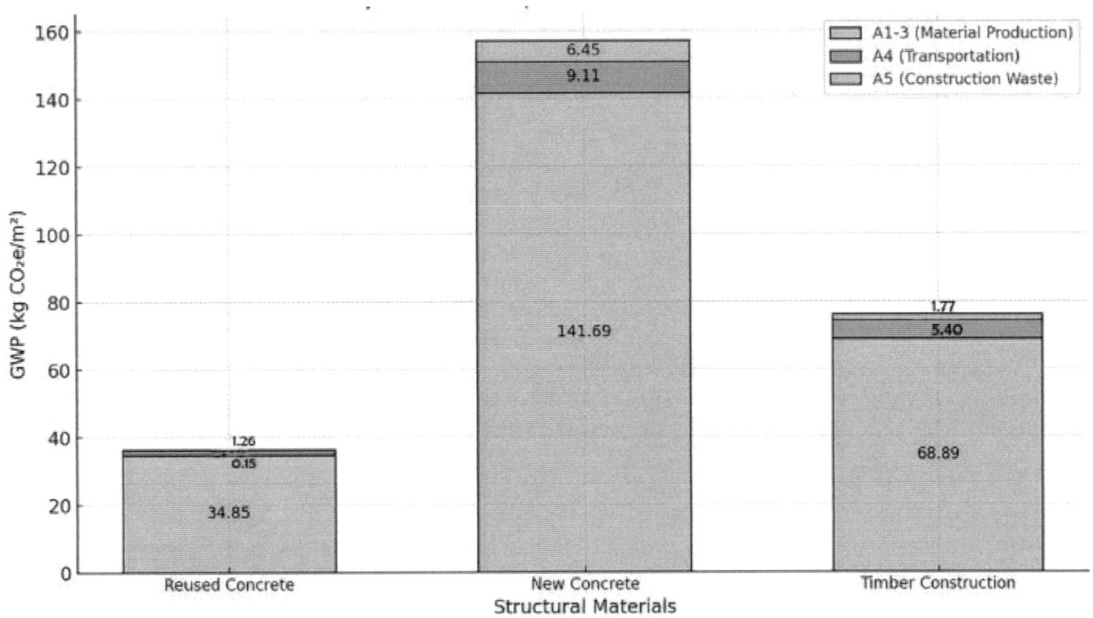

Figure 2. Global warming (GWP) from row house construction using reclaimed concrete elements, conventional virgin concrete and light timber across the lifecycle phases A1-A5

The study confirms that reusing structural concrete significantly lowers environmental impact, outperforming both traditional concrete and timber. Conventional concrete, with its high 157 kg CO_2 e/m^2 emissions, remains the least sustainable due to cement production's carbon intensity. Cement manufacturing is among the most polluting industrial processes, contributing heavily to CO_2 emissions through limestone calcination and high energy demand. These findings align with national standards, such as the Swedish benchmarks for small residential structures (Boverket, 2017), reinforcing the urgency of adopting alternative structural materials.

Although timber is widely regarded as a sustainable building material, its comparative impact depends on long-term carbon storage and sustainable forestry. Conventional GWP assessments often exclude biogenic carbon storage, affecting timber's relative performance. While timber construction contributes to emission reduction goals, its effectiveness depends on responsible forest management and material longevity (Andersen et al., 2022).

This study strongly supports reclaimed concrete elements may become a low-carbon alternative for structural applications. Its substantial GWP reduction underscores its role in circular construction while maintaining structural integrity. Additionally, localized reuse strategies and optimized transportation further enhance environmental benefits.

3.2 LCA STAGE CONTRIBUTIONS AND SENSITIVITY ANALYSIS

Accurate quantification of embodied carbon emissions and identification of mitigation potential relieve a comprehensive understanding of contributions from various life cycle stages. This section examines the proportional GWP impact through material manufacturing (A1–A3), transportation (A4), and construction waste management (A5). The percentage distribution of GWP over each scenario's various stages of the lifespan is illustrated in the figure. A thorough life cycle stage analysis further emphasizes the critical impact of material production (A1–A3) on total emissions. It can be found that the conventional concrete demonstrates a pronounced concentration of greenhouse gas (GHG) emissions in early life cycle stages (A1-A3), contributing 141.69 kg CO_2 e/m^{2}of its total emissions from raw material extraction and production. This highlights its reliance on carbon-intensive virgin resource processing. In comparison, the lumber derives 68.89 kg CO_2 e/m^2 of emissions from A1-A3, reflecting energy demands in forestry operations and sawmill processing, while the reclaimed concrete shows a markedly lower A1-A3 share at 34.85 kg CO_2 e/m^2, as recycling bypasses resource extraction and reduces manufacturing energy. The 72.8 kg CO_2 e/m^2 reduction in A1-A3 emissions for reclaimed concrete versus conventional concrete directly correlates with avoided virgin material use. This supports circular economy principles by demonstrating that reusing structural materials minimizes upstream impacts.

https://doi.org/10.52202/080684-0009

In every case, the findings indicate that stage A1–A3 (life cycle stage of product stage) primarily contributes to greenhouse gas emissions, highlighting its significant influence on overall environmental performance outcomes.

In the A4 (transportation) stage; the distances in sourcing materials play a crucial role in transport-related emissions, as Figure 3 illustrates. Compared to 5.8 % for new concrete and 2.3 % for timber, reused concrete has the lowest transport emissions at 0.4%, highlighting the enhanced carbon efficiency of localized material reuse. The study's material reuse strategy defines transportation distances as "short" or "medium." Initially, transportation emissions were assessed for an inspection station only 150 meters from the construction site in the case with relocated concrete elements, indicating a "short" distance. To evaluate the sensitivity of emissions to increased transport requirements for the reclaimed concrete elements, an alternative scenario considered an inspection station located 25 kilometers away ("medium"). The sensitivity analysis results (Figure 4) indicate that in the case of a reused concrete structure, a medium transport distance can lead to more than ten times the A4 emissions compared to a short transport distance. This finding emphasizes the urgent need for implementing regionally optimal sourcing policies to reduce the environmental impact of transportation in supply chains for building materials. Although A4 emissions are secondary to those from material manufacturing (A1–A3), their overall contribution to embodied carbon remains significant, particularly for goods transported over long distances. The results highlight that achieving the best carbon reduction outcomes in building projects depends on proximity to reuse locations, effective logistical planning, and minimized reliance on transportation. In the A5 (construction and installation) stage, emissions primarily arise from on-site energy consumption, equipment operation, and construction waste management. While reassembling and deconstructing recycled concrete components does generate emissions, these amounts are still significantly lower than those resulting from cement production in traditional concrete. The clear environmental benefits of material reuse greatly enhance circular building practices compared to the exploitation and processing of virgin resources. Further quantitative studies on on-site emission reduction strategies, energy-efficient deconstruction and reassembly processes, and improved waste disposal methods will improve the efficiency of A5 operations. Addressing these issues is crucial for optimizing circular building methods, lowering embodied carbon, and promoting sustainable material reuse systems. The findings reinforce the necessity of combining localized reuse strategies with efficient construction techniques to enhance GWP reductions during the A4 and A5 life cycle phases.

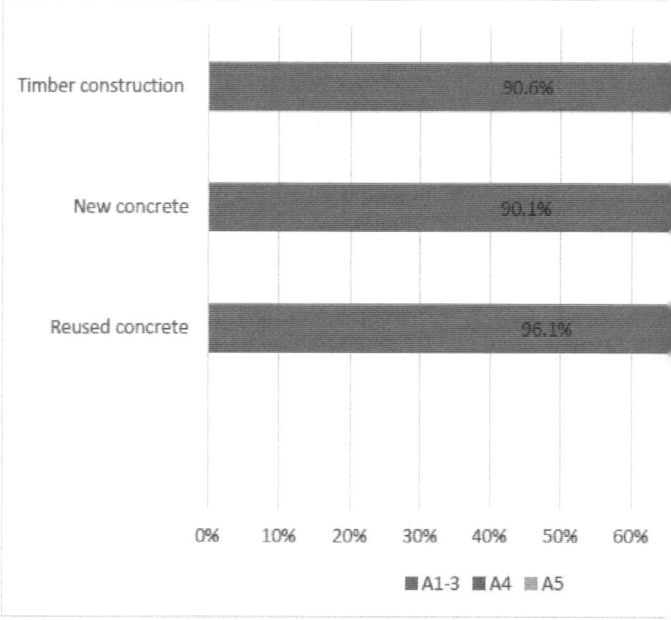

Figure 3: Proportion of GWP (kg CO_{2e}/m^2) for structures based on reclaimed concrete elements, conventional virgin concrete and light timber. Blue stands for material production, red for transportation and green for construction waste.

https://doi.org/10.52202/080684-0009

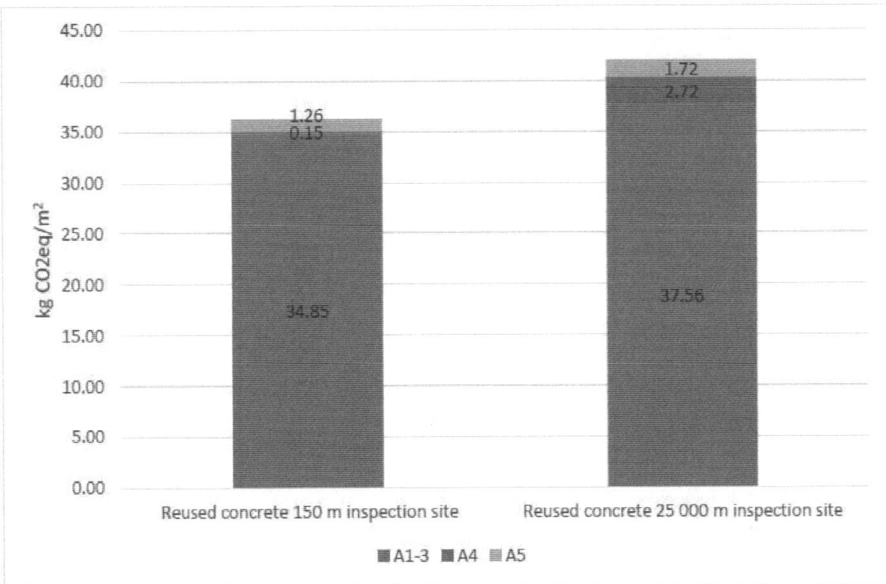

Figure 4: Sensitivity analysis of emissions with increased transport, assuming a 25 km inspection station ("medium" distance).

Carefully constructed assumptions and calculations ensured methodological consistency and validity. When accurate donor building data was unavailable, assumptions were derived from industry standards, past LCA assessments, and legal requirements. BIM models, architectural drawings, and structural inventories were utilized to estimate material quantities, ensuring the precision of resource utilization. While project-scale data predicted energy consumption and emissions during construction (A5), project-specific logistics and typical industry norms helped determine transport distances. Variations in transit lengths, construction energy consumption, and material processing impacts were evaluated using sensitivity studies, thereby assessing the robustness of the assumptions. In compliance with EN 15804 and EN 15978 criteria, the OneClick LCA tool was used for lifetime computations to ensure methodological accuracy. All assumptions and data limitations were documented clearly to enhance transparency, thereby improving the reliability and repeatability of the research. The assumption was made that the cement and concrete industry is highly localized globally. The average travel distance for in-situ concrete is 16km, while the average distance for concrete's raw materials is 48km (ICE,2023). To be economically competitive, localized sourcing of reuse material is crucial. When transportation distances were raised to "medium" for the relocated concrete, the transport suddenly accounted for twice as much. A "long" distance would have significantly impacted the reclaimed concrete case and shows that proximity in transportation is a critical factor when assessing construction waste (A5) and material reuse initiatives.

4 CONCLUSIONS AND RECOMMENDATIONS

This study compares the environmental performance of three structural solutions—reused concrete elements, new cast-in-place concrete, and timber—through a life cycle assessment (LCA) perspective. The findings demonstrate that reusing concrete elements significantly reduces embodied carbon emissions, positioning it as a key strategy for sustainable construction. Very few examples of reusing concrete elements exist so far, probably because building with virgin concrete is cheap. However, by avoiding the carbon-intensive cement production process, reclaimed concrete can lower the Global Warming Potential (GWP) of building frames by approximately 75%, saving over 100 kg CO_2 e/m² of floor area. Therefore, this potential should be investigated more as a viable low-carbon alternative for the construction industry.

Timber construction is widely recognized for its environmental benefits, offering a 52% reduction in GWP compared to conventional concrete. While timber is renewable and sequesters carbon, its processing emissions and durability limitations impact its overall sustainability. Although timber performs better than virgin concrete, reused concrete emerges as the most effective option for reducing embodied emissions, reinforcing the importance of circular economy strategies in construction.

Achieving sustainable construction requires context-specific solutions that balance carbon reduction with practical considerations such as cost, structural performance, workability, and material availability.

https://doi.org/10.52202/080684-0009

Expanding material reuse faces logistical barriers, including transportation distances, infrastructure limitations, and concerns over the long-term strength of reclaimed materials. To overcome these challenges, future research should explore metrics beyond GWP, such as durability and economic feasibility, strengthening reuse infrastructure as well as innovating design and engineering methods to facilitate material reuse in structural applications.

The results align with EU climate targets by supporting decarbonization in the built environment and promoting sustainable material management. Reclaimed concrete not only reduces waste but also advances low-carbon, resource-efficient construction, making it a fundamental strategy for climate-resilient building design. By prioritizing material reuse and minimizing embodied emissions, the construction sector can take significant steps toward carbon neutrality and a circular economy.

ACKNOWLEDGEMENTS

The authors thank the Swedish Energy Agency [Grant No. P2023-01473] for financial contributions to this study.

REFERENCES

International Energy Agency (IEA). (2022). Buildings: A source of enormous untapped efficiency potential. Retrieved from https://www.iea.org

Boverket. (2023). Miljöindikatorer för bygg- och fastighetssektorn. Retrieved from https://www.boverket.se

Küpfer, C., Bastien-Masse, M., & Fivet, C. (2023). Reuse of concrete components in new construction projects: Critical review of 77 circular precedents. Journal of Cleaner Production, 383, 135235.

Nikolakopoulos, A., Steriotis, T., Charalambopoulou, G., Karagiannakis, G., Dimitrakis, D., Michalis, V., & Katsiotis, M. (2024). Reducing carbon emissions in cement production through solarization of the calcination process and thermochemical energy storage. Computers & Chemical Engineering, 180, 108506

Abubakar, I. R., Maniruzzaman, K. M., Dano, U. L., AlShihri, F. S., AlShammari, M. S., Ahmed, S. M. S., ... & Alrawaf, T. I. (2022). Environmental sustainability impacts of solid waste management practices in the global South. International journal of environmental research and public health, 19(19), 12717

Minunno, R., O'Grady, T., Morrison, G. M., & Gruner, R. L. (2021). Investigating buildings' embodied energy and carbon: A systematic literature review and meta-analysis of life cycle assessments. Renewable and Sustainable Energy Reviews, 143, 110935.

Al-Najjar, A., & Malmqvist, T. (2025). Embodied carbon saving of reusing concrete elements in new buildings: A Swedish pilot study. Resources, Conservation and Recycling, 212, 107930.

Küpfer et al., (2023).Reuse of concrete components in new construction projects: critical review of 77 circular
precedents. J. Clean. Prod., 383 (2023), Article 135235, 10.1016/j.jclepro.2022.135235

Küpfer et al., (2024), Reuse of cut concrete slabs in new buildings for circular ultra-low-carbon floor designs
J. Clean. Prod., 448 (2024), Article 141566, 10.1016/j.jclepro.2024.141566

Miljöbyggnad. (2023). Miljöbyggnad 4.0. Swedish Green Building Council.

BREEAM. (2023). BREEAM SE V6. Swedish Green Building Council.

M. Erlandsson, "Blå Jungfrun version 2017 med nya cement," IVL Svenska Miljöinstitutet, 2017.

Boverket, "Limit values for climate impact from buildings and an expanded climate declaration," Swedish National Board of Housing, Building and Planning, 2023

Sbahieh, S., Serdar, M. Z., & Al-Ghamdi, S. G. (2023). Decarbonization strategies of building materials used in the construction industry. Materials Today: Proceedings.

Norouzi, M., Chàfer, M., Cabeza, L. F., Jiménez, L., & Boer, D. (2021). Circular economy in the building and construction sector: A scientific evolution analysis. Journal of Building Engineering, 44, 102704.

United Nations Environment Programme (UNEP). (2022, November 9). 2022 Global Status Report for Buildings and Construction. Retrieved April 23, 2024, from
https://www.unep.org/resources/publication/2022-global-status-report-buildings-and-construction

United Nations Environment Programme (UNEP). (2022). 2022 Global Status Report for Buildings and Construction: Towards a Zero-emission, Efficient and Resilient Buildings and Construction Sector. Nairobi, Kenya: UNEP.
n Energy, 95, 389-40

Low Carbon Concrete Group 2022. https://www.emccement.com/pdf/ICE_Low-Carbon-Concrete-Routemap_27.04.22.pdf, Accessed 2025-01-30

Bertin, I., Saadé, M., Le Roy, R., Jaeger, J.-M., & Feraille, A. (2022). Environmental impacts of design for reuse practices in the building sector. Journal of Cleaner Production, 349, 131228. https://doi.org/10.1016/j.jclepro.2022.131228

Andersen, J. H., Rasmussen, N. L., & Ryberg, M. W. (2022). Comparative life cycle assessment of cross-laminated timber building and concrete building with special focus on biogenic carbon. Energy and Buildings, 254, 111604. https://doi.org/10.1016/j.enbuild.2021.111604

Proceedings of the Circular Building Sector Conference 2025, Lund, Sweden, 1-3 June 2025

IMPACT OF REASSEMBLY ON THE MECHANICAL PROPERTIES OF STRUCTURAL FLOOR ELEMENTS MADE OF INDUSTRIAL WOOD RESIDUES

Sara Khanalizadehtaromi[1*] and Karin Sandberg[1]

[1]RISE Research Institutes of Sweden, Wood Technology Research

*Corresponding Author: sara.khanalizadehtaromi@ri.se

ABSTRACT

Background and aim. Considering the significant amount of timber constructions that end up in landfills or are incinerated, promoting efficient and circular use is essential. Designing structural elements for dis- and reassembly can extend their lifespan. However, uncertainties remain about these elements' material properties and functional performance after being disassembled, and whether they meet technical requirements for structural building products. This study investigates the impacts of using industrial wood residues to produce I-beams and multiple disassembly cycles on the mechanical properties of floor elements.

Methods and Data. The E-modulus and bending strength of elements were measured with bending tests performed according to EN 408:2010. The effects of dis- and reassembly on flooring elements made from a combination of graded sawn timber and industrial wood residues in the form of ungraded sawn timber offcuts were tested and evaluated after repeated cycles and compared to reference values. Initially, six elements were disassembled once or twice, and three elements were tested until failure to be considered as reference elements.

Findings. Two different types of reassembly processes were considered for the elements. The first reassembly type resulted in a decrease in both bending strength and E-modulus mean values. In contrast, the second reassembly type led to an approximately 78% increase in bending strength and a slight 9% decrease in E-modulus.

Theoretical / Practical / Societal implications. Using industrial wood residues in the form of ungraded sawn timber offcuts and graded sawn timber to produce load-bearing systems increases industrial wood residue utilization in structural elements. Studying the mechanical properties of elements after one or two dis- and reassembly processes ensures the user of the quality of elements after disassembly and increases the reuse rate and carbon storage time. The study shows that new end-of-life scenarios can be defined for flooring elements and industrial wood residues.

KEYWORDS: Circularity, Design for Disassembly, Experiment, Flooring Systems, Wood Residues.

1 INTRODUCTION

The construction industry's resource dependency and consumption, global greenhouse gas (GHG) emissions, and waste generation are massive (Munaro et al., 2021). The implementation of the circular economy concepts is recognized as the main solution to the existing environmental impacts of the construction sector and its transition to a more sustainable industry (Çimen, 2021). Studies on strategies to reduce the construction sector's embodied carbon emissions mention using materials with low embodied energy, such as timber, better design practices, reduction, reuse, and recovery of construction materials, refurbishment of existing buildings, and increased use of local materials. (Pomponi & Moncaster, 2016; Akbarnezhad & Xiao, 2017).

While timber is recognized as one of the most sustainable construction materials, increased demand and use of timber results in increased volumes of wood processing residues (Saal et al., 2017). Although, it is known that industrial wood residues are mostly incinerated or used in the production of engineered wood products such as chipboards; Saal et al. (2017) mention the utilization of these residues as a question that needs further analysis due to unknown available quantities, no clear internal or external consumption extents, and a few available studies

https://doi.org/10.52202/080684-0010

on utilization scenarios. Apart from sawn timber residues at the material processing phase, the significant amount of construction timber lost at landfills at their end-of-life phase or incinerated cannot be neglected either.
A construction project's linear life cycle starts with material extraction, processing, and manufacturing of components. It continues with the building assembly and use phase and ends with the demolition and waste creation stages (Crowther, 2005). The transition of this linear life cycle to a real cyclic one needs defining alternative end-of-life scenarios also known as closing the material loop. In the proposed cyclic life cycle model of a built environment the demolition stage is replaced with deconstruction and alternative end-of-life scenarios are defined as relocation or reuse of the entire building, reuse of components in a new building, reuse of material in production of new components, and recycling new material to produce new material. (Crowther, 2005).
Deconstruction as an alternative end-of-life scenario is defined by Rios et al. (2015) as salvaging material from a dismantled structure for reuse or recycling. Deconstruction has both opportunities and challenges. Opportunities existing in deconstruction can be categorized as environmental, social, economic, and other benefits (Rios et al., 2015). Deconstruction's challenges can be disregarding elements or materials that are damaged during deconstruction as they are not usable any longer.
Uncertainties also remain regarding the material properties and performance of elements after disassembly, and whether they meet technical requirements for structural building products (Rios et al., 2015). Jockwer et al. (2020) mention the lack of existing methods to evaluate the performance of the dismantled elements before reuse as one of the reasons that the circularity concepts are not yet effectively established in timber buildings. This can also be due to considering buildings to be long-lasting and not anticipating disassembly and reuse of their elements (Jockwer et al., 2020).
Design for deconstruction (DfD) refers to the importance of considering deconstruction as the end-of-life scenario in the design stage of structures (Densley Tingley, 2013). Designing structural elements for easier disassembly, and reuse can extend their lifespan and enhance future circular use. Cristescu et al. (2020) summarized novel design concepts for deconstruction and reuse of timber buildings in a state-of-the-art with a focus on Design for Deconstruction and Reuse (DfDR) in low-rise timber structures (Cristescu et al., 2020).

1.1 RESEARCH AIM

This study aims to investigate the impact of multiple dis- and reassembly cycles on the mechanical properties of I-beams for floor elements and the impact of using industrial wood residues in the I-beams' flange production.
In this experimental research, flooring elements that were designed for deconstruction with I-beams made of a combination of graded timber and industrial wood residues in the form of ungraded timber offcuts were studied. The number of elements received from the producer to be tested was limited. Two research questions were defined:

1) How will the combination of offcuts and graded timber affect the material properties of flooring elements?
2) How will the material properties of these flooring elements change after one or more dis- and reassembly processes?

2 MATERIAL AND METHOD

2.1 FLOORING ELEMENTS

The structure of the load-bearing elements investigated in this study was a section of flooring systems built by Masonite Beams AB in Sweden. The width and length of these flooring systems' sections were 150 *mm* and 4800 *mm*, respectively. All studied sections were built with 10 I-beams connected with 9 noggings, chipboard on top, and batten at the bottom. A drawing of an element can be seen in Figure 1.

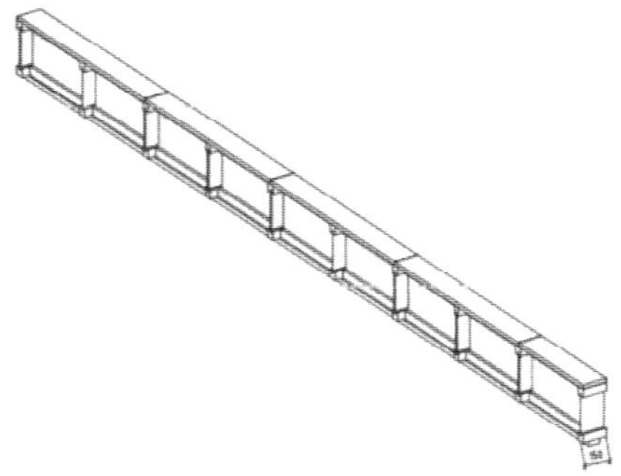

Figure 1: Drawing of an element investigated. Source: Masonite Beams AB.

The I-beams and noggings were made of H300s beams from Masonite Beams AB production where the total height of the beams was 300 *mm*, with 47×47 *mm* flanges, and 10 *mm* web. The chipboard thickness was 22 *mm*, and the width was 150 *mm*. The width of the battens used at the bottom of the elements was 70 *mm*, and the height was 34 *mm*. The flanges of the I-beams used in these elements were produced with finger jointing industrial wood residues in the form of ungraded timber offcuts with a minimum length of 150 *mm* and graded timber with strength class C30. To use the industrial wood residues, a new finger joint machine was added to the production line that could combine pieces with a minimum length of 150 *mm*. Different properties of these new finger-jointed pieces had to be tested before being used in the production of I-beams' flanges.

2.1.1 Labeling system

A total number of nine elements were studied. The elements' labels include a letter followed by two numbers separated by a dot. The letter indicates the group to which the element belongs to. The groups were called A, B, and R. Groups A and B included elements that experienced the dis- and reassembly processes twice and once, respectively. Group R refers to the reference elements. The first number refers to the number of the element within its group, and the second number indicates the number of times the element was tested. As an example, the element labelled A2.3 was the second element in group A tested for the third time.

2.2 METHOD

To answer both research questions defined earlier in this study, the mechanical properties of elements must be investigated. The European Standard EN 408:2010 includes laboratory methods to determine the mechanical properties of structural-size timber. In this study, the Swedish national version of EN 408:2010 that is SS-EN 408:2010+A1:2012 was used to investigate the mechanical properties of flooring elements. In accordance with this standard, the displacement (w) of elements was measured at the centre of the elements' span under the four-point bending test. Figure 2 shows a flooring element under the four-point bending test setup.

Figure 2: One of the flooring elements under the four-point bending test setup.

The global modulus of elasticity in bending, $E_{m,g}$, in N/mm^2, was determined based on equation (1).

$$E_{m,g} = \frac{3al^2 - 4a^3}{2bh^3(2\frac{w_2 - w_1}{F_2 - F_1} - \frac{6a}{5Gbh})} \quad (1)$$

where a = distance between a loading position and the nearest support, in mm, b = width of cross-section, in mm, and h = depth of cross-section, in mm. Defining $F_{max,est}$ as the estimated maximum load, in N, $F_2 = 0.4F_{max,est}$ and $F_1 = 0.1F_{max,est}$. The displacement values corresponding to F_2 and F_1 are w_2 and w_1, respectively. G = shear modulus. Here, based on the recommendations from standard EN 408:2010, G was considered infinite. The bending strength of beams was calculated according to Equation (2).

$$f_m = \frac{3Fa}{bh^2} \quad (2)$$

where a, b, and h were defined same as Equation (1). f_m = bending strength, in MPa and F = load, in N.

2.3 TEST STEPS

The steps taken to test the elements were different based on the group they were labelled as. The test was performed at RISE's laboratory located in Skellefteå, Sweden.

2.3.1 Reference group

Three of the nine elements, labelled group R, were tested until failure occurred under a four-point bending test following SS-EN 408:2010+A1:2012. The aim was to investigate the mechanical properties of elements built with I-beam flanges produced from a combination of industrial wood residues and graded sawn timber.

2.3.2 Dis- and reassembled groups

The other six elements, from groups A and B, were built with the same I-beams and noggings as group R. Moreover, they were designed for easier future disassembly leading to less damage to the materials by using screws and glue instead of nails and glue to attach the batten at the bottom to the I-beams and noggings in a 976 mm length, where the disassembly of elements was planned. The producer provided the instruction plans for the dis- and reassembly of elements.

The effects of deconstruction on the mechanical properties of these two groups were studied by testing them under a four-point bending test up to a certain load level, disassembling, reassembling, and bending the elements afterward. This cycle was done once or twice. Two different reassembly processes, type 1 for group A and type 2 for group B, were implemented for the dis- and reassembly of elements. In other words, the type of the dis- and reassembly processes performed on the elements was the classification factor for elements in groups A and B. The following subsections describe the disassembly process and the type of reassembly for each group of elements.

2.3.3 First-time disassembly for groups A and B

Both groups A and B, were designed in a way that they could be disassembled into two unequal parts in terms of size for easier handling and transportation from the first to the second location of use. An example of a disassembled element can be seen in Figure 3. The disassembly process included five steps as follows:

1) Removing the screws of the batten from underneath.
2) Removing the glued batten using a crowbar.
3) Removing the screws connecting the I-beam to its adjacent nogging.

https://doi.org/10.52202/080684-0010

4) Cutting the chipboard from the top of the I-beam's flange.
5) Taking two parts of the element apart.

Figure 3: An example of a disassembled element after its first four-point bending test.

2.3.4 First-time reassembly, type 1 for group A

It should be mentioned that based on producer's instruction plan, this reassembly type is recommended if the surface of the nogging's flange was destroyed less than 50% after the first disassembly and has enough surface for gluing back the batten. The reassembly process type 1 had four steps including:

1) Adding a 45×45 *mm* piece of timber on the upper part of the cross-section cut, between the nogging and the I-beam. The piece can be glued and nailed or glued and screwed.
2) Connecting the chipboard on top to the added 45×45 *mm* piece of timber with glue and screw.
3) Adding screws connecting the I-beam to its adjacent nogging.
4) Attaching the batten underneath with screws and glue.

2.3.5 First-time reassembly, type 2 for group B

This reassembly type had five steps. It should also be mentioned that this reassembly type is recommended by the producer if the nogging's flange surface was destroyed for 50% or more during the first disassembly and does not have enough surface for gluing back the batten. The mentioned recommendation does not rule out the use of this reassembly type if the nogging's flange surface was destroyed for less than 50%. The steps included:

1) Adding a 45×45 *mm* piece of timber on the upper part of the cross-section cut, between the nogging and the I-beam. The piece can be glued and nailed or glued and screwed.
2) Connecting the chipboard on top to the added 45×45 *mm* piece of timber with glue and screw.
3) Adding screws connecting the I-beam to its adjacent nogging.
4) Four pieces of 34×70×200 *mm* timber screwed and glued to both sides of two I-beams in the middle.
5) Attaching two parts of 28×70×976 *mm* battens underneath the element. The battens were laterally shifted and were glued and screwed.

Figure 4 shows a view of the elements from underneath with both types of reassemblies.

Figure 4: View of the elements from underneath with reassembly types 1, group A, (on the top) and type 2, group B, (at the bottom) after the first dis- and reassembly.

2.3.6 Second-time disassembly for group A

Disassembling the elements of group A for the second time had 5 steps similar to the first-time disassembly. The difference can be seen in step 4 where the section to cut the chipboard changes from the vicinity of the I-beam's flange to the vicinity of the 45×45 *mm* piece added during the reassembly process.

1) Removing the screws of the batten from underneath.
2) Removing the glued batten using a crowbar.
3) Removing the screws connecting the I-beam to its adjacent nogging.
4) Cutting the chipboard from the top close to the added 45×45 *mm* piece of timber.
5) Taking two parts of the element apart.

2.3.7 Second-time reassembly, type 1 for group A

Before running the four-point bending test for the third time on elements in group A, they were reassembled once again under the following process including five steps:

1) Adding another 45×45 *mm* piece of timber on the upper part of the cross-section cut beside the 45×45 *mm* piece added to the element on the first reassembly process. The piece can be glued and nailed or glued and screwed.
2) Connecting the chipboard on top to the 45×45 *mm* piece added in step 1 with glue and screw.
3) Adding screws connecting the I-beam to its adjacent nogging.

4) Two pieces of 34×70×200 *mm* timber screwed and glued to one side of two I-beams in the middle.
5) Attaching one part of 34×70×976 *mm* batten underneath the element.

3 FINDINGS

This section presents the results from the four-point bending tests on all the tested elements. In all tables, w_{max} (*mm*) is the displacement value when reaching the maximum force F_{max} (*kN*), $E_{m,g}$ (*N/mm²*) and f_m (*MPa*) are the E-modulus and bending strength values, respectively.
Table 1 presents the results of the reference elements R1-3.

Table 1: Results of testing reference elements under four-point bending test until failure and corresponding E-modulus and bending strength values.

Element	F_{max} (*kN*)	w_{max} (*mm*)	$E_{m,g}$ (*N/mm²*)	f_m (*MPa*)
R1.1	15. 9	33.8	8250	15.2
R2.1	15.2	33.8	7398	14.5
R3.1	16.3	33.0	8019	15.6
Mean	15.8	34.9	7889	15.1

Table 2 presents the results related to elements A1-3 before disassembly. Tables 3 and 4 provide the results of elements A1-3 after their first and second dis- and reassembly processes, respectively.
The results indicated a decrease in all the mean values after each cycle of dis- and reassembly processes. Compared to the mean values related to elements tested before disassembly, maximum force (F_{max}) and correlatively bending strength after the first and second dis- and reassembly processes decreased by 33%, and 41%, respectively. Modulus of elasticity also showed around 11% decrease in values after both dis- and reassemblies compared to the state before disassembling elements.
Looking at the dis- and reassembly steps related to this group, these lower values can be explained by the impact of the two disassembly processes on the integrity of the elements by cutting the chipboard, unscrewing, and screwing back the I-beam to their adjacent noggings. All these factors lead to lower strength and enable more deflections under lower applied stress in the elements.

Table 2: Results of testing elements A1-3 under four-point bending test before disassembly and corresponding E-modulus and bending strength values.

Element	F_{max} (*kN*)	w_{max} (*mm*)	$E_{m,g}$ (*N/mm²*)	f_m (*MPa*)
A1.1	7.0	16.8	6514	6.7
A2.1	7.1	17.4	6068	6.8
A3.1	6.0	16.5	4764	5.8
Mean	6.7	16.9	5782	6.4

Table 3: Results of testing elements A1-3 under four-point bending test after one dis- and reassembly process and corresponding E-modulus and bending strength values.

Element	F_{max} (*kN*)	w_{max} (*mm*)	$E_{m,g}$ (*N/mm²*)	f_m (*MPa*)
A1.2	1.7	6.3	5123	1.7
A2.2	5.5	17.7	5367	5.2
A3.2	6.2	18.6	4943	6.0
Mean	4.5	14.2	5144	4.3

Table 4: Results of testing elements A1-3 under four-point bending test after two dis- and reassembly processes and corresponding E-modulus and bending strength values.

Element	F_{max} (*kN*)	w_{max} (*mm*)	$E_{m,g}$ (*N/mm²*)	f_m (*MPa*)
A1.3	3.8	11.2	5163	3.7
A2.3	3.9	10.2	5359	3.7
A3.3	4.0	11.4	4770	3.8
Mean	3.9	10.9	5098	3.7

Tables 5 and 6 present the results of elements B1-3 before and after their dis- and reassembly processes, respectively. With the reassembly type 2, the elements' F_{max} and correlatively the bending strength increased by around 78%, while a slight decrease of 9% was shown in the modulus of elasticity.

Table 5: Results of testing elements B1-3 under four-point bending test before disassembly and corresponding E-modulus and bending strength values.

Element	F_{max} (*kN*)	w_{max} (*mm*)	$E_{m,g}$ (*N/mm²*)	f_m (*MPa*)
B1.1	5.3	11.6	6624	5.0
B2.1	5.1	13.0	6268	4.8
B3.1	5.4	12.0	6988	5.1
Mean	5.2	12.2	6627	5.0

Table 6: Results of testing elements A1-3 under four-point bending test after one dis- and reassembly process and corresponding E-modulus and bending strength values.

https://doi.org/10.52202/080684-0010

Element	F_{max} (*kN*)	w_{max} (*mm*)	$E_{m,g}$ (*N/mm²*)	f_m (*MPa*)
B1.2	9.1	23.8	6015	8.8
B2.2	9.8	27.1	5530	9.4
B3.2	9.0	23.0	6640	8.6
Mean	9.3	24.7	6062	8.9

The differences between the values of elements A1-3 and B1-3 before their first disassembly presented in Tables 2 and 5, respectively, can be interpreted by different factors impacting the mechanical properties and quality of timber elements. As the beams are made from a combination of graded timber and industrial residues in the form of ungraded offcuts, the impacting factors can be named as the number of ungraded offcuts and consequently, the amount of glue used in finger jointing in the production of elements.

The impact of the two different types of reassemblies can be seen in the values presented for A1-3 and B1-3 after their dis- and reassembly processes. While the disassembly processes impact the integrity of elements, reassembly type 2 showed to have a more compensating impact on the B1-3 elements' properties. Although the elements showed to experience higher deflections under lower applied stress leading to lower E-modulus, the impacts of added timber reinforcements to the sides and under the elements with screws and glue can be seen in the increased bending strength.

Compared to both groups of elements A1-3 and B1-3, reference elements had higher mean values. The decrease in the values of elements in groups A and B compared to the values from the reference elements can be explained by the fact that while the battens in the reference elements were one piece nailed and glued under the elements, the battens underneath the floor elements in groups A and B were cut in 976 *mm* length in the section that was planned for disassembly and were glued and screwed. The change caused a lower strength in bending and more flexibility in the elements.

4 CONCLUSIONS

In this research, the use of industrial wood residues in combination with graded sawn timber in the production of structural flooring elements and the effects on the mechanical properties of these elements was studied. To extend the lifespan of these elements and increase the carbon storage time they were also designed for easier disassembly. Existing uncertainties regarding the effects of dis- and reassembly on elements were also investigated by studying the mechanical properties of flooring systems after one or two disassembly cycles.

Looking at the values from the reference elements and comparing them to the elements that were designed for disassembly the reference elements have both higher E-modulus and bending strength. Although this can be interpreted as a need to improve elements that are designed for disassembly; compared to the values of the flooring systems made with only graded sawn timber from the same producer, these elements' properties are still within an acceptable range before disassembly.

The decrease witnessed in the mechanical properties of group A, enables using the elements in the structures with lower requirements after the first and second dis- and reassembly processes. For group B, although there was a small decrease in E-modulus value, the bending strength was increased significantly.

The results of this study emphasize that the production of structural elements from both industrial residues and graded sawn timber leads to an increase in industrial wood residues utilization rate in load-bearing systems and ensures the quality of structural elements after dis- and reassembly when the right reassembly type is chosen.

It is worth mentioning, that using industrial wood residues with a minimum 150 *mm* length required the manufacturer to add a new machine to the production line and test different properties of the finger-jointed pieces before producing I-beams' flanges. Although uncertainties about the available quantities of industrial wood residues and their consumption scenarios still exist, the results of this study highlight the possibility of defining new end-of-life scenarios for both industrial wood residues and the produced flooring elements.

In the built environment and construction industry, the results can emphasize the existing possibilities in defining alternative end-of-life scenarios for buildings and their elements, increased use of reused structural elements, and establishment of more circular transition concepts in this industry.

For future studies, performing the same tests with dis- and reassembly processes on more elements or computational simulations can verify the results presented in this study.

ACKNOWLEDGEMENTS

The research for this article was carried out under the project DUET- Circular Design and Use of Wood Building Elements within the Bioeconomy in the North program and financed in Sweden by Vinnova, Sweden's Innovations Authority.

The authors acknowledge the contribution of Tommy Persson and Mikael Thors from Masonite Beams AB and Urban Häggström from RISE for all their support and efforts in performing the tests at RISE's laboratory in Skellefteå.

REFERENCES

Akbarnezhad, A., & Xiao, J. (2017). Estimation and minimization of embodied carbon of buildings: A review. Buildings, 7(1), 5. https://doi.org/10.3390/buildings7010005

Çimen, Ö. (2021). Construction and built environment in circular economy: A comprehensive literature review. Journal of cleaner production, 305, 127180. https://doi.org/10.1016/j.jclepro.2021.127180

Cristescu, C., Honfi, D., Sandberg, K., Shotton, E., & Walsh, S. J. (2020). Design for deconstruction and

reuse of timber structures–state of the art review. https://doi.org/10.23699/bh1w-zn97
Crowther, P. (2005). Design for disassembly—themes and principles. RAIA/BDP Environment Design Guide, Australia.
Densley Tingley, D. (2013). Design for deconstruction: an appraisal (Doctoral dissertation, University of Sheffield).
Jockwer, R., Goto, Y., Scharn, E., & Crona, K. (2020, November). Design for adaption–making timber buildings ready for circular use and extended service life. In *IOP conference series: earth and environmental science* (Vol. 588, No. 5, p. 052025). IOP Publishing. http://dx.doi.org/10.1088/1755-1315/588/5/052025
Munaro, M. R., Tavares, S. F., & Bragança, L. (2022). The ecodesign methodologies to achieve buildings' deconstruction: a review and framework. *Sustainable Production and Consumption*, *30*, 566-583. https://doi.org/10.1016/j.spc.2021.12.032
Pomponi, F., & Moncaster, A. (2016). Embodied carbon mitigation and reduction in the built environment–What does the evidence say? Journal of environmental management, 181, 687-700. https://doi.org/10.1016/j.jenvman.2016.08.036
Rios, F. C., Chong, W. K., & Grau, D. (2015). Design for disassembly and deconstruction- challenges and opportunities. Procedia engineering, 118, 1296-1304. https://doi.org/10.1016/j.proeng.2015.08.485
Saal, U., Weimar, H., & Mantau, U. (2019). Wood processing residues. *Biorefineries*, 27-41. http://dx.doi.org/10.1007/10_2016_69
SIS. (2012). SS-EN 408:2010+A1:2012 Timber structures – Structural timber and glued laminated timber – Determination of some physical and mechanical properties. Swedish Standards Institute.

https://doi.org/10.52202/080684-0010

Proceedings of the Circular Building Sector Conference 2025, Lund, Sweden, 1-3 June 2025

THREE OBJECTIVES FOR A HOLISTIC CIRCULAR ECONOMY POTENTIAL ASSESSMENT OF BUILT ASSETS

Goran Sibenik[1*], Rudi Stouffs[2], and Pieter Herthogs[1]

[1]Singapore-ETH Centre, Future Cities Lab Global Programme, CREATE campus, Singapore

[2]National University of Singapore, College of Design and Engineering, Department of Architecture, Singapore

[*]Corresponding Author: goran.sibenik@sec.ethz.ch

ABSTRACT

Background and aim. Built assets can contribute to the circular economy (CE) in several ways, implying there are multiple CE objectives to consider. Existing measurement methods often yield suboptimal results for objectives that are not their focus. We are developing a holistic method for measuring the CEP of built assets; the present paper proposes three key CE objectives essential for a holistic CEP calculation.

Methods and Data. A holistic set of CE objectives encompasses all relevant end-of-use and end-of-life strategies for the CEP of a built asset. We analyse existing circularity quantifications, identify challenges that prevent a holistic assessment, and determine the requirements for a novel set of CE objectives. Furthermore, we propose and verify the novel set of objectives using the CEP framework and three illustrative use cases.

Findings. The three objectives for a holistic CEP assessment are longevity, reusability, and transformability. Longevity implies that a resource remains in place; reusability refers to outflows that retain a similar form and embodied value; and transformability involves the outflows that change their form. Through validation we demonstrate that these objectives apply to previously identified CEP influencing factors and all use cases.

Theoretical / Practical / Societal implications. The novel system of objectives provides a foundation for more accurate measurement of the CEP in the building sector. The proposed set is sufficient for navigating the complex landscape of CE assessments, strategies and parameters. A quantification method encompassing all objectives and reliably reflecting real-world performance would incentivise more circular design of built assets.

KEYWORDS: adaptability, building configuration, circularity, flexibility, quantification.

1 INTRODUCTION

1.1 CIRCULAR ECONOMY IS A PARADIGM ENCOMPASSING A RICH SET OF PRACTICES

The growing environmental impact of the construction industry is increasingly recognized in both research and practice (Bertino et al., 2021). A potential solution to reduce this unsustainable influence lies in approaches encapsulated within the concept of the circular economy (CE). The principles of CE are diverse and encompass a broad set of recommendations: eliminating waste and pollution, circulating products and materials, and regenerating nature (Ellen MacArthur Foundation, 2021). However, only certain approaches are directly relevant to the construction industry. While it is widely accepted that transitioning to more sustainable practices is essential, and that the construction industry plays a pivotal role in this transition, it remains challenging to quantify and compare the effects of different practices. The construction industry faces specific challenges and requirements for achieving this transition, which have been the topic of multiple literature reviews. These reviews focus on multiple aspects, such as multiple phases of the building lifecycle (Cimen, 2021), strategies influencing CE practices (Eberhardt et al., 2022), obstacles to implementation (Charef et al., 2021), or company-specific implementation (Nunez-Cacho et al.,

2018). Consolidating this extensive set of influencing factors into a measurable set of performance indicators has to our knowledge not been achieved.

1.2 MEASURING BUILDING CE POTENTIAL IS A CRITICAL OBSTACLE

The inherent CE potential (CEP) of built assets refers to the various ways in which built assets and their sub-parts can support or hinder CE, some of which are challenging to measure. We use the term "built asset" as defined in ISO 19650 (ISO, 2018), as it encompasses both infrastructural constructions and carries an economic meaning. However, terms like "construction asset", "built object" or "built structure" are also appropriate. Prior studies have identified two critical obstacles to effectively leveraging CE principles: a lack of a holistic perspective on CE performance (Ossio et al., 2023) and a lack of practical methods for its measurement (Eberhardt et al., 2022; Hossain et al., 2020). In our previous work, we explored multiple quantification methods for the CEP which consider the configuration of built assets and its relation to the CE (Sibenik et al., 2025). The ultimate goal of our research is to develop a holistic digital tool capable of measuring the CEP of built assets. By doing so, we aim to cover both methodological and technological gaps, which are identified as critical barriers to advancing CE principles in the construction industry (Gasparri et al., 2023). In this way, our work aims to contribute to the four most important sub-clusters within the gaps as identified by Gasparri et al. (2023): design, policies/standards, assessment method and digitalisation.

Assessing the value of a built asset at its end of life (EoL) or end of use (EoU) is particularly complex. We refer to EoL and EoU as described in Murakami et al. (2010); however due to the more complex user arrangements of built assets compared to other commodities, we will adapt these terms for the construction industry in the discussion (section 5). Built asset lifetimes are generally long and unpredictable, demand for construction materials and components depends on the broader urban context, and there is often insufficient information about disassembly procedures, as well as the condition and performance of constituents. Moreover, different sub-parts of a built asset have different lifespans. Although various measurement and certification methods exist, they are typically limited on specific objectives and EoL scenarios, resulting in a lack of holistic assessment. These methods promote best practices for a certain CE objective, while other practices relevant for the CE are often excluded.

1.3 STRUCTURING THIS PAPER

This paper proposes a set of three key objectives that together holistically and effectively represent the ways in which built assets can influence the CE: longevity, reusability and transformability. Effectively, in this context, means that numerous end-of-use and end-of-life strategies relevant for the CE are represented by three objectives that consider a limited set of factors affecting the CEP for built assets.

We introduce these objectives in the following sections. The background section outlines existing CE objectives and the main challenges associated with applying CE principles in the construction industry. The methodology section presents how we developed and tested the novel holistic set of objectives for CEP assessment of built assets in this study. In the findings section, we formally present our novel system centred on the three objectives for implementing CE principles in built assets. Finally, the discussion connects the findings to our future research directions.

2 BACKGROUND

2.1 CE STRATEGIES FOR BUILT ASSETS ARE INCONSISTENT WITH EACH OTHER

The literature reveals various CE goals for built assets and their constituents. These goals are sometimes referred to as dimensions, types, aspects, or actions, but they all describe strategies articulating CE practices, particularly at the EoU of the built asset. For clarity, we will refer to the various goals described in literature as CE strategies, to distinguish them from our new proposal consisting of a set of three objectives. Durmisevic (2015) identifies three dimensions of transformation by design that determine strategies across different asset tiers of a building: building transformation (adaptability on building level), system transformation (reconfiguration, reuse, repurpose), and material dimension (upcycling/ downcycling). Durmisevic (2015) also highlights "design quality" or "prevention by design" as a key factor influencing the future building value, emphasizing how psychological factors and the importance of place influence the sustainability of buildings. The Circularity Gap Report for Holland identifies four groups of CE strategies, referred to as "scenarios", each with a different impact on the labour market (Circle Economy & Metabolic, 2022). The scenario most relevant to the design of built assets includes three strategies: "design to reduce", "design for cyclability" and "design to last". The scenarios focus on different effects on the labour market; therefore, the strategies are not necessarily mutually exclusive or exhaustive across the scenarios.

The EU waste hierarchy (Directive 2008/98/EC, 2008) outlines five waste management strategies: prevention, preparing for reuse, recycling, recovery, and disposal. The hierarchy recognizes construction and demolition waste as the priority resource stream, and aims to reduce its disposal. The EC has developed a tool to measure CE principles in buildings, called Level(s) (Directorate-General for Environment, 2025). This tool includes indicators for efficient and circular material life cycles, such as the use of: (a) bill of quantities, materials and lifespans, (b) construction and demolition waste and materials, (c) design for adaptability and renovation, and

https://doi.org/10.52202/080684-0011

(d) design for deconstruction, reuse, and recycling resource. Another EU agency publication emphasises importance of reuse of the existing building stock, listing three circular renovation actions: increasing lifespans, reducing material consumption, and making use of new generation materials (European Environment Agency, 2022). According to Hakaste et al. (2024), durability, adaptability, and reusability (including deconstruction) are crucial CE strategies influencing a building's lifecycle performance. These three strategies contribute to the CE in different ways. Reusability and deconstruction include three additional strategies: ease of disassembly, ease of reuse, and ease of recycling. Each of these strategies requires specific design considerations and dispositions.

Bertino et al. (2021) identify four EoL strategies for buildings: maintenance, refurbishment, demolition, and deconstruction. Additionally, they outline four deconstruction strategies for a built asset: reuse of the entire building (relocating it to a new site), components reuse in other buildings, material reprocessing, and material recycling. Similarly, Marsh et al. (2022) propose CE strategies for concrete, including design for durability, component reuse, and material recycling.

From this diverse body of literature, it is evident that the CE strategies identified across various works lack a unified structure and are not consistently aligned.

2.2 CE IN CONSTRUCTION INDUSTRY SHOWS CHARACTERISTIC CHALLENGES

This subsection lists how we summarized characteristic challenges of implementing CE principles in the construction industry from diverse research work. The list presented here is not exhaustive; more comprehensive literature reviews on the barriers and gaps in the CE application can be found in sources such as Charef et al. (2021) or Gasparri et al. (2023). The challenges highlighted in this subsection are specifically relevant to the quantification of CEP in the construction sector. The following challenges, drawn from the literature (more detailed reasoning can be found in Sibenik et al. (2025)), will guide the novel proposal of CE objectives presented in this work:

Current practices:

- Destructive demolition: common practice, often arbitrary, but faster and cheaper (Bertino et al., 2021)
- Current building stock: most existing buildings were not designed with EoL scenarios in mind (Bertino et al., 2021; European Environment Agency, 2022)

CE flows:

- Avoiding demolition: retaining structures has significant potential for greenhouse gas (GHG) saving (European Environment Agency, 2022; Moisio et al., 2024)
- Building components configuration: simpler, reusable, and deconstruct-able constituents are preferred (Bertino et al., 2021)
- Building adaptability: this is a widely recognized CE strategy (Eberhardt et al., 2022)
- CE flow of components: reuse in their original form is desirable (Ossio et al., 2023)
- CE flow of materials: recycling is the most common practice (Marsh et al., 2022)

Quantification:

- Prioritizing powerful CE strategies: strategies beyond recycling and recovery should be prioritized for greater impact (Morseletto, 2020)
- LCA calculations: these often focus on downscaling and may not capture the full CEP (Ossio et al., 2023)
- Transportation performance at EoL: evaluating transportation impacts requires numerous assumptions, complicating assessments (Moisio et al., 2024)

3 METHODS AND DATA

The objectives of CEP assessment represent various approaches to evaluating the value of building constituents after their primary use has ended and their purpose changed. This study posits three objectives for a holistic CEP assessment and demonstrates their application. The objectives are informed by reviews, case studies, and reports, as well as by the CEP framework previously proposed by Sibenik et al. (2025). Additionally, we outline factors that influence objective-based performance of built assets, including essential considerations for future calculation.

A previously conducted review of quantification methods (Sibenik et al., 2025) serves as the starting point. The CEP framework it introduced comprises three key elements that collectively shape the CEP of a built structure.:

1) Design strategies: the interdependencies between constituents, particularly combinations of their geometrical and topological properties, significantly influence the CEP and the methods used for its measurement.
2) Asset tiers: the hierarchical partitive relationships within a built structure, from the entire built asset to the material tier. By considering multiple tiers, individual and combined performance calculations determine the overall CEP.
3) CE flows: transitions between uses and lifecycles, meaning changes in primary use or location, involve diverse strategies for new purpose of all constituents of built asset. These strategies have varying environmental performance, particularly regarding their GHG emissions.

In addition, we incorporated additional analyses focused on works specifically determining the CEP of built assets. Such works often address the topic without necessarily providing a quantification method or naming the objectives of the potential to be measured.

By critically reviewing existing trends in the literature and employing deductive reasoning, the present study proposes three CE objectives. Initially, the system of three objectives was tested through examples related to the design strategies of built assets, each tier of the asset, and various end-of-use strategies. This test evaluated the system's scope of application and its holistic nature.

Subsequently, the objectives were tested in three descriptive test cases:

(1) Adaptation potential: focuses on quantifying the potential to adapt building configuration; the adaptation potential test case is identified from the literature review (Sibenik et al., 2025),
(2) Temporary work components: this use case investigates the CEP of temporary work components such as formwork; it is explored as part of our research project and rarely addressed in the existing literature (Tizzani et al., 2023).
(3) Materials passports: this use case uses the set of objectives with the intention of creating a materials passport; it is a widely recognized approach used to calculate and document the materials and components within a building (Honic et al., 2019).

These illustrative use cases were selected to represent a range of applications and challenges within the CE framework. The examples were discussed and analysed through meetings and workshops within the with the authors and one other CE researcher, providing insights into the applicability and robustness of the proposed objectives. The team of this project, called Circular Future Cities, explores different aspects of implementing CE in the building sector (ETH Zurich, 2025).

4 FINDINGS

4.1 CE OBJECTIVES ARE LONGEVITY, REUSABILITY AND TRANSFORMABILITY

We observe that while different CE objectives are frequently applied together, existing methodologies typically focus on specific aspects, such as materials passports or LCA. As a result, current CE calculations often fail to encompass all critical circularity aspects of a built asset, especially the challenges listed in subsection 2.2. Our proposed set of three CE objectives consolidates methods for different design strategies and asset tiers into a unified system.

We set the boundary conditions for the novel system unifying the CEP objectives, which needs to be:

- Holistic – all common CE flows are considered by the system and it can be applied to any type of built asset and its sub-parts.
- Measurable – all objectives will provide a numerical value describing the performance of any building or its sub-parts.
- Simultaneous – objectives can be implemented simultaneously within a single built asset, as constituents often intended for different CE flows.
- Compatible – results of objectives can be quantified individually, but then they can be combined into a single comprehensive result.
- Inclusive – not limited to high-performing elements or those specifically designed to meet a particular objective. Instead, all building constituents can be assessed, with their circularity performance reflected as a positive or negative value.

The integrative approach considering various CE objectives represents a novel contribution. Although this research paper cannot be separated from the framework and literature review (Sibenik et al., 2025) and the CEP calculation tool currently under development, it establishes the link between the two and explains the main solution that addresses the challenges identified in the assessment methods. The findings lead to the three objectives that should be targeted during the design of a built asset, but can also be used to evaluate assets not explicitly following the objectives. These three CE objectives and their relationships are depicted in Figure 1 and defined and elaborated upon below. Examples of typical assets, across all asset tiers, which can be calculated with specific objectives are represented in Table 1.

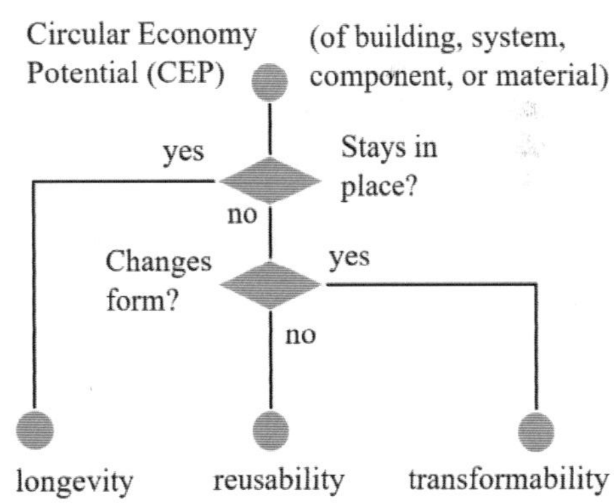

Figure 1 Graphical diagram representing relations between the CEP objectives based on their features

4.1.1 Longevity

Longevity is the objective to keep constituents of a built asset where they are currently located in space, maximizing potential timespan, while continuing to meet existing or changing performance requirements. This objective applies to all constituents of a built asset intended to remain in one place over time. It is closely associated with design strategies such as adaptability, maintainability, and flexibility. Buildings with open and

https://doi.org/10.52202/080684-0011

flexible floor plans often align with this objective. The constituents following the longevity objective may be considered a fixed asset in economic terms, however, with a significantly longer lifespan than one year. Achieving longevity requires specific design considerations to ensure buildings, systems, components or materials are durable, easily accessible and can meet changing future needs. Longevity is assessed individually for each component; however, the assessment must also be performed at the system and building tiers to provide overall results.

Table 1 Examples of "built structure units" that are likely to perform well for particular objectives

	Longevity	Reusability	Transform-ability
Building	Keeping a building	Moving a building	Demolishing a building
System	Over-dimensioned structural system	Kit-of-parts for partitioning walls	HVAC system
Component	Concrete column	Steel beam	Heat Pump
Material	Concrete	Gravel for surface cover	Coolant

4.1.2 Reusability

Resources that do not remain in place (i.e. outflows into the CE) are distinguished according to their further use. Outflows that retain their form and embodied value are represented with the reusability objective. Outflows that are transformed into another form follow the transformability objective.

Reusability is the objective to maximally retain the original functional performance of a building sub-part after extracting it from its location in space. In economic terms, reusability can be considered a movable asset. The reusability objective focuses on constituents created for dispositions like design for deconstruction, modularity, prefabrication, and the use of standardized constituents adaptable to various contexts. Constituents that score well in reusability should meet the following criteria: enable damage-free deconstruction and detachment, be easily accessible, maintain their value and functionality after deconstruction, requiring minimal repair or improvement, retain generic functional properties suitable for reuse, and exclude significant changes in embodied emissions for new use, aside from transportation and storage impacts. Reusability assessment can be calculated for individual components and is less dependent on the system and building tiers; however, the reusability still needs to include assessments of accessibility and deconstruct-ability. Some external factors like market value or technical requirements might also affect reusability; assessing such factors are out of our scope as they rely on more speculative methods.

4.1.3 Transformability

Transformability is the objective to minimize the additional GHG emissions when a constituent changes its state and is transformed in such a way that it remains a part of the CE, even if changing its intended use. The transformability objective addresses outflows that undergo changes in form and embodied GHG value. In economic terms, transformability can be seen as aggregate asset. It includes widely practiced circularity practices such as recycling, incineration, upcycling or downcycling, and even biodegradation. The effectiveness of these practices varies significantly and is seldom quantified and compared with other transformability options. To achieve a high score in transformability, energy or GHG emissions invested must be kept low. The calculation compares input and output GHG levels for a likely EoU transformation and considers the state-change activities. Relevant design strategies at the component tier include ease of deconstruction and accessibility, similar to reusability, which also require consideration of higher asset tiers. A typical example of transformability at the building tier is demolishing a building, where the resulting GHG values are compared to the initial ones, accounting for all activities in the process. This objective at the building tier resembles material flow analysis calculations.

4.2 OBJECTIVES CONSIDER ALL ELEMENTS OF THE CEP FRAMEWORK

While the proposed objectives should encompass all elements of the CEP framework presented by Sibenik et al. (2025), their application yields varied results depending on the context. To enhance understanding, the objectives are tested with all elements of the framework and use cases.

The design strategies outlined as circular building adaptability determinants by Hamida et al. (2023) are listed and aligned with the proposed objectives, as illustrated in Table 2. These strategies have varying relevance on the objectives, highlighting their interdependencies. Design strategies for each construction asset tier are analysed for their suitability for specific objectives, revealing that some tiers align more closely with certain objectives while others are better suited to alternative objectives.

These objectives are also contextualized within lifecycle '9R' CE strategies described by Circle Economy & Realdania (2025). The CE strategies show a hierarchical order with "refuse" being the most circular and "recover" the least circular. We evaluate CE EoU strategies within the context of the objectives, aiming to encompass all types of circular strategies. The strength of their alignment is detailed in Table 4.

https://doi.org/10.52202/080684-0011

Table 2 Probability of relevance of design strategies for CE objectives (A-always, S-sometimes, R-rarely, rest N.A.)

Design strategy	Asset tier	Longevity	Reusability	Transform-ability
Flexibility	building	A	R	
	system	A	A	
	component			
	material			
Deconstruct-ability	building		A	R
	system	S	A	S
	component	S	A	S
	material	S	S	S
Multi-usability	building	S		
	system			
	component			
	material			
Regularity	building	A	S	S
	system	A	A	S
	component	S	A	S
	material	R	A	S
Convertibility	building	A	S	S
	system	A	S	S
	component	R	R	S
	material			S
Reversibility	building			A
	system			A
	component			A
	material			A
Maintain-ability	building	A	A	R
	system	A	A	R
	component	A	A	R
	material	A	A	R
Recovery	building	S	S	
	system	S	S	
	component			R
	material			R
Scalability	building	A	A	R
	system	A	A	R
	component	A	A	R
	material	A	A	R
Refit-ability	building	S	R	R
	system	S	A	R
	component	S	A	R
	material			

Table 3 Strength of alignment of CE EoU strategies with particular CE objectives for built assets

EoU strategy	Longevity	Reusability	Transform-ability
Refuse	Strong	Weak	Weak
Rethink	Strong	Strong	Weak
Reduce	Strong	Strong	Strong
Reuse	Weak	Strong	Weak
Repair	Strong	Strong	Weak
Refurbish	Strong	Strong	Weak
Remanufacture	Weak	Weak	Strong
Repurpose	Weak	Weak	Strong
Recycle	-	-	Strong
Recover	-	-	Strong

4.3 USE CASES COMBINE MULTIPLE OBJECTIVES

Applying our three proposed objectives to three use cases (adaptive reuse, temporary work components, and materials passports) reveals that all cases involve multiple objectives (Figure 2). By applying our proposed objectives to these three use cases, we demonstrate the applicability and coverage of our set of objectives. As this paper reports on the conceptual development of our CE objectives, the use cases are purely descriptive.

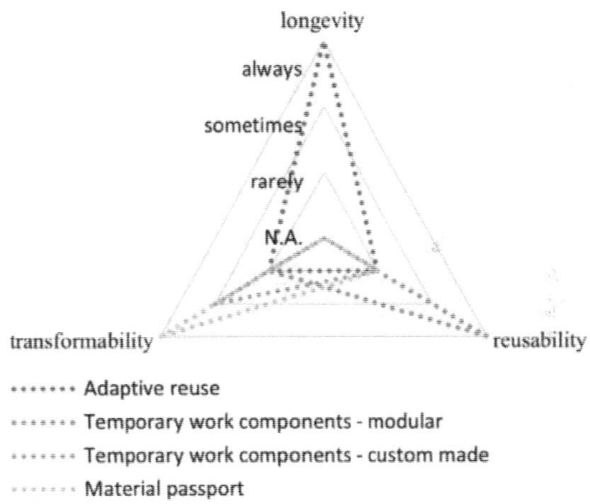

Figure 2 Relevance of objectives for use cases represented in a radar chart

4.3.1 Adaptive reuse

Adaptive reuse has been examined in multiple studies, with additional details in our previous work (Sibenik et al., 2025). This approach assesses the extent to which an entire building can be adapted, focusing on prolonging its lifespan as a whole. However, adaptive reuse is not limited to the longevity objective alone. The asset tiers considered include systems, components, and materials, besides the whole built asset.

A typical case of adaptive reuse is that certain systems align with the longevity objective, such as structural or façade systems. However, other components are rarely accounted for or quantified. In the proposed set of three objectives, all parts of the built asset—including those that do not follow the longevity objective—are considered. This means that a building is not quantified for adaptive reuse as a whole, but separately for each objective. All constituents are calculated for a specific objective and have their own CEP score. Some systems will be reused in the same place, therefore score best with longevity, while others elsewhere, and calculated for reusability or transformability objectives. The removable constituents also contribute with their CEP score to the final score of the asset. Depending on the objective, different calculation methods are applied, generally combining elements of existing methods from Sibenik et al. (2025) and considering different design strategies for different objectives.

This comprehensive approach provides a CEP for both existing and novel buildings, surpassing current adaptive reuse methodologies by considering all constituents, including the ones following reusability and transformability objectives.

4.3.2 Temporary work components

Temporary work components, such as concrete formwork, are considered in this use case (Tizzani et al., 2023). Two common types of formwork construction are on one side modular systems that are used in similar ways multiple times, and on the other side custom-made formwork made on site and discarded after use. The most common material for modular formwork is aluminium, while the custom-made formwork often consists of wooden planks and panels. There are also realizations which combine these two types of construction. Although these realizations of temporary work components are considered typical, it is difficult to measure their CEP with current measuring tools.

Aluminium systems are created with an intention to be used multiple times in the initial form, combined with other elements of the system. They are characterized by a modular and standardized design. Although they can be recycled and they could potentially score positively with the transformability objective, their primarily objective and better score is expected with the reusability objective. Therefore, their score is calculated based on the design strategies relevant to reusability, such as deconstruct-ability and refit-ability.

Wooden formwork created on site corresponds better to the transformability objective, as custom made, and it is commonly used in another form after the EoU. Therefore, the system can be calculated for the CE strategy which is likely to happen, such as incineration or recycling, where the difference of GHG before and after the process indicates its CE score.

In the system including both reusability and transformability, both types of construction can be modelled and calculated, results compared, and the better performing construction selected. Additionally, the CEP score of temporary work components can be added to the CEP score of the entire structure. Existing CEP measurement tools generally do not allow such detailed calculations for formwork or other temporary work components.

4.3.3 Materials passports

Materials passports document the quantities of materials within a built asset, focusing primarily on the material tier of a construction asset. Other tiers are typically overlooked, and configurational properties are considered only in a limited capacity, such as certain properties on the component level. While materials can be evaluated using the proposed objectives, only the transformability objective and material tier are deemed relevant.

Compared to the novel set of three objectives, materials passports focus on the transformability of materials but do not fully address the objectives of longevity and reusability, nor do they account for all construction asset tiers that retain higher embodied GHG value. To calculate a materials passport using the proposed set of objectives, it is necessary to consider the transformability objective and, in some cases, reusability for specific types of materials or components. The results will align in a similar direction, although our calculation and the data used do not fully reflect a materials passport. Relying solely on such calculation is not advised, as it does not consider potentially more sustainable options, such as the reuse of components in their original form (Ossio et al., 2023) and structural durability (Marsh et al., 2022).

5 DISCUSSION

The proposed holistic set of objectives for calculating the CEP of built assets accommodates all elements (design strategies, asset tiers, CE flows) of the CEP framework and supports our three test use cases, some that have not been previously considered in other evaluation methods (temporary work components, removable components of adaptable buildings). The synthesis of the objective-based calculation and elements of the CEP framework is represented in Figure 3. While similar objectives have been widely discussed separately in literature, aligning them within a single set of objectives respecting the boundary conditions of being holistic, measurable, simultaneous, compatible, and inclusive makes a significant step towards a holistic CEP calculation. This integrated approach aims on enabling the comparison of objectives, prioritization of strategies, and a robust assessment of the actual CE behaviour of built assets. The system's adaptability could allow for tailored prioritization depending on regional urban plans—for instance, emphasizing reusability in one area while focusing on longevity in another. However, it is crucial to recognize the interrelated nature of the CE objectives; excluding certain objectives from the assessment could overlook better-performing options.

https://doi.org/10.52202/080684-0011

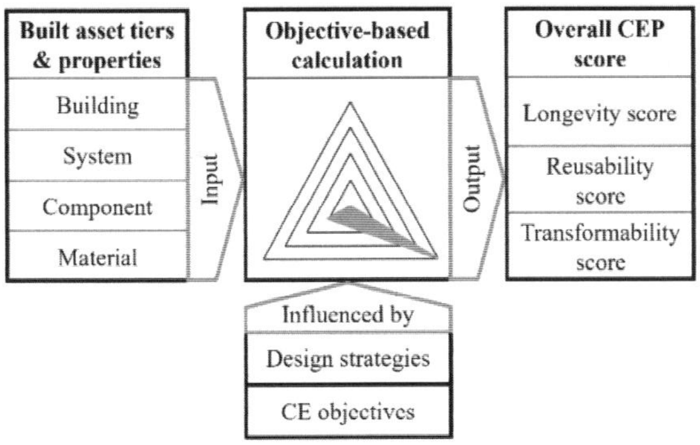

Figure 3 CEP assessment considers all built asset tiers, and it is influenced by design strategies and potential EoU scenarios. It provides results per objective, all contributing to the total CEP score.

All elements of the CEP framework are addressed by at least one objective. The challenges identified in the literature are considered for the novel system of objectives with special attention given to configurational properties of built assets, adaptive reuse with the retention of built assets and outflow of constituents, evaluation of both positive and negative impacts of state transformation, and accounting for both high-performing and low-performing elements.

The set of objectives allows for adapted definitions of EoU and EoL for the built environment. In this context, EoL refers to the point when any asset tier ceases to follow their intended objective, while EoU indicates a change of its role within a higher tier. Therefore, assets following reusability are expected to have multiple EoU points, whereas longevity and transformability have EoU and EoL occurring at the same point in time. Scenarios such as adaptive reuse, following multiple objectives, do not reach EoU or EoL on the building tier, however their lower tiers do.

While promising, this study has limitations. Numerical calculations for the proposed framework are still under development and are not included in this publication. Incorporating and interrelating all the necessary methods could present challenges for some influencing factors of the CEP assessment. Initial tests yielded encouraging results, but acquiring reliable data remains a significant obstacle, as observed in other methodologies like LCA and materials passports. We use available databases for GHG properties, environmental product declarations, and expert assessments to perform the calculation, especially those used by established methods and research when applicable. While we aim to include existing indicators where possible, detailed investigation into data reliability is not part of this study. However, areas in need of improvement will be indicated. A broader range of tests, supported by detailed calculations, is necessary to fully assess the tool's applicability.

Moreover, practical implementation of the approach has not been achieved yet, and the circularity objectives, derived through deductive reasoning, may not be sufficient for all circularity cases. Although the performed tests did not reveal shortcomings, more robust testing could uncover unaddressed issues. These risks will be addressed when the CEP assessment tool reaches higher development level with a more elaborated and diverse set of case studies. Additionally, presenting the framework to the audiences beyond the research team is necessary to assess its adoptability and usability. Receiving feedback from the community actively advancing CE principles in the building sector is particularly valuable.

Future research will focus on developing calculation methods based on the three objectives. This includes creating a tool to measure the CEP and establishing case studies as a proof of concept. All three use cases have been successfully supported by the proposed objectives, and the next steps involve incorporating exact data and realization of a measurement method. This method will subsequently be realized as a decision-support tool working with BIM models. Leveraging BIM-authoring software tools and models is a key goal to expedite the measurement process and identify requirements for a fully automated CEP assessment.

6 CONCLUSIONS

This work proposes a structured organization of CE objectives in a holistic system for built assets, addressing multiple challenges identified in the literature and existing quantification methods. The proposed set of objectives aims to enable a holistic evaluation of the CEP, providing a basis for comparing built assets or their tiers (systems, components, materials), designed with varying objectives or without any. Unifying the objectives relevant for the construction assets is a prerequisite for the development of a new tool. The novelty of this work lies in its comprehensive set of three objectives, which encompasses all relevant design strategies and CE strategies for EoU and EoL, while also considering various asset tiers. Although it is still a work in progress, the solution currently follows all identified crucial boundary conditions of being holistic, measurable, simultaneous, compatible, and inclusive, which, in combination, are still not available in practice.

Developing this objective-based concept as a digital tool will offer a fast and efficient method for assessing the CEP. By unifying approaches that are currently fragmented or separately addressed, the objectives ensure a comprehensive evaluation and allow for the calculation of diverse CE performance metrics within the built environment.

ACKNOWLEDGEMENTS

This research was conducted at the Future Cities Laboratory (FCL) Global at Singapore-ETH Centre (SEC). FCL Global is funded and supported by the National Research Foundation (NRF) Singapore under its Campus for Research Excellence and Technological Enterprise (CREATE) programme and ETH Zurich, and

https://doi.org/10.52202/080684-0011

includes collaborations with the National University of Singapore (NUS), Nanyang Technological University (NTU) Singapore and the Singapore University of Technology and Design (SUTD).

REFERENCES

Bertino, G., Kisser, J., Zeilinger, J., Langergraber, G., Fischer, T., & Osterreicher, D. (2021). Fundamentals of Building Deconstruction as a Circular Economy Strategy for the Reuse of Construction Materials. *Applied Sciences*, *11*(3), 939. https://doi.org/10.3390/app11030939

Charef, R., Morel, J.-C., & Rakhshan, K. (2021). Barriers to Implementing the Circular Economy in the Construction Industry: A Critical Review. *Sustainability*, *13*(23), 12989. https://doi.org/10.3390/su132312989

Cimen, O. (2021). Construction and built environment in circular economy: A comprehensive literature review. *Journal of Cleaner Production*, *305*, 127180. https://doi.org/10.1016/j.jclepro.2021.127180

Circle Economy & Metabolic. (2022). *The Circularity Gap Report: Built Environment, the Netherlands*. Circle Economy.

Circle Economy & Realdania. (2025). *Case Studies in the Circular Built Environment: The 9R Strategies in Action*. Circle Economy.

Directive 2008/98/EC. (2008). *Directive 2008/98/EC of the European Parliament and of the Council of 19 November 2008 on waste and repealing certain Directives (Text with EEA relevance)*. Official Journal of the European Union L 312/3.

Directorate-General for Environment. (2025). *Level(s)—European framework for sustainable buildings*. Environment. https://environment.ec.europa.eu/topics/circular-economy/levels_en

Durmisevic, E. (2015). Buildings as Material Banks. In M. Roso Popovac, A. Idrizbegovic-Zgonic, S. Klaric, N. Rustempasic, & A. Causevic (Eds.), *The 3rd International Conference: The Importance of Place*. CICOPBH.

Eberhardt, L. C. M., Birkved, M., & Birgisdottir, H. (2022). Building design and construction strategies for a circular economy. *Architectural Engineering and Design Management*, *18*(2), 93–113. https://doi.org/10.1080/17452007.2020.1781588

Ellen MacArthur Foundation. (2021). *Completing the Picture: How the circular economy tackles climate change*. Ellen MacArthur Foundation.

ETH Zurich. (2025). *[CFC] Circular Future Cities*. Future Cities Laboratory. https://fcl.ethz.ch/research/cycles-and-districts/circular-future-cities.html

European Environment Agency. (2022). *Linking circular economy and climate change mitigation in building renovation* (Briefing no. 12/2022). doi: 10.2800/416573

Gasparri, E., Arasteh, S., Kuru, A., Stracchi, P., & Brambilla, A. (2023). Circular economy in construction: A systematic review of knowledge gaps towards a novel research framework. *Frontiers in Built Environment*, *9*, 1239757. https://doi.org/10.3389/fbuil.2023.1239757

Hakaste, H., Hakkinen, T., Lahdensivu, J., & Saarimaa, S. (2024). *Lifecycle Performance in Promoting the Longevity of Buildings* (2024:23). Ministry of the Environment. http://urn.fi/URN:ISBN:978-952-361-187-0

Hamida, M. B., Jylha, T., Remoy, H., & Gruis, V. (2023). Circular building adaptability and its determinants – A literature review. *International Journal of Building Pathology and Adaptation*, *41*(6), 47–69. https://doi.org/10.1108/IJBPA-11-2021-0150

Honic, M., Kovacic, I., Sibenik, G., & Rechberger, H. (2019). Data- and stakeholder management framework for the implementation of BIM-based Material Passports. *Journal of Building Engineering*, *23*, 341–350. https://doi.org/10.1016/j.jobe.2019.01.017

Hossain, Md. U., Ng, S. T., Antwi-Afari, P., & Amor, B. (2020). Circular economy and the construction industry: Existing trends, challenges and prospective framework for sustainable construction. *Renewable and Sustainable Energy Reviews*, *130*, 109948. https://doi.org/10.1016/j.rser.2020.109948

ISO. (2018). *Organization and digitization of information about buildings and civil engineering works, including building information modelling (BIM)—Information management using building information modelling—Part 1: Concepts and principles* (ISO 19650-1:2018).

Marsh, A. T. M., Velenturf, A. P. M., & Bernal, S. A. (2022). Circular Economy strategies for concrete: Implementation and integration. *Journal of Cleaner Production*, *362*, 132486. https://doi.org/10.1016/j.jclepro.2022.132486

Moisio, M., Salmio, E., Kaasalainen, T., Huuhka, S., Räsänen, A., Lahdensivu, J., Leppänen, M., & Kuula, P. (2024). Towards urban LCA: Examining densification alternatives for a residential neighbourhood. *Buildings & Cities*, *5*(1). https://doi.org/10.5334/bc.472

Morseletto, P. (2020). Targets for a circular economy. *Resources, Conservation and Recycling*, *153*, 104553. https://doi.org/10.1016/j.resconrec.2019.104553

Murakami, S., Oguchi, M., Tasaki, T., Daigo, I., & Hashimoto, S. (2010). Lifespan of Commodities, Part I: The Creation of a Database and Its Review. *Journal of Industrial Ecology*, *14*(4), 598–612. https://doi.org/10.1111/j.1530-9290.2010.00250.x

Nunez-Cacho, P., Gorecki, J., Molina-Moreno, V., & Corpas-Iglesias, F. A. (2018). What Gets Measured, Gets Done: Development of a Circular Economy Measurement Scale for Building Industry. *Sustainability*, *10*(7), 2340. https://doi.org/10.3390/su10072340

Ossio, F., Salinas, C., & Hernández, H. (2023). Circular economy in the built environment: A systematic literature review and definition of the circular construction concept. *Journal of Cleaner Production*, *414*, 137738. https://doi.org/10.1016/j.jclepro.2023.137738

Sibenik, G., Sanchez, B., Herthogs, P., & Stouffs, R. (2025). *Review of measuring methods for built structure configurations influencing circular economy flows*. ETH Zurich. https://doi.org/10.3929/ethz-b-000728295

Tizzani, F., Herthogs, P., & Stouffs, R. (2023). Developing a cost-control and project-planning based implementation of circular construction in temporary works: A framework of core supportive technologies. In F. Biondini & D. M. Frangopol, *Life-Cycle of Structures and Infrastructure Systems* (1st ed., pp. 3578–3585). CRC Press. https://doi.org/10.1201/9781003323020-438

https://doi.org/10.52202/080684-0011

CITIES AS LOCAL GOVERNMENTS FOSTERING THE CIRCULAR ECONOMY IN THE CONSTRUCTION SECTOR – UNRAVELLING THE KEY ACTIONS AND OPERATIONAL ROLES OF CITIES THROUGH A MULTIPLE-CASE STUDY

Lauri Alkki[1*], Johanna Alakerttula[1], and Leena Aarikka-Stenroos[1]

[1]Tampere University, Unit of Industrial Engineering and Management, Finland

*Corresponding Author: lauri.alkki@tuni.fi

ABSTRACT

Background and aim. Climate change and environmental issues have driven cities to adopt more sustainable practices, with the circular economy seen as a solution. Cities, as built environments are responsible for 75% of global resource use and over 70% of greenhouse gas emissions, play a critical role in the circular transition. Harnessing the potential of circular construction is a key means for cities as local governments to achieve sustainability goals and reduce negative environmental impacts in the built environment. However, research largely describes circular actions in construction at general and conceptual levels. In addition, although the city's role in the circular transition within construction is recognised, few studies identify specific actions enabling cities to contribute to this transition.

Methods and data. To address these gaps, we conducted a qualitative multiple-case study of four Finnish construction cases engaging cities – the construction of an eco-industrial park and a circular city district; construction waste management through mass coordination; and a circular construction research project – based on primary (n=11 interviews and ethnographic observation) and secondary (n=over 100 documents) data.

Findings. We recognised that cities as local governments play an important role in fostering circular construction through 26 key actions categorised as facilitate collaboration, govern and monitor, develop, and operate. We also identified that cities actions manifest through two operational roles, actor and platform, where actions tend to have identifiable characteristics of both roles, depending on the action and its implementation.

Theoretical / Practical / Societal implications. This study contributes to research on circular cities and construction by highlighting cities' potential in the circular transition of the construction sector. It also provides practical guidance for city-level managers and policymakers on circular decision-making at the local level.

KEYWORDS: circular actions; circular economy; circular city; circular construction; operational roles.

1 INTRODUCTION

Given the tension between the planet's resources and economic growth, the circular economy (CE) has been proposed as a solution to replace linear pollutive and waste-generating actions (Kirchherr et al., 2017; Reike et al., 2018). Implementing more circular actions enables a move towards more sustainable consumption and production while simultaneously limiting environmental impacts such as climate change and biodiversity loss by maintaining the value embedded in products and resources for longer (Ghisellini et al., 2016; Kirchherr et al., 2017).

Cities can be seen as a core part of the CE transition, from both a built environment and local government perspectives (Prendeville et al., 2018; Petit-Boix & Leipold, 2018; Christensen, 2021; Paiho et al., 2021; Hürlimann et al., 2022). Cities, as built environments, are not only significant consumers of raw materials and energy but also hotspots of innovations, policy action, capital, data, talents, and resources (Ellen MacArthur Foundation [EMF], 2022). Cities account for approximately 60% of global gross domestic product,

https://doi.org/10.52202/080684-0012

60% of the resources used, and 70% of global carbon emissions (United Nations [UN], 2022). In addition, it is expected that by 2050 almost 70% of the world's population will live in cities (World Bank Group, 2023). While we acknowledge that the cities as built environments have significant environmental and economic impacts, in this study we focus on the perspective of the cities as local governments and their actions to facilitate the transition toward more sustainable and circular actions.

Consequently, it is acknowledged that cities, as local governments, have a crucial governmental role to play regarding the built environment: cities are responsible, for example, for land-use planning (Turcu & Gillie, 2020; Williams, 2019) and street, water, and waste infrastructures and their maintenance (Caragliu et al., 2011). Cities are also landowners and owners of several properties, such as schools, daycare facilities, and hospitals. Cities issue building permits and oversee the construction and demolition of buildings (Ministry of the Environment, 2021) as well as take care of municipalities' waste management (Christensen, 2021). Therefore, understanding how a city, as a local government, can contribute to and foster circular construction is critical in reducing the harmful environmental impacts of cities as built environments. While focusing on circular construction we refer to actions that maintain construction materials, buildings, and infrastructure in use and circulation by reducing, sharing, reusing, refurbishing, repairing, and recycling in all lifecycle phases (Pomponi & Moncaster, 2017; Ghaffar et al., 2020; Benachio et al., 2020; Dams et al., 2021).

The construction sector is among the key areas of focus for cities seeking to meet their sustainability goals and reduce negative climate impacts (Paiho et al., 2021; Rios et al., 2022). As construction accounts for 36% of global final energy use and 39% of energy-related carbon emissions, while construction and demolition waste (CDW) accounts for 36% of all waste generated in the European Union (UN Environment, 2018; European Commission, 2019; Eurostat, 2020). In addition, the construction sector is one of the largest consumers of natural resources, accounting for more than half of the total materials used globally (Organisation for Economic Co-operation and Development [OECD], 2018). As a result, the construction sector plays a crucial role in ensuring future sustainability and preserving biodiversity.

Based on the construction sector's huge potential in the circular transition and the critical role of cities in this regard, this study focuses on key actions taken by cities, as local governments, to foster circularity in the construction sector. We look at the actions taken by the cities through two operational roles – the *city as a platform* (e.g., Tukiainen et al., 2015; Anttiroiko, 2016; Bollier, 2016; Haveri & Anttiroiko, 2021), in which the city is seen as an enabler of different actors' actions; and the *city as an actor* (e.g., Acuto et al., 2020), in which the city is considered to take an active role in planning, implementing, and operating. However, these roles are not mutually exclusive. Cities must often act as both platform and actor simultaneously, depending on how the action is implemented, to enable the CE transition.

Previous studies of circular cities have identified certain key actions that cities can take when implementing circular strategies to foster circularity across their functions (Alhola et al., 2018; Lakatos et al., 2021; Bonoli et al., 2021). However, many studies have been either product- or policy-oriented (Bonoli et al., 2020) or presenting more general strategies and actions that a city can follow to implement CE and sustainability goals (Prendeville et al., 2018). Concrete circularity plans and actions remain scarce (Paiho et al., 2021), and little is known about the practical implications and applications of what cities can do regarding the CE in the construction sector (Caragliu et al., 2011; Turcu & Gillie, 2020; Williams, 2021). In addition, while studies have identified various roles that can be played by public actors and cities (von Malmborg, 2004; Frantzeskaki et al., 2016; Kronsell & Mukhtar-Landgren, 2018; Uusikartano et al., 2020; 2021). Less emphasis has been given on the city's fundamental operational roles—whether as an actor or a platform—that serve as the foundation for identifying specific actions and responsibilities. Overall, research on circular construction has highlighted the need for more empirical studies to determine which actions can effectively drive the transition to a circular economy in the construction sector (Adams et al., 2017; Munaro et al., 2020; Guerra & Leite, 2021) and how cities can contribute to this shift (Girard & Nocca, 2019).

By focusing on city-level solutions in circular construction, our study complements a wider discussion of circularity in the construction sector. It contributes to an understanding of how cities as local governments can be involved in a wide range of circular construction projects and promote, facilitate, enable, and manage circularity in the built environment. Accordingly, this study aims to clarify how cities, as local governments, can actively implement circular initiatives in construction—either through direct own action or by facilitating the efforts of other actors and stakeholders. In particular, this study aims to answer two research questions: i) *what are the key actions a city can take to foster circular construction?* and ii) *how do the operational roles of the city manifest in actions fostering circular construction?*

To meet our research aim, we conducted a qualitative multiple-case study (including 11 interviews, ethnographic follow-up, and over 100 secondary sources) in the Finnish context focusing on four different circular construction cases engaging cities with different set of actions enabling and fostering circular construction. Therefore, the cases provide an insightful empirical setting to explore how a city as a local government can

https://doi.org/10.52202/080684-0012

implement its circular strategy in the construction sector by delving into the key actions of cities and their operational roles in fostering circular construction.

Our study contributes to circular city research by providing a categorisation of how cities can foster circularity through construction with empirically based examples of actions. Moreover, it deepens the understanding of the roles of public actors, thus not only contributing to circular city research but also to circular construction research. Furthermore, our research provides insights for decision-makers in city organisations and construction companies by providing a comprehensive understanding of how a city can contribute to and foster circular construction in its own organisations and public–private partnerships.

2 LITERATURE OVERVIEW

In this section, we provide an overview of cities' importance in the CE transition and describe why focusing on cities in the circular construction context is crucial.

2.1 CIRCULAR ECONOMY IN CITIES

Rapid urbanisation brings social, economic, technical, and environmental challenges, such as how to provide affordable housing, well-connected transport systems, water, energy, and waste infrastructure, basic services, and a safe environment for all citizens (Paiho et al., 2021; World Bank Group, 2023). Cities, as built environments, are the hotspots where environmental problems and challenges arise, but at the same time, they are cradles and ecosystems where sustainability challenges are solved and nurtured (Henrysson et al., 2022; UN Environment, 2018). Thus, the transition towards the CE and more sustainable practices is gaining attention within cities (OECD, 2020; Paiho et al., 2021; Prendeville et al., 2018).

Importantly, the CE offers new tools for cities to respond to climate change and resource challenges by rethinking how to use, reuse, recycle, and sustain the value of materials, products, and assets (Sodiq et al., 2019) in collaboration with other actors, such as citizens, companies, and researchers (Prendeville et al., 2018; Williams, 2021). Seven key CE sectors have been identified in the city context: construction, food, waste, procurement, water, transport, and energy (Paiho et al., 2021; Rios et al., 2022). These sectors are not entirely separate; rather, they are closely linked, as cities are complex ecosystems of public and private actors, innovation cultures, business networks, infrastructures, and resources (Paiho et al., 2020; Henrysson et al., 2022).

The actions identified in the literature on how cities can foster the CE indicate that cities can use their policy tools as catalysts for circular change and define and coordinate their CE actions through CE strategies and roadmaps (European Investment Bank [EIB], 2018; Prendeville et al., 2018). To add to this, the CE in cities is also driven by different levels of policy instruments (e.g., national- and EU-level), providing different tools to promote the CE. For example, the waste framework directive (2008/98/EC) and circular economy action plan (COM(2020)98) that enhances sustainability requirements, promote waste prevention as well as utilisation and recycling of waste, and encourage resource-efficient practices. As key components of the EU Green Deal, they collectively support the transition to a circular and environmentally responsible built environment. City administrations can lead by example, for instance, by offering, procuring, and tendering circular solutions and services and accelerating circular business development (e.g., EIB, 2018; Prendeville et al., 2018; Alhola et al., 2018). Economic support is identified as a way to promote the circular transition, as it can include indirect financial incentives as well as direct economic help, such as financing for sustainability projects (e.g., Uusikartano et al., 2021). Cities can include circular principles in their technical infrastructure and services, such as energy, water, transport, infrastructure, and education (e.g., Rios et al., 2022). In addition, cities can promote circular awareness and create a culture of collaboration among city organisations, citizens, companies, and other organisations (e.g., Paiho et al., 2020; Henrysson et al., 2022). Cities may call themselves circular cities; nonetheless, it remains challenging for them to define which combinations of CE initiatives will result in the most environmentally friendly performance (ICLEI, 2021; Paiho et al., 2021; Lakatos et al., 2021).

In addition to the actions identified by cities as fostering the CE transition, different roles for cities and public actors have been identified, such as innovator, operator, coordinator, organiser, financer, enabler, catalyser, influencer, supporter, policymaker, and regulator (e.g., von Malmborg, 2004; Frantzeskaki et al., 2016; Kronsell & Mukhtar-Landgren, 2018; Uusikartano et al., 2020; 2021). Circular actions taken by cities have also been studied from intra-urban and interurban perspectives. Intra-urban studies focus on one city, either the whole city or a limited area within it, such as a district or an even smaller unit, such as an area, neighbourhood, or household, and the actors and their interactions inside it (Bork-Hueffer, 2014). Common to all identified public actors' roles is that cities cannot be seen as isolated entities. Rather, they need to work with their surrounding environment in the CE transition and must therefore leverage the actions they take on their own as well as those taken in cooperation with other actors and stakeholders.

However, in this study, we dive deeper into the overarching starting point of the role division; that is, we focus on the nature of the actions taken as part of the operational roles of a city as an actor and a platform. While the previous literature has identified the actions that cities can take to foster the CE, most remain at a very conceptual level, and more empirical evidence is needed

(Girard & Nocca, 2019; Lakatos et al., 2021; Isoaho & Valkama, 2024). By bringing cities' operational roles and key actions together in the context of the CE, we provide a more comprehensive way to understand how cities can foster the transition toward CE.

2.2 THE IMPORTANCE OF CITIES FOR CIRCULAR CONSTRUCTION

An unanimously agreed and comprehensive definition of circular construction (Benachio et al., 2020) has yet to be established. One oft-quoted definition provided by Pomponi and Moncaster (2017, p. 711) of the CE in the built environment is '*building that is designed, planned, built, operated, maintained, and deconstructed in a manner consistent with CE principles'*. Based on this definition, Benachio et al. (2020, p. 5) refined the definition of the CE in the construction sector as '*the use of practices, in all stages of the life cycle of a building, to keep the materials as long as possible in a closed loop, to reduce the use of new natural resources in a construction project'*. An essential element of both definitions is the mention of circular practices at different lifecycle stages. On the other hand, Dams et al. (2021, p. 1) explained that '*the concept of circular construction requires that a building should not be merely a static, physically whole entity, but instead should be a changing, evolving combination of functions and processes and be able to adapt to changing societal or functional requirements over long periods of time'*. From our point of view, the need for a combination of processes and actions is interesting, as is the constant development that comes from reacting to external stimuli. Moreover, Ghaffar et al. (2020) observed that '*in circular construction, buildings and infrastructure will be designed according to circular principles*,' bringing in the perspective that circular construction considers more than just buildings. In addition, the EMF (2023) sets out the following three CE principles: to eliminate waste and pollution, circulate products and materials (at their highest value), and regenerate nature. Based on these definitions, we refer to circular construction as *actions that aim to maintain construction materials, products, buildings, and infrastructure in use and circulation at their highest value by reducing, sharing, reusing, refurbishing, repairing, and recycling in all lifecycle phases*.

Cities are among the key actors in the CE transition in the construction sector, as they are not only major centres of the built environment but also involved in construction projects in one way or another (cf. Campbell-Johnston et al., 2019; Christensen, 2021). For example, cities are pivotal in advancing the construction sector towards the CE as economic hubs. That is, urban areas can achieve cost savings through the reduction, reuse, and recycling of materials while also driving innovation and technological advancements in sustainable building practices (Joensuu et al., 2020; Christensen, 2021). City governments have the authority to implement policies and regulations that promote circular construction, and cities can engage communities to foster a culture of sustainability (Isoaho & Valkama, 2024). Circular construction in cities can significantly mitigate climate change by reducing the greenhouse gas emissions associated with building materials. Overall, cities' influence on resource use, economic activities, policymaking, and community engagement makes them essential in the CE transition. (e.g., Campbell-Johnston et al., 2019; Lakatos et al., 2021; Hürlimann et al., 2022) In addition, cities' (circular) actions are influenced by external policy instruments, providing frameworks, boundary conditions, and guidelines for more sustainable ways of operating. For example, the Environmental Product Declaration (EPD), the revision of the Construction Products Regulation (CPR 2024), along with the Waste Framework Directive (2008/98/EC), Energy Performance of Buildings Directive (EPBD), and the New Circular Economy Action Plan (COM(2020)98), establishes a strong regulatory foundation for circular construction in the EU and driving the transition towards more circular and sustainable ways of operating in the construction sector, and along that providing cities the opportunity utilize them in their own operations.

While the role of cities in fostering circular construction is recognised, there is a scarcity on the actions identified in previous studies. However, previous studies have identified some actions on how cities can foster circular construction, for example, the maintenance of existing infrastructure reduces the need for new materials, while promoting consumer practices and services aligned with the CE encourages sustainable consumption (e.g., Caragliu et al., 2011; Lakatos et al., 2021; Hauashdh et al., 2022). Land-use planning ensures efficient use of space and resources, and public procurement criteria prioritise sustainable materials and methods (e.g., Turcu & Gillie, 2020; Williams, 2019). By optimising industrial structures, cities can enhance resource efficiency and reduce waste. The use of local renewable resources minimises environmental impact (e.g., Lin & Kao, 2020), and industrial symbiosis allows industries to share resources and by-products (e.g., Joensuu et al., 2020). Establishing eco-industrial parks (EIPs) fosters collaboration and innovation towards sustainability (e.g., Uusikartano et al., 2021). Additionally, the utilisation of industrial waste in building materials and effective demolition and waste management practices ensure that materials are reused and recycled, closing the loop in construction processes (e.g., Joensuu et al., 2020; Lin & Kao, 2020; Junli et al., 2021). Through these actions, cities can significantly contribute to a more sustainable and circular construction sector as well as built environment.

Previous studies are still lacking a more comprehensive understanding on the actions that cities can take in relation to circular construction (Wang et al., 2018; Joensuu et al., 2020; Benachio et al., 2020; Christensen, 2021). Moreover, most of the actions identified for cities to foster

https://doi.org/10.52202/080684-0012

circularity remain at a rather general level and do not address the operational role of the city. One explanatory factor is that the construction sector is at an early stage of the CE transition and is seeking ways to contribute to it (Adams et al., 2017; Çimen, 2021).

3 METHODOLOGY

In this section, we explain the methodological choices of our qualitative multiple-case study.

3.1 RESEARCH DESIGN AND CASE SELECTION

Our study follows a qualitative case study research strategy, as it enables us to study empirically real-life contexts and obtain in-depth information on real-life phenomena (Yin, 2009), here, of cities', as local governments, circular construction actions. We conducted a multiple-case study of four different circular construction cases engaging cities in Finland. By employing purposeful sampling, we selected four cases from Finland's largest cities: Helsinki, Tampere, and Turku. The aim was to select cases with diverse circular construction characteristics to ensure variation within the data (Palinkas et al., 2015) and thus allow us to identify similarities and differences in how cities foster circular construction.

We chose to study cases from the construction sector in which the city is engaged and which are different in nature. The cases studied are (see Table 1) the construction of an EIP, representing circular business development; the construction of an urban area, representing urban planning; mass coordination, representing construction waste management; and an (EU-funded) research project focusing on reusing old concrete elements, representing innovation towards circularity.

In all the selected cases, cities play important roles, for example, by facilitating, coordinating, supervising, funding, or supporting circular construction innovation, development, and operation. Common to all the cases is the drive to minimise waste and maximise resource efficiency.

Table 1: Case-specific details.

	Description	Background	Circular construction aspects
Case 1	Establishment of an EIP (ECO3)	ECO3 is an EIP located in the Tampere region of Finland. The core of ECO3's bio and circular business activity is companies that develop businesses based on construction waste management, nutrient cycle, the wood-based CE, bioenergy and fuel, and technical cycles. ECO3 is an initiative of the City of Nokia's development company.	ECO3 provides an area reserved for companies engaged in bio- and circular business; therefore, many construction material and waste processing companies are located in the area. The city can deliver its construction waste to ECO3 companies or use secondary construction products that are processed in the area.
Case 2	Development of a CE city district (Hiedanranta)	Hiedanranta is a new city district located in the Tampere region of Finland. It is a former industrial area that the City of Tampere is developing into a new city district based on CE principles.	The city has taken into consideration carbon neutrality and CE principles in planning and building the city district. The carbon neutrality of buildings, recycling of construction materials, excavated soil coordination, space sharing, CE projects, and the piloting and use of digitalisation are all observed in the city district.
Case 3	Circulation of excavated soils (mass coordination)	Finland's main cities (Helsinki region, Turku, and Tampere) are coordinating the excavation of soils to promote the reuse and recycling of infra-construction waste. The aims of the mass coordination are to proactively predict soil mass flows, monitor and guide the design and construction of future applications, and maintain up-to-date data.	The primary function of mass coordination is to direct excavated soils directly from their place of origin to the next destination of use, namely, from one site to another, and to improve soil mass economy and material efficiency. This coordination promotes the reuse of excavated soils; in that sense, CE construction may decrease carbon emissions since excavated soils do not normally need to be transported far.
Case 4	Pilot project of a novel construction approach (reuse of old concrete elements)	ReCreate is a Horizon2020 project addressing circular construction (particularly concrete element reuse) with reuse pilots located in Tampere, Finland.	ReCreate aims to develop the deconstruction of intact precast structural concrete elements from old buildings and their reuse in new buildings. The main objective is to keep concrete in circulation as a high-value product and reduce energy consumption and the carbon footprint compared to virgin production, aggregate recycling, and backfilling.

3.2 DATA GATHERING AND ANALYSIS

Our research data consist of primary and secondary sources (see Table 2). Among the primary data sources are semi-structured interviews and ethnographic observations, while secondary sources include media data, city reports, and web pages.

Table 2: Data types analysed in the study.

Data types	Case 1	Case 2	Case 3	Case 4
Semi-structured interviews	N=2 CEO in a city owned development company (10/2021) Senior university research fellow (11/2021)	N=3 KAM in the consultant company (10/2021) DM in a city-owned company (10/2021) Senior university research fellow (9/2021)	N=3 Mass coordinator in the city (three mass coordinators in different cities) (7-10/2021)	N=3 BDM & PM in a construction company (10/2021) Housing and development manager & PM in city organisation (10/2021) Senior university research fellow (10/2021)
Ethnographic observation	Attending consortium/project meetings, seminars, and site visits focusing on the cases (N=76 ethnographic observation situations)			
Secondary data	Media data, new, web pages, seminar presentations, and city reports and documents focusing on the cases (N=157 secondary sources)			
Abbreviations	CEO =Chief executive officer; KAM = Key account manager; (B)DM = (Business) Development manager; PM = Project manager			

Our abductive research approach focused on identifying comprehensively different actions cities can take to foster circular construction through various data sources. The data analysis was conducted as a thematic analysis. Interviews were recorded, transcribed, and saved to ATLAS.ti data analysis software, where two researchers coded all relevant text excerpts on how cities can contribute to circular construction. Data from observations, minutes, reports, and secondary sources were analysed and saved to Excel and combined with the interview data from ATLAS.ti. In the data analysis, we sought excerpts related to actions taken by cities to foster circular construction either directly or indirectly (by helping other actors) after which all the coded parts were reanalysed and categorised. Lastly, we further analysed our findings by deleting and combining the overlapping results. By using multiple tactics and tools in the data analysis, we gained a comprehensive overview of how cities can foster circular construction through two operational roles (actor and platform). Two researchers collected and analysed the data, which increased the data triangulation and the reliability of the results (Flick, 2004).

4 RESULTS

In this section, we present our multiple-case study's results of the key actions cities can take to foster circular construction and how their operational roles, that is, actor and platform, manifest.

4.1 KEY ACTIONS TAKEN BY CITIES TO FOSTER CIRCULAR CONSTRUCTION

Based on our analysis, we identified that the key actions cities as local governments take to foster circularity in construction occur in four different categories – facilitate collaboration, steer and monitor, develop, and operate (see Table 3). However, the city's governmental role in fostering CE in construction is complex and not always reduceable to easily categorisable actions; rather, actions fall within and between the identified categories. It is also clear that the city may contain different layers and different levels of bodies depending on the size of the city (organisational aspect of the city). In our analysis, we have taken into account that a city may, for example, contain different units (city government, city council, zoning, building control, licensing and permitting, in-house companies owned by the city, etc.) that include individuals whose own perceptions and attitudes affect their decision-making. Overall, a city can be seen as a larger entity, and each city is hence unique, as its list of actions is also strongly influenced by its location, size, and structure and their subsequent impacts (external factors).

The actions in the first category, ***facilitate collaboration***, are linked to the systemic nature of the CE as well as to the fact that cities, as built environments, link actors. Thus, the city has a critical role to play in facilitating, supporting, and maintaining cooperation between actors and stakeholders (i.e., between actors but also within the city organisation) to enable actions fostering the CE. The second category, ***steer and monitor***, focuses on the opportunities offered by the city's obligations. Actions in this category are taken as part of the city's responsibilities and obligations and seek, in particular, to understand how cities' duties can be used to foster the CE transition. The third category, ***develop***, focuses on the opportunities for the city to be involved in innovation and development projects as either an active driver or a participant/enabler. The actions in the last category, ***operate***, focus on different ways a city can actively contribute to fostering circularity in the construction sector within the city area and more generally in society. Actions in this category are not mandatory for cities but can be critical in promoting the CE transition. In addition, the actions in this category can most strongly intersect with the characteristics of the other categories. Table 3 presents the categories and their descriptions.

https://doi.org/10.52202/080684-0012

Table 3: Categorisation of the key actions cities can take to foster circular construction.

Category	Description	Key actions identified
Facilitate collaboration	Actions stemming from the systemic nature of the CE and the fact that cities, as built environment, are hotspots for actors to operate and collaborate.	- Participate in CE actions and projects to promote the adoption of circular solutions, products, and services - Support the sharing economy to facilitate the uptake of circular solutions - Foster and maintain public–private partnerships to enhance the implementation of circular solutions, products, and services - Enable platforms for collaboration and the development of innovative circular solutions - Pilot initiatives to promote circular solutions, products, and services - Collaborate with research institutions to increase understanding and implementation of circular solutions, products, and services - Join networks to share best practices and enhance understanding of circular solutions, products, services, and implementation
Steer and monitor	Actions stemming from the city's responsibilities and obligations, i.e., how to utilise cities' duties to foster the CE.	- Align city policy and strategy with circular and sustainability objectives to support the transition towards CE - Establish acquisition and CE procurement criteria to promote circular solutions in city projects and create markets for them - Make city investments to support the transition towards CE - Building control services (supervision) supporting circular projects and solutions - Align licensing and permits with CE objectives for new projects focusing on circular principles
Develop	Actions stemming from the city's involvement in innovation and development towards the CE.	- Develop an EIP to enhance material efficiency and collaboration - Utilize city-owned land to develop areas for CE (city as a landowner) - Planning and zoning of different areas that supports the transition towards CE - Market and commercialize CE to create markets for circular solutions, products, and services - Participate in developing national CE tools to understand the impact and comparability of circular solutions - Develop and coordinate regional databases and platforms to use collected data in decision-making supporting CE - Recruit, educate, and train staff to increase the city's competence in circular solutions and implementation - Use research and survey results to create circular concepts and plans - Align city-owned companies with the city's circular objectives
Operate	Actions stemming from the non-mandatory opportunities and possibilities to foster circularity that are open to cities.	- Align city properties and assets with CE objectives - Use voluntary agreements (e.g., Green deals) to support CE transition - Implement knowledge-based management from research and surveys to enhance CE actions - Use existing calculation methods (e.g., LCA, carbon footprint) to compare CE solutions - Coordinate processes within the city and with other organizations to support the transition towards CE

Facilitate collaboration: A city can play a vital role in advancing the CE in the construction sector through different facilitating actions. First, by participating in, facilitating, and contributing to regional and national CE roadmaps and strategies, a city can align its policies and practices in collaboration with other cities and the government to best respect the CE principles (*Case 1*). A city can also enable and accelerate the sharing economy, for example, by promoting the more efficient use of common spaces and buildings and by creating platforms for industrial symbioses and the more efficient use of material and energy flows between companies (*Cases 1 and 2*). Moreover, a city can facilitate cooperation and innovation among different actors and stakeholders, such as by establishing an EIP for bio and CE companies, cooperating with research institutes to gain data for urban development, and developing information exchanges and best practices between different cities (*Cases 1, 2, 3, and 4*). A city can also support areas where landmasses and recycled materials can be stored and processed and provide financing and enabling functions for pilot projects supporting circular construction (*Cases 1, 2, 3, and 4*). Furthermore, a city can build a brand and raise awareness of the benefits and opportunities of the CE in the construction sector by organising and participating in CE competitions, seminars, events, and conferences nationally and internationally (*Cases 1 and 2*). By doing so, a city can help facilitate the systemic transition to the CE in the built environment.

Steer and monitor: A city can foster the CE in the construction sector through different steering and monitoring actions. By implementing circular construction policies in its roadmaps and strategies, a city can set a clear vision and direction for the transition to a more sustainable and resource-efficient built environment (*Cases 1, 2, and 3*). A city can also define CE criteria for more sustainable procurement and acquisition and use its purchasing power to create markets and new business opportunities in line with circular construction (*Cases 1, 2, and 3*). For example, a city may increase the use of recycled materials, such as in infrastructure projects, by adopting special procurement criteria and mass coordination (*Cases 2 and 3*). A city can also take into account CE criteria in land donation and plot allocation and encourage the development of circular buildings and neighbourhoods (*Cases 2 and 4*). Moreover, a city government needs to be committed to developing circular practices over the long term and provide support and incentives for companies to develop their own CE actions (*Cases 1, 2, and 3*). In addition, the city can implement circular policies and regulations, such as considering the principles of the CE and the reduction and recycling of waste in construction and demolition permits and environmental permits and licensing as well as promoting and enabling circular design and construction practices through construction supervision (*Cases 1, 2, 3, and 4*). By adopting more circular ways of steering and monitoring, the city can harness the potential of circularity and benefit from circular construction in terms of reducing carbon footprint, saving energy, enhancing durability and longevity, and increasing economic value. By doing so, a city can enable and promote CE innovation and contribute to a more resilient and sustainable urban environment.

Develop: A city can foster circular construction by implementing various actions related to development that aim to reduce the environmental impact and resource consumption of the construction sector. One such action is to develop an EIP, which is a planned area where businesses cooperate to optimise the use of materials, energy, and water and minimise waste and emissions (*Case 1*). A city can act as a landowner and planner to facilitate the development of EIPs, city districts, and areas in accordance with CE principles to foster circular construction (*Cases 1, 2, and 3*). Moreover, a city can use sustainable zoning to allocate areas for land and resource recycling where materials from demolition and renovation projects can be collected, sorted, and reused or recycled (*Case 3*). A city can also improve the engagement of different actors and stakeholders, such as developers, contractors, architects, and customers, by developing guidelines to ensure the benefits of circular construction are realised and advising on the best practices of successful projects (*Cases 1, 2, and 3*). Cities can contribute to and participate in the development of national CE calculation tools and methods, which can help measure and monitor the circularity performance of buildings and materials (*Cases 1, 2, 3, and 4*). In addition, cities can develop and coordinate regional and national databases and platforms to share collected CE data for regional and national use and develop operating models to collect and use CE data in decision-making and planning and maintain these models by compiling databases (*Cases 1, 2, and 3*). However, the city organisations must have sufficient knowledge of the CE and sustainable decision-making, which is why a city can also recruit CE experts into its organisations and develop its internal CE knowledge and expertise through education and training (*Cases 1, 2, and 3*). Furthermore, a city can conduct its own surveys and engage in concept building to support circular construction, such as assessing the carbon neutrality or CE potential of districts and neighbourhoods (*Cases 1 and 2*). Besides carrying out their own surveys, cities can participate in research projects and utilise the research data gained in their own urban development, such as by identifying the barriers and enablers of circular construction and testing new circular products, services, and solutions. In addition, a city can establish or support companies that foster the CE, such as those that offer product-as-a-service, sharing platforms, or product life extension models (*Cases 1, 2, and 4*). By coordinating its internal processes, such as mass coordination, land use, and construction supervision, a city can ensure the effective implementation of circular construction actions.

Operate: A city can foster the CE in the construction sector through different actions and modes of operation. On its own properties, a city may support the recovery and reuse of materials, products, and elements (*Cases 1 and 2*). This can reduce the demand for new materials and extend the lifespan of existing ones as well as lower the environmental impact of demolition and disposal. On the other hand, through voluntary agreements, a city can participate in Green Deals to foster the CE in various domains, such as construction, waste management, and mobility (*Cases 1, 2, and 3*). Green Deals can help a city share best practices, access funding, and create a supportive regulatory framework for the CE. Additionally, the transition towards more circular actions can generate profits and savings for the city by enabling it to use recovered and recycled materials more efficiently and instead of virgin raw materials (*Case 3*). This can lower the costs of procurement, transportation, and disposal and reduce the dependency on external suppliers and vulnerability to price fluctuations. Furthermore, a city can use the services and products produced in the EIP as well as the industrial symbiosis for public–private partnership to support the development and implementation of innovative solutions for urban challenges (*Case 1*). The EIP can provide a city with access to cutting-edge technologies, expertise, and networks that can help foster the CE in the construction sector. As well as operating in circular projects as an active member or by offering the project ground for development, a city can benefit by co-developing new solutions, creating new business opportunities, and enhancing public–private partnerships (*Cases 1, 2, and 4*). Moreover, in a city-level context, data (e.g., information on material flows, waste generation, energy consumption, carbon emissions, and economic indicators) are generated that can be used in decision-making on the implementation of circular actions, for example, planning how to handle the material flows in the area (*Cases 1, 2, and 3*). Thus, it is important that the city can collect, process, analyse, and maintain CE data. These data can be used to monitor and evaluate the progress and impact of the city's CE actions as well as to communicate and engage with various actors and stakeholders, such as citizens, businesses, and policymakers. In addition, calculation tools and metrics, such as LCA and carbon footprint calculations, can be used in decision-making, evaluating the overall sustainability of a project, comparing different options and scenarios, identifying hotspots and improvement areas, and optimising the environmental performance of construction projects. By taking these actions, a city can foster the CE in the construction sector and contribute to the global goals of climate action, resource efficiency, and social inclusion.

4.2 CITIES' OPERATIONAL ROLES IN FOSTERING CIRCULAR CONSTRUCTION

In this study, we focused on the two operational roles through which a city can foster circular construction – the city as an actor and the city as a platform. As shown by the findings on the key actions cities can take, a city can be an active actor, for example, in planning city districts, funding CE projects, establishing new companies focusing on developing areas and districts as well in line with CE principles, educating their own staff, recruiting CE professionals, and operating EIPs. In addition, a city can apply CE criteria in procurement and thus actively

https://doi.org/10.52202/080684-0012

foster circular construction and markets for circular solutions and products. On the other hand, as a platform, a city can facilitate industrial symbiosis and enable the circulation of construction materials. It can also offer (e.g., through plot donation and zoning) areas and spaces to other organisations to test and pilot new circular construction solutions (e.g., sustainable dismantling, forms of the sharing economy, or new construction methods). Such a policy will enable the sharing economy for citizens and companies and integrate CE principles into roadmaps and politics, both as cities' own roadmaps and as policies setting targets for circular construction (although the operational side often requires close cooperation with companies and other stakeholders). Finally, such strategies will allow cities to contribute to regional and national-level CE projects, for example, by streamlining bureaucracy. Thus, based on our analysis, we have compiled the dynamism of cities fostering circular construction into Figure 1. This serves as a starting point to understand how cities can foster the CE in the construction sector in each unique case.

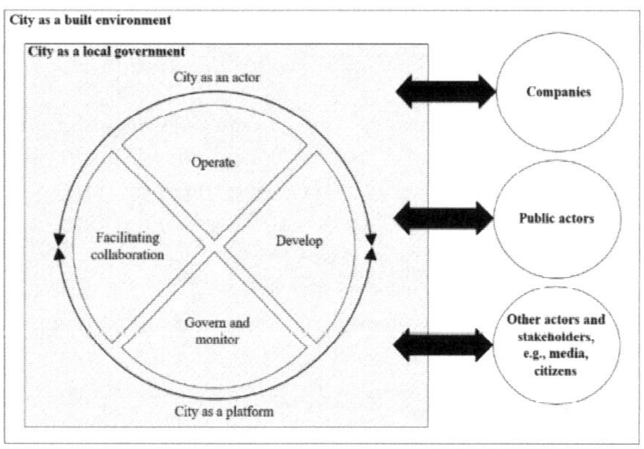

Figure 1: Cities fostering circular construction in collaboration with others.

In *case 1*, the establishment of an EIP, the city took various actions to foster circular construction at different stages in the process of setting up the EIP. The city can be seen as an actor because it has a central role in land use planning, environmental permitting, infrastructure building, concept building, and facilitating the operation within the area. However, the city can be seen as a platform as well since it also provides an area where companies can jointly develop bio and CE businesses and create new industrial symbioses.

In *case 2*, the city focused on building an entire circular city district (Hiedanranta) according to CE principles. Doing so required the city to have extensive knowledge and expertise from a wide range of areas since a circular city district seeks to maximise its use of CE solutions and methods. From the perspective of circular construction, the city can be thought of as playing the roles of both actor and platform. The Hiedanranta city district is being constructed with a deliberate emphasis on all aspects of the CE, including in the construction itself. Since the city developed the idea and proposed its implementation, it has played a central actor role in the project, especially in the planning and construction phases. At the same time, however, it can be seen as a platform, as it has offered an area where a circular city district can be piloted and tested, and citizens can inhabit a built environment while engaging in circular practices such as the sharing economy.

In *case 3*, the circulation of excavated soils, the cities focus on mass coordination as their internal function. The involved cities aim to promote the recycling of landfill, concrete and brick aggregate, and demolition materials by planning and coordinating demolition and new construction projects so that resources obtained in one project are efficiently used in another. Since the city is responsible for the operation of mass coordination, it plays an active and key role in its success. Therefore, the city's role in actions to foster circular construction in case 3 is principally that of an actor.

Lastly, in *case 4*, a city participates in a pilot project focusing on a novel construction approach, and its role is to take actions that enable and streamline the process. This pilot project focuses on concrete element reuse, an almost completely new method for all the actors involved. The city is involved in the development of the process whereby precast concrete elements can be reused, learning how to enable this and determining what it can do to facilitate this reuse. Consequently, the city's actions revolve around supporting the reuse of precast concrete elements. To ensure the smooth operation of the process, the city is discovering how the new approach will be implemented. Therefore, the city's role in fostering circular construction in case 4 is principally that of a platform.

However, we found that dividing cities' actions to foster circular construction according to their operational roles is not the most meaningful outcome of this role division. Rather, it is more valuable to understand what is gained from the different features of the operational roles taken in each action, that is, understanding that actions can have very different effects depending on how they are implemented and that actions can be approached in more depth by delving into their mechanism and logic. Indeed, it is crucial to understand which role the city is perceived to be playing at any given time (especially within the city but also by other actors) to reflect on which actions are optimal in fostering the CE in the current situation. It is also worth noting that the same action can have characteristics of both the roles played by a city, largely depending on the situation and desired outcome. Thus, there is no absolute division between the actions taken when the city is playing the operational role of an actor and those taken when it is acting as a platform.

5 DISCUSSION AND CONCLUSIONS

In this section, we synthesise our key findings and provide our theoretical and practical contributions as well as our study's limitations and ideas for future research.

5.1 KEY FINDINGS

Cities, as local governments, are complex entities that have internal duties as well as possibilities to foster circular construction through their departments, governmental inputs, and in-house companies. However, as cities also support and interconnect companies, organisations, and citizens, through which they have a large impact on circular decisions, products, and services. Overall, cities have a major role in creating markets for the circular construction.

Focusing on the key actions taken by cities to foster the CE in construction, we identified a total of 26 actions categorised into four different categories, *facilitate collaboration, steer and monitor, develop, and operate,* a city can take to foster circular construction. In addition, analysing these actions enabled us to delve into two operational roles played by the city, the city as an actor and the city as a platform, through which the actions were implemented. However, the outcome is affected by the situation (e.g., what action is considered; how the action is implemented) and its consequent constraints, as well as by other actors in the industry (such as companies, stakeholders, and authorities). Overall, the city's circular construction actions are a summation of all the actions presented in the study. Each case presents a unique combination of these actions, which is why our results provide an excellent starting point for thinking in different situations and suitable combination of actions on how a city can foster circular construction.

However, the division of the city's role into actor and/or platform is not absolute, as the city almost always has identifiable characteristics of both roles, which may vary according to the point of view and the situation. Our study also reveals that when the city acts as an actor, it can itself take concrete actions to foster circular construction. In contrast, when it acts as a platform, its actions are more directed to enabling and supporting other actors' circular construction actions. Moreover, our analysis indicates that the scale of the case under consideration notably affects the role of the city: in large-scale projects, the city's role as actor and platform is more easily identifiable, while in more focused and smaller projects, its role is often identified as that of either an actor or a platform. Although our study identified a wide range of actions that cities could take to contribute to circular construction, they require the desire to operate in a circular manner to gain the full potential of CE.

5.2 THEORETICAL CONTRIBUTIONS AND PRACTICAL IMPLICATIONS

In particular, our study contributes to the literature streams on circular cities (e.g., Prendeville et al., 2018; Petit-Boix & Leipod, 2018; Wang et al., 2018; Girard & Nocca, 2019; Christensen, 2021; Paiho et al., 2021; Williams, 2021) and circular construction (e.g., Adams et al., 2017; Benachio et al., 2020; Çimen, 2021) by linking them and deepening the understanding of cities' impact on the construction sector in the context of the CE and providing a list of actions that cities can take to foster circular construction with empirical examples. In addition, our view of the operational roles of cities, that is, as an actor (Acuto et al., 2020) and as aplatform (Tukiainen et al., 2015; Anttiroiko, 2016; Bollier, 2016; Haveri & Anttiroiko, 2021), in fostering circularity complements the discussion of the roles of cities and public actors in the CE transition (von Malmborg, 2004; Frantzeskaki et al., 2016; Kronsell & Mukhtar-Landgren, 2018; Uusikartano et al., 2020; 2021) by providing an overarching starting point for understanding these role(s) in relation to actions and how actions can relate to different roles depending on the situation. In addition, our study provides a comprehensive empirical-based categorisation of key actions taken by cities to foster circular construction and implement their own circular strategies. Thus, it answers the need identified in previous studies on circular cities, circular construction, and the CE in general for more empirical-based evidence on the actions promoting the CE (e.g., Adams et al., 2017; Paiho et al., 2021).

Our study's practical implications are twofold. It deepens cities' understanding of how to foster the CE in the construction sector through different key actions and operational roles. It also reveals how cities can help other actors in the construction sector to understand the impact of cities in the CE transition. Our research provides guidance on how to engage cities and how cities can contribute to fostering the CE in construction projects and the built environment. City organisations and cities' in-house company managers, in particular, are given an overview of how a city can foster and contribute to the development of circular construction. More understanding of cities' different operational roles is provided, guiding cities to develop circular construction more holistically. Practically, our results provide a comprehensive categorisation of the key actions cities can take to foster circular construction to eliminate waste and pollution, better enhance the circulation of products and materials (at their highest value), and regenerate nature. Thus, our results help not only city organisations but also various actors in the construction sector (i.e., companies and other stakeholders) more comprehensively reach circularity and sustainability objectives. In addition, our study provides examples for city organisations of situations in which different key actions can be applied through our empirical multiple-case study setting.

https://doi.org/10.52202/080684-0012

5.3 LIMITATIONS AND FURTHER RESEARCH

Based on our methodological choices and research setting, our study has some limitations. To gain a comprehensive understanding of cities' key actions and operational roles, we focused on easily accessible cases from the Finnish context, making our qualitative multiple-case study geographically limited. Thus, future research initiatives could expand the geographical focus and undertake a regional comparison of how cities' circular construction actions converge and differ.

Moreover, although we selected four different cases for our multiple-case study, all of which are linked to circular construction engaging cities, other cases may reveal certain key actions that did not emerge in our study. This realisation is, naturally, influenced by the increased understanding of circular actions taken by cities, companies, and stakeholders. Consequently, future studies could look at other circular construction cases and determine how the actions and roles identified in this study emerge and whether new actions can be identified.

Cities, as local governments, have numerous opportunities to take independent action (internally) but also collaborate with various actors and stakeholders (externally) to drive the transition toward CE. Our research serves as a strong foundation for examining ecosystem revenue—identifying key partners for cities and understanding how they should collaborate to unlock the full potential of circular construction. Future cities will not be mere centralised and rule-driven bodies that only decide on strictly city-related issues; rather, they will be enablers and facilitators of innovations and CE business through collaboration. Thus, cities will not only operate in isolation. Therefore, more information is needed about a city's different actions and roles over time and how they develop in collaboration with other actors.

ACKNOWLEDGEMENTS

This work is supported by the European Union's Horizon 2020 research and innovation program (ReCreate) research funding (grant ID 958200), the Academy of Finland's 'Circular Economy Catalysts: From Innovation to Business Ecosystems (CICAT2025)' research funding (grant IDs 320194 and 346626), and the European Regional Development Fund through the projects '6Aika: ILPO – Climate positive business parks and value chains' (grant ID A75909). Alkki's work was also supported by research grants by The Finnish Foundation for Technology Promotion as well as the Jenny and Antti Wihuri foundation. The authors are also grateful to the CITER research group for their valuable comments on developing the paper.

REFERENCES

Acuto, M., Kosovac, A., Pejic, D. and Jones, T.L. (2020). The city as actor in UN frameworks: Formalizing 'urban agency' in the international system? Territory, Politics, Governance, 11(3), 1–18. https://doi.org/10.1080/21622671.2020.1860810

Adams, K. T., Osmani, M., Thorpe, T., & Thornback, J., (2017). Circular economy in construction: Current awareness, challenges and enablers. Proceedings of the Institution of Civil Engineers – Waste and Resource Management, 170(1), 15–24. https://doi.org/10.1680/jwarm.16.00011

Alhola, K., & Nissinen, A. (2018). Integrating cleantech into innovative public procurement process – evidence and success factors. Journal of Public Procurement, 18(4), 336–354. https://doi-org.libproxy.tuni.fi/10.1108/JOPP-11-2018-020

Anttiroiko, A. V. (2016). City-as-a-platform: The rise of participatory innovation platforms in Finnish cities. Sustainability, 8(9), 1-31, 922. https://doi.org/10.3390/su8090922

Benachio, G. L. F., Freitas, M. D. C. D., & Tavares, S. F. (2020). Circular economy in the construction industry: A systematic literature review. Journal of Cleaner Production, 260(1), 1-17, 121046. https://doi.org/10.1016/j.jclepro.2020.121046

Bollier, D. (2016). The city as platform: How digital networks are changing urban life and governance. The Aspen Institute.

Bonoli, A., Zanni, S., & Serrano-Bernardo, F. (2021). Sustainability in building and construction within the framework of circular cities and European New Green Deal. The contribution of concrete recycling. Sustainability, 13, 2139. https://doi.org/10.3390/su13042139

Bork-Hueffer, T. (2014). Intra-urban and interurban quality of life approaches. In A. C. Michalos (Ed.). Encyclopedia of quality of life and well-being research. Springer, 3371–3375. https://doi.org/10.1007/978-94-007-0753-5_4065

Campbell-Johnston, K., Cate, J., Elfering-Petrovic, M., & Gupta, J. (2019). City level circular transitions: Barriers and limits in Amsterdam, Utrecht and The Hague. Journal of Cleaner Production, 235(20), 1232-1239. https://doi.org/10.1016/j.jclepro.2019.06.106

Caragliu, A., Del Bo, C., & Nijkamp, P., (2011). Smart cities in Europe, Journal of Urban Technology, 18(2), 65–82. https://doi.0.1080/10630732.2011.601117

Christensen, T. (2021). Towards a circular economy in cities: Exploring local modes of governance in the transition towards a circular economy in construction and textile recycling. Journal of Cleaner Production, 305, 127058. https://doi.org/10.1016/j.jclepro.2021.127058

Çimen, Ö. (2021). Construction and built environment in circular economy: A comprehensive literature review. Journal of Cleaner Production, 305, 127180. https://doi.org/10.1016/j.jclepro.2021.127180

COM (2020)98. https://environment.ec.europa.eu/strategy/circular-economy-action-plan_en (accessed 20.3.2025)

CPR, (2024). European Commission. https://single-market-economy.ec.europa.eu/sectors/construction/construction-products-regulation-cpr_en (accessed 20.3.2025)

Dams, B., Maskell, D., Shea, A., Allen, S., Driesser, M., Kretschmann, T., Walker, P., & Emmitt, S. (2021). A circular construction evaluation framework to promote designing for disassembly and adaptability. Journal of Cleaner Production, 316, 128122. https://doi.org/10.1016/j.jclepro.2021.128122

European Investment Bank (EIB) (2018). The 15 circular steps for cities. https://www.eib.org/en/publications/circular-economy-15-steps-for-cities

Ellen MacArthur Foundation (EMF) (2022). Cities and the circular economy. https://archive.ellenmacarthurfoundation.org/explore/cities-and-the-circular-economy

Ellen MacArthur Foundation (EMF) (2023). Circular economy introduction. https://ellenmacarthurfoundation.org/topics/circular-economy-introduction/overview

European Commission (2019). Accelerating the transition to the circular economy. https://ec.europa.eu/info/publications/accelerating-transition-circular-economy_en

Eurostat (2020). Waste statistics. https://ec.europa.eu/eurostat/statistics-explained/index.php/Waste_statistics#Total_waste_generation

Flick, U. (2004). Triangulation in qualitative research. A companion to qualitative research (pp. 178–183). SAGE Publications.

Frantzeskaki, N., Dumitru, A., Anguelovski, I., Avelino, F., Bach, M., Best, B., ... & Rauschmayer, F. (2016). Elucidating the changing roles of civil society in urban sustainability transitions. Current Opinion in Environmental Sustainability, 22, 41-50. https://doi.org/10.1016/j.cosust.2017.04.008

Ghaffar, S., Burman, M., & Braimah, N. (2019). Pathways to circular construction: An integrated management of construction and demolition waste for resource recovery. Journal of Cleaner Production, 244, 118710. https://doi.org/10.1016/j.jclepro.2019.118710

Ghisellini, P., Cialani, C., & Ulgiati, S. (2016). A review on circular economy: The expected transition to a balanced interplay of environmental and economic systems. Journal of Cleaner Production, 114, 11–32. https://doi:10.1016/j.jclepro.2015.09.007

Girard, L., & Nocca, F., (2019). Moving towards the circular economy/city model: Which tools for operationalizing this model? Sustainability 11(22), 6253. https://doi.org/10.3390/su11226253

Guerra, B. C., & Leite, F., (2021). Circular economy in the construction industry: An overview of United States stakeholders' awareness, major challenges, and enablers. Resources, Conservation and Recycling, 170, 1-14, 105617. https://doi.org/10.1016/j.resconrec.2021.105617

Hauashdh, A., Jailani, J., Rahman, I. A., & Al-Fadhali, N. (2022). Strategic approaches towards achieving sustainable and effective building maintenance practices in maintenance-managed buildings: A combination of expert interviews and a literature review. Journal of Building Engineering, 45, 2352–7102, https://doi.org/10.1016/j.jobe.2021.103490

Haveri, A., & Anttiroiko, A.V. (2021). Urban platforms as a mode of governance. International Review of Administrative Sciences, 89(1). https://doi.org/10.1177/00208523211005855

Henrysson, M., Papageorgiou, A., Björklund, A., Vanhuyse, F., & Sinha, R. (2022). Monitoring progress towards a circular economy in urban areas: An application of the European Union circular economy monitoring framework in Umeå municipality. Sustainable Cities and Society, 87, 1-17, 104245. https://doi.org/10.1016/j.scs.2022.104245

Hürlimann, A. C., Warren-Myers, G., Nielsen, J., Moosavi, S., Bush, J., & March, A. (2022). Towards the transformation of cities: A built environment process map to identify the role of key sectors and actors in producing the built environment across life stages. Cities, 121, 1-7, 103454. https://doi.org/10.1016/j.cities.2021.103454

ICLEI (2021). Circular city actions framework. https://circulars.iclei.org/wp-content/uploads/2021/10/Circular-City-Action-Framework_V2.pdf

Isoaho, K., & Valkama, P. (2024). Understanding circular city policies as a discontinuation strategy: Policy insights from circular construction. Environmental Innovation and Societal Transitions, 52, 100897. https://doi.org/10.1016/j.eist.2024.100897

Joensuu, T., Edelman, H., & Saari, A. (2020). Circular economy practices in the built environment. Journal of Cleaner Production, 276(10), 1-20, 124215. https://doi.org/10.1016/j.jclepro.2020.124215

Junli, L., Wie, S., Huaming S., Ruiping L., & Junqing H. (2021). Toward the construction of a circular economy eco-city: An energy-based sustainability evaluation of Rizhao city in China. Sustainable Cities and Society, 71, 102956. ISSN 2210-6707, https://doi.org/10.1016/j.scs.2021.102956

Kirchherr, J., Reike, D., & Hekkert, M., (2017). Conceptualizing the circular economy: An analysis of 114 definitions. Resources, Conservation and Recycling, 127, 221–232. https://doi.org/10.1016/j.resconrec.2017.09.005

Kronsell, A., & Mukhtar-Landgren, D. (2018). Experimental governance: The role of municipalities in urban living labs. European planning studies, 26(5), 988-1007. https://doi.org/10.1080/09654313.2018.1435631

Lakatos, E., Yong, G., Szilagyi, A., Clinci, D., Georgescu, L., Iticescu, C., & Cioca, L-I. (2021). Conceptualizing core aspects on circular economy in cities.

 https://doi.org/10.52202/080684-0012

Sustainability, 13(14), 7549. https://doi.org/10.3390/su13147549

Lin, G., & Kao, C. (2020). Contesting eco-urbanism from below: The construction of 'zero-waste neighborhoods' in Chinese cities. International Journal of Urban and Regional Research, 44,(1), 72-89. https://doi.org/10.1111/1468-2427.12813

Ministry of the Environment (2021). The national building code of Finland. https://ym.fi/en/the-national-building-code-of-finland

Munaro, M. R., Tavares, S. F., & Bragança, L., (2020). Towards circular and more sustainable buildings: A systematic literature review on the circular economy in the built environment. Journal of Cleaner Production, 260(1), 1-25, 121134. https://doi.org/10.1016/j.jclepro.2020.121134

OECD (2018). Global material resources outlook to 2060. Economic drivers and environmental consequences. https://www.oecd.org/environment/raw-materials-use-to-double-by-2060-with-severe-environmental-consequences.htm 5.2.2020

OECD (2020). The circular economy in cities and regions. Synthesis report. https://www.oecd-ilibrary.org/sites/10ac6ae4-en/index.html?itemId=/content/publication/10ac6ae4-en&_csp_=1ebb6c64661c2ec985d31fb93fe18274&itemIGO=oecd&itemContentType=book

Palinkas, L., Horwitz, S., Green, C., Wisdom, J., Duan, N., Hoagwood, K., Horwitz, S., Olin, S., & Hoagwood, K., (2015). Purposeful sampling for qualitative data collection and analysis in mixed method implementation research. Administration and Policy in Mental Health and Mental Health Services Research, 42(5), 533–544. https://doi.org/10.1007/s10488-013-0528-y

Paiho, S., Mäki, E., Wessberg, N., Paavola, M., Tuominen, P., Antikainen, M., Heikkilä, J., Rozado, C., & Jung, N., (2020). Towards circular cities—Conceptualizing core aspects. Sustainable Cities and Society, 59, 102143. https://doi.org/10.1016/j.scs.2020.102143

Paiho, S., Wessberg, N., Pippuri-Mäkeläinen, J., Mäki, E., Sokka, L. Parviainen, T. Nikinmaa, M., Siikavirta, H. Paavola, M. Antikainen, M., Heikkilä, J., Hajduk, P. & Laurikko, J. (2021). Creating a circular city–An analysis of potential transportation, energy and food solutions in a case district. Sustainable Cities and Society, 64, 102529. https://doi.org/10.1016/j.scs.2020.102529

Petit-Boix, A., & Leipold, S. (2018). Circular economy in cities: Reviewing how environmental research aligns with local practices. Journal of Cleaner Production, 195(10), 1270-1281. https://doi.org/10.1016/j.jclepro.2018.05.281

Pomponi, F., & Moncaster, A. (2017). Circular economy for the built environment: A research framework. Journal of Cleaner Production, 143, 710–718. https://doi.org/10.1016/j.jclepro.2016.12.055

Prendeville, S., Cherimb, E., & Bocken, N. (2018). Circular cities: Mapping six cities in transition. Environmental Innovation and Societal Transitions, 26, 171–194. https://doi.org/10.1016/j.eist.2017.03.002

Reike, D., Vermeulen, W. J. V., & Witjes, S. (2018). The circular economy: New or refurbished as CE 3.0? — Exploring controversies in the conceptualization of the circular economy through a focus on history and resource value retention options. Resources, Conservation and Recycling, 135, 246–264. https://doi.org/10.1016/j.resconrec.2017.08.027

Rios, F., Panic, S., Grau, D., Khanna, V., Zapitelli, J., & Bilec, M. (2022). Exploring circular economies in the built environment from a complex systems perspective: A systematic review and conceptual model at the city scale. Sustainable Cities and Society, 80, 103411. https://doi.org/10.1016/j.scs.2021.103411

Sodiq, A., Baloch, A., Khan, S., Sezer, N., Mahmoud, S., Jama, M., & Abdelaal, A. (2019). Towards modern sustainable cities: Review of sustainability principles and trends. Journal of Cleaner Production, 227, 972–1001. https://doi.org/10.1016/j.jclepro.2019.04.106

Tukiainen, T., Leminen, S., & Westerlund, M., (2015). Cities as collaborative innovation platforms. Technology Innovation Management Review, 5(10), 16–23. http://doi.org/10.22215/timreview/933

Turcu, C., & Gillie, H. (2020). Governing the circular economy in the city: Local planning practice in London. Planning Practice & Research, 35(1), 62–85. https://doi.org/10.1080/02697459.2019.1703335

UN Environment (2018). Global Status Report – Towards a zero-emission, efficient, and resilient buildings and construction sector. GLOBAL STATUS REPORT 2018. https://globalabc.org/sites/default/files/2020-03/2018_GlobalAB_%20Global_Status%20_Report%20_English.PDF

United Nations (UN) (2022). Goal 11: Make cities inclusive, safe, resilient and sustainable. https://www.un.org/sustainabledevelopment/cities/

Uusikartano, J., Väyrynen, H., & Aarikka-Stenroos, L. (2020). Public agency in changing industrial circular economy ecosystems: Roles, modes and structures. Sustainability, 12(23), 10015. https://doi.org/10.3390/su122310015

Uusikartano, J., Väyrynen, H., & Aarikka-Stenroos, L., (2021). Public actors and their diverse roles in eco-industrial parks: A multiple-case study. Journal of Cleaner Production, 296(10), 1-10, 126463. https://doi.org/10.1016/j.jclepro.2021.126463

Von Malmborg, F. (2004). Networking for knowledge transfer: towards an understanding of local authority roles in regional industrial ecosystem management. Business Strategy and the Environment, 13(5), 334-346. https://doi.org/10.1002/bse.419

Wang, N., Lee, J., Zhang, J., Chen, H., & Li H. (2018). Evaluation of urban circular economy development: An empirical research of 40 cities in China. Journal of

Cleaner Production, 180, 876–887. https://doi.org/10.1016/j.jclepro.2018.01.089
The Waste Framework Directive 2008/98/EC. https://eur-lex.europa.eu/eli/dir/2008/98/oj/eng (accessed 20.3.2025)
Williams, J., (2021). Circular cities: What are the benefits of circular development? Sustainability, 13(10), 5725. https://doi.org/10.3390/su13105725
World Bank Group (2023). Urban development. https://www.worldbank.org/en/topic/urbandevelopment/overview
Yin, R. K. (2009). Case study research: Design and methods (Vol. 5). sage.

https://doi.org/10.52202/080684-0012

DEVELOPING A CIRCULAR, TECHNOLOGY-ENRICHED TESTBED-ON-WHEELS FOR TEMPORARY AND SHARED USE

Marjo Kenttälä[1,2*], Tuuli Jylhä[1], and Suvi Nenonen[3]

[1]Aalto University, marjo.kenttala@aalto.fi

[2]The City of Helsinki

[3]University of Helsinki

*Corresponding Author: marjo.kenttala@aalto.fi

ABSTRACT

Background and aim. In this fast-paced world, making choices for permanent learning environments that combine physical and digital environments can be complex due to the rapid development of educational technology (EdTech). One circular solution is to utilise modular buildings that offer flexibility, temporality, shared resources, and lower construction and maintenance costs while supporting the learning experience. This paper aims to analyse the mobilization and use phases, including the required servitization models, of a relocatable, shareable and circular classroom that also functions as a testbed for EdTech startups. This testbed-on-wheels, named the Mobile Testbed Tekla, operates in the City of Helsinki, Finland.

Methods and Data. The paper presents an ongoing case study utilising action research methodology on the Mobile Testbed Tekla, which is relocatable, sharable, flexible, multifunctional, and adaptable within urban structure. The data is collected through observations, project documentation, and an expert interview. Tekla functions as both a classroom and a testbed, moving from school to school every 2-4 weeks in Helsinki, Finland.

Findings. Action research with the iterative cycles provides learning points related to physical, digital and social structure of the testbed-on-wheels. The structural, logistic, technical, and functional elements are described in the process of co-creation and co-use of this new learning environment.

Theoretical / Practical / Societal implications. The academic contributions of the paper highlight the use of modular buildings to address temporal demands. Practical implication is valuable for stakeholders on the demand and supply side of learning environments, which explore connectivity and use of new technology.

KEYWORDS: Educational technology, relocatable classroom, shared use, temporality, testbed.

1 INTRODUCTION

While modern educational technology innovations are making education more adaptive, interactive, and student-centered, pedagogical practices and the physical learning environment need to be aligned. The challenges that digitalization is causing for schools can be solved by new pedagogical practices and integrated learning environments in schools. However, in this fast-paced world, making choices for an effective learning environment can be complex. This is why it is useful to use modular elements that not only provide flexibility and upgradability, but also reduce costs, shorten construction time and facilitate maintenance and replacement of parts (Galal El Deen, 2017). According to Kyrö *et al.* (2019), modular buildings support circularity and enhanced usability through features such as flexible ownership arrangements, adaptability, including multifunctionality and elasticity, and seamless integration into existing urban environments. Modular buildings also reflect key principles of the circular economy (Circular Economy, 2017.), such as the reusability of components, the extension of service life through approaches that allow used parts to be taken back for reuse, and the use of

https://doi.org/10.52202/080684-0013

business models that connect products with services to foster aligned objectives and added value.

To proceed with small-scale steps in providing an integrated learning environment and training, new pedagogical practices one can start with explorative pilots. Instead of choosing one technology one can try different solutions. Additionally, instead of bringing technology to existing classrooms at school one can develop a modular and movable classroom which can be shared by diverse school communities. At the same time, this movable classroom can offer start-ups and SMEs (small and medium-sized enterprises) in the EdTech (educational technology) sector the opportunity to test and co-develop their solutions with end-users, so that the mobile classroom becomes an EdTech testbed-on-wheels. EdTech Testbeds offer the opportunity to connect learning technology and schools in two ways: they help EdTech companies to develop their products based on real user data and support teachers' continuous professional learning, and by participating in testing, teachers and learners can influence the development process and adopt new technologies with greater confidence (Vanbecelaere, S. et al. 2023). This realization of an integrated learning environment as a shared resource is also an exploration of more effective resource use. Although relocatable container-based solutions for learning environments exist, research on processes and practices to realize shared use and movable classrooms that can be moved every 2-4 weeks are still rare. There is no research on movable EdTech testbeds at all, as there have been none.
This paper aims to analyse the mobilization and use phases, including the required servitization models, of a movable classroom that also functions as a testbed for EdTech startups. This testbed-on-wheels, named the Mobile Testbed Tekla, operates in the City of Helsinki, Finland. As the research is ongoing, the current results are provisional and will be specified throughout the research process.

The paper is structured as follows. First, the theory section discusses relocatable and temporary (modular) schools, educational technology and testbeds. Next, the action research methodology is presented, followed by empirical data and results on the Mobile Testbed Tekla. Finally, the conclusions outline both the practical and scientific contributions of this research, as well as the topics for further study.

2 THEORY: MOBILE, TECHNOLOGY-ENRICHED LEARNING AND TESTING ENVIRONMENT

Movable, temporary learning spaces have existed for over half a century. Already in the 1970s, it was recognized that relocatable classrooms could provide schools with much-needed additional space (e.g. Baas, 1973), and they could bring specialized educational content to different locations (Erickson, 1971). By the 1980s, relocatable classrooms had become a growing business (Sylvester, 1988), and were particularly utilized as a solution to temporary space shortages caused by, for example, fluctuations in population (Wilson & Schneider, 1989) or urgent space demands (Allison, 1988) like an unexpected influx of refugee children (Silva, 1985). Over the decades, relocatable classrooms have evolved to serve a wide range of needs, and they have become more attractive and versatile.

The concept of relocatable modular school (RMS) has been investigated e.g. by Nguyen et al. (2023). An effective RMS building addresses temporary classroom requirements, responds to the changing in educational delivery programs and, at the same time, provides a pleasant, safe, secure, accessible, well-illuminated, well-ventilated, and aesthetically pleasing learning environment. Like the ordinary school, RMS includes not only the physical structure but also a variety of building systems such as mechanical, plumbing, electrical and power, telecommunications, security, and fire suppression. RMS facilities can allow students to learn in a unique environment that is distinct from their usual classroom, which can stimulate creativity and promote learning by making the educational experience more engaging and interesting. The classrooms can be designed to be visually appealing, incorporating natural light and greenery, or technology-integrated, which can have a positive impact on students' mental and physical well-being and lead to better learning outcomes (Nguyen et al., 2023).

Blazy et al. (2024) have conducted the “Green Classrooms” project, which responds to the growing demand to improve the quality of educational space and increase school space by providing additional mobile classrooms. These can be used as classrooms or as multi-functional spaces, such as a library or a study or break area. The design is based on a modular system that can be easily adapted to the existing site conditions. A key element of the facilities is innovative ecological solutions, such as stormwater retention systems and renewable energy installations.

As schools strive to integrate digital tools, they need environments that support this process without significant investments in permanent infrastructure. A co-usable and mobile learning environment offers a resource-efficient and scalable solution that can be adapted to different educational contexts in teaching. The development of embedded, integrated, or hybrid learning environments includes the merge of physical and virtual spaces as well as the integration of formal and informal spaces in order to stress the need to overcome disciplinary and organizational boundaries. Space matters, but not just physical space, the process of co-creation has an important role too (Ninneman et al. 2020).

https://doi.org/10.52202/080684-0013

While EdTech has the potential to revolutionize learning, it often fails to deliver the expected impact. EdTech companies often encounter difficulties in gaining access to schools for testing and refining their innovations. (Batty et al., 2019). As EdTech becomes more accessible to teachers, governments are emphasizing the need for educators to stay current with its advancements and provide feedback on their experiences to support its continuous improvement (Vanbecelaere et al., 2023). It is important to be able to combine both the needs of EdTech companies for testing with end-users and, on the other hand, the need for teachers to see and try out the latest solutions.

In recent years, various EdTech Testbeds have been launched globally (Vanbecelaere et al., 2023). It was not until 2022 that The Global EdTech Testbed Network (GETN) was established (Globaledtech.org). As Batty et al. (2019) define, EdTech Testbed is "an environment to test and experiment with EdTech in a real-world setting". The City of Helsinki established one of the earliest EdTech Testbeds worldwide (Nordic EdTech Group, 2024). It allows the practical testing and co-development of new technologies and learning solutions while giving teachers the opportunity to provide user feedback and become familiar with advanced tools and digital learning environments, and providing learners with engaging learning experiences, and the chance to develop e.g. their transversal competencies (Kenttälä, 2020). GETN recognizes that the broader concept of EdTech Testbeds encompasses a variety of resources, objectives, and roles (Vanbecelaere et al. 2023).

In Finland, information and communication technology (ICT) is integrated into all grades of basic education, being applied in different subjects and multidisciplinary learning modules, where students are taught various ICT applications and their practical uses (FNAE 2014). The Education Policy Report (Finnish Government 2021) emphasizes leveraging new technologies and digitalization to support, advance, and improve the accessibility of learning, while also developing digital skills and media literacy. The goal is for Finland to become a global leader in developing and utilizing sustainable educational digitalization by 2027 (Ministry of Education and Culture, Finland 2023).

Schools across Finland are provided with the necessary IT equipment and internet access (UNESCO, 2024). Teachers have significant autonomy over the learning materials they use in their teaching (Nordic EdTech Group 2024). Despite these resources, digital technology is still infrequently utilized in Finnish lower secondary schools. Its use is generally limited and focuses mainly on information retrieval, editing, and storage (Oinas et al. 2023). Alternative and low-barrier methods to promote the use of educational technology are thus welcome in the field of education. This would also encourage those teachers who are hesitant to try educational technologies, a group often identified as the late majority and laggards in Rogers' Diffusion of Innovation (2003).

As the Finnish National Sustainable Development Certification of Educational Establishments (Okka Foundation, 2024) states, to develop high-quality learning environments for a sustainable future, educational institutions are encouraged to form partnerships to create, among other things, demonstration environments for new technologies. These environments can be utilized in training, and companies are offered opportunities to visit them. Learning environments can facilitate the creation of new innovations and support sustainable entrepreneurship. (Okka Foundation, 2024)

3 METHODS

This paper adopts an action research approach. This approach, as noted by Tripp (2005) and McNiff (2013), enables the cycle of action inquiry (Figure 1) — including diagnosing, action planning, action taking, evaluating, and specifying learning (Susman and Evered, 1978) — to be repeated throughout the development process, permitting learning not only from theory but also the development of nascent theory. It also allows for the integration of various data sources and methods.

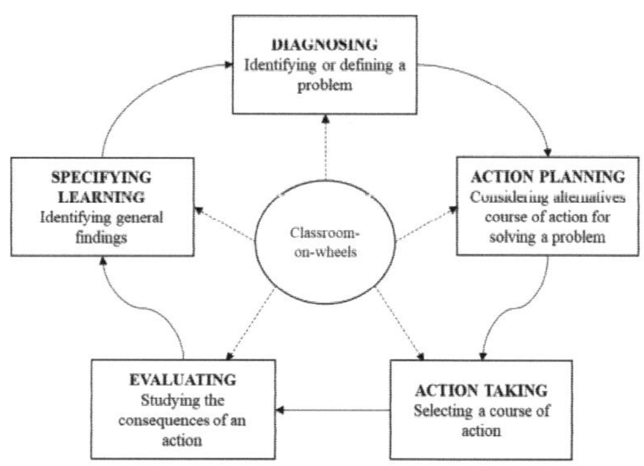

Figure 1: Action research cycle (Susman and Evered, 1978).

In this research, the main data collection methods are observations and project document analysis, including, for example, meeting notes, presentations, collected statistics, personal notes, and published material of Tekla. Additionally, the experiences of the other project member were collected via in-depth interview, which was used as secondary data to verify and strengthen the interpretation of the primary data.

The testbed-on-wheels experiment is ongoing and is estimated to continue until June 2026. This paper focuses on the first part of the study, and it documents one iteration of the action research including five phases (timeline in Figure 2). Each phase included several action cycles, which are presented in the results section. This cyclical nature is crucial, as within action research, ongoing

https://doi.org/10.52202/080684-0013

reflection and assessment enable modifications to be made as the project evolves (Koshy, 2005), ensuring that the study remains flexible and responsive. The results are presented in alignment with the actions taken, since "action research is about two things: action (what you do) and research (how you learn and explain what you do)" (McNiff & Whitehead, 2009). Meanwhile, the testbed-on-wheels concept is continuously being developed based on user experiences, and more action cycles are being undertaken.

To summarize, the project started in April 2023 with Identifying the problem, continued to action planning and designing and converting the container to testbed-on-wheels. Ten months after the start of the project the container was ready for the pilot, which lasted until the end of the spring semester 2024, a total of four months (Figure 2). In 2023 the project team consisted of three people: the project manager, who started in April 2023, and the project specialist, who began a month later. Their task in the project was to develop the container as a learning and testing environment. In May 2023, a part-time project coordinator responsible for project bookkeeping, reporting, and archiving joined the project for 6 months. From January to mid-March 2024, a subsidized part-time employee assisted the container host with the pilot. Later, this employee was hired as a part-time assistant from April to June 2024 to support teachers in using the loanable technologies in regular classrooms. In March 2024, another project specialist joined the project, assisting with start-up collaboration, agreement drafting, and the development of the operational model.

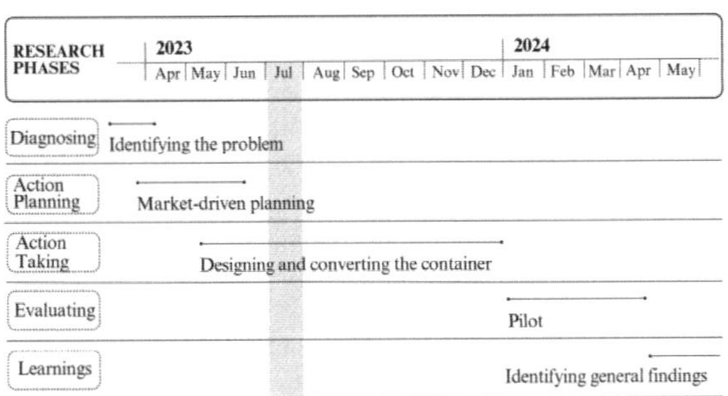

Figure 2: Timeline of the research phases.

4 RESULTS

The results are structured in five phases (Figure 2).

4.1 IDENTIFYING THE PROBLEM

The problem addressed in this research is the lack of a shareable, pedagogically and digitally rich learning environment that supports diverse stakeholders, including learners, teachers, educational technology start-ups and SMEs, and city organizations (e.g., the divisions of the City of Helsinki). Despite various approaches to developing technology-enriched learning environments, a unified definition of a mobile learning environment for temporary and shared use and clear guidelines for its design and implementation in diverse contexts are still missing. It is challenging to develop a learning environment that addresses the diverse needs of multiple stakeholders and supports shared use effectively. Additionally, there is limited knowledge about best practices, key development stages, and the challenges involved in creating such environments service models. This research seeks to address these gaps by providing a clear development framework and proposing a model that can guide the design and implementation of similar learning environments, which can serve also as a testbed, testing and co-developing platform.

4.2 ACTION PLANNING BASED ON MARKET ALTERNATIVES

The action planning consisted of four tasks (Figure 3). This was done to form alternatives to build the testbed-on-wheels.

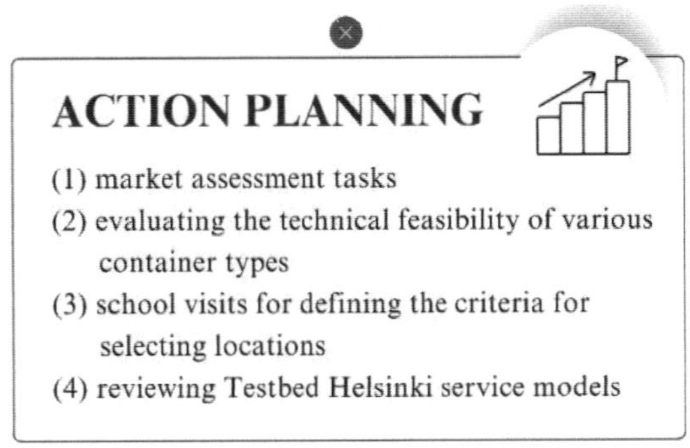

Figure 3: Tasks in action planning.

In the market assessment tasks, a systemic market study was conducted to compare and benchmark the alternatives. These alternatives were benchmarked to identify exemplary practices for adaptation and to avoid less effective practices. Initially, various physical space options were considered, such as a variety of containers, an event trailer, a library bus, a van, and a mini house on wheels. More specifically, eight environments used in education or other services in Finland in terms of shared spaces, tools, and technologies, as well as movability and facilitation were also analyzed as benchmarks. The benchmarking was based on information collected through site visits and observations, interviews with the benchmarking case representatives, and public materials.

The benchmarking identified four clusters of concepts for supplying educational technology to learners: (1) Placing the technology in permanent locations to be shared by multiple users across various organizations. (2) Making the technology available for loan to teachers at schools. (3) Moving the technology from one classroom to another, with facilitation included. (4) Integrating the

https://doi.org/10.52202/080684-0013

facilitated workshops into a movable learning environment that visits a few cities in Finland for an extended period during warm weather.

Based on benchmarking, a solution that allows educational technology to be shared both inside and outside the classroom was sought. The aim was to create an environment that facilitates product development for companies in collaboration with end-users, supports the use of technology, and provides low-barrier access for teachers and learners. This would facilitate its use during regular teaching sessions and by external educators working with learners and teachers. Based on this, containers were concluded to be the best solution for developing testbed-on-wheels.

In the second task, the technical feasibility of various container types was evaluated. This was to ensure that the chosen container model would be feasible for the City of Helsinki: that it could operate all year round, that a group of learners can work there, and that it would allow low-threshold participation in workshops enriched with educational technology. The project team visited a total of three container rental companies, including providers of traditional sea containers of various sizes, office containers, and different kinds of glass containers built for trade fairs. The visits provided information on container handling and weights, ventilation, heating and cooling alternatives, window and door locations, window protection, lighting, electrical wiring, power supply to the container, wall materials, and customization options.

While comparing the alternatives, twenty critical design issues were identified, including electricity demands and supply, vandalism protection of the container, and estimating the fitness-for-use of the container for providing technology-enriched workshops. For some of these issues, more expert knowledge was required, necessitating broader collaboration with specialists from the City of Helsinki. For example, to address the electricity demand and supply issues, consultation with the city's technical experts was needed.

The school visits in Helsinki helped define the criteria for selecting the container's visiting locations. The most important criterion identified was the ability to provide electricity from the school to the container. The previous service models of EdTech Testbed Helsinki were also reviewed.

As a result of the exploratory work resulted in the following inputs: (1) the technical and functional requirements for the container, (2) the criteria for selecting schools to participate in the experiment, and (3) information for the call to invite EdTech companies to join the experiment and further develop their products within the container (Figure 4).

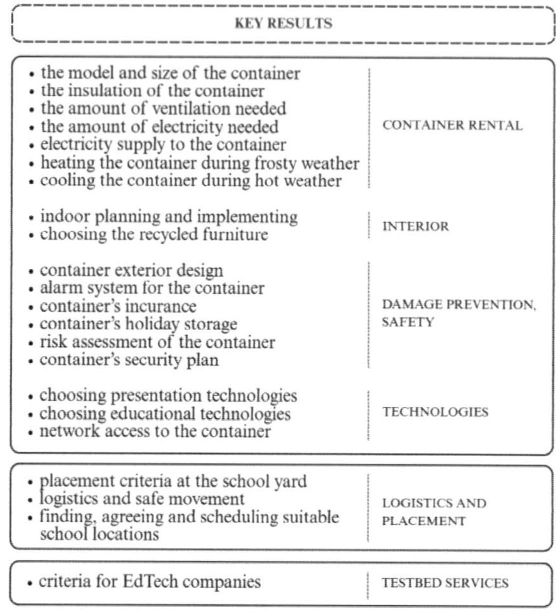

Figure 4: Key results in action planning.

4.3 ACTION TAKING DESIGNING AND CONVERTING THE CONTAINER INTO TESTBED-ON-WHEELS

In action taking, seven tasks were taken to design and convert the container into testbed-on-wheels (Figure 5).

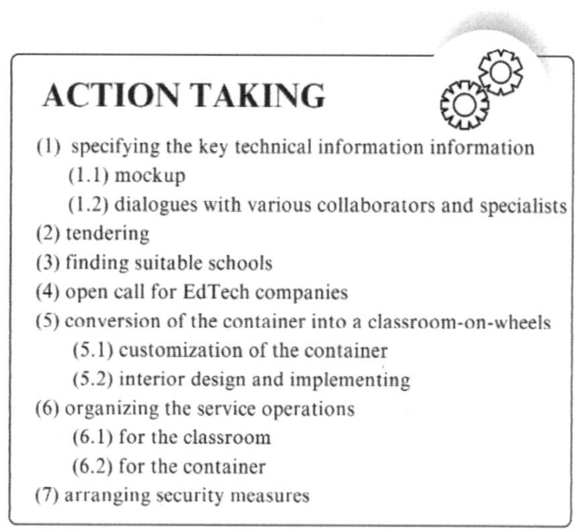

Figure 5: Tasks in action taking.

First, a mockup of the container to visually test how it would function as a classroom was created. It was built in a meeting room at their office, roughly the container's size. Tables and chairs were arranged, and the positions of the door and windows were given to illustrate walking routes, the layout of the teaching lesson, and the practicalities of entering and leaving the classroom. Through this mockup, it was concluded that the container could only accommodate half a class of students at a time, which consequently impacted the operating model of the container (A in Figure 6).

Second dialogues with collaborators and specialists were conducted between May-September 2023 to define the critical technical details for the testbed-on-wheels. The transportation of the container from school to school was planned to be done by the students of the logistics program of a vocational school. Via the dialogue with the teacher, a significant limitation was found (B in Figure 6): the crane of the school, which would be used to lift the container onto the platform, could lift only about 3000 kg. Therefore, the weight of the container became the most important technical detail in the selection of the container. Additionally, the weight limit required the inclusion of a van in the transport arrangements to carry loaned equipment and loose furniture when the container is moved from one school to another.

Furthermore, discussions with technical experts helped determine the adequate ventilation required per person (liters/second/person), and it was concluded that the container would need mechanical ventilation (C in Figure 6). Another group of technical experts assessed the necessary amount of electricity and related infrastructure. Their calculations indicated that the container would require a 32-ampere power plug for its power supply. With the ICT experts, the necessary presentation technology and other equipment, such as computers and tablets, needed in the container to facilitate teaching were determined. Similarly, the technological equipment to be brought into the classrooms was defined. The media experts told what Wi-Fi options could be installed in the mobile space. In addition, based on the dialogue, the facility services of the Education Division of the city offered existing furniture from schools to be re-used in the container.

After identifying the key information on the technical requirements for the container and the limitations it brings, based on the mockup and dialogues with various collaborators and specialists, the final decisions on the technical details were made. This was essential in defining the tender request to rent the container published in September 2023 and the tender request for the interior design in October 2023.

Third, the school visits were continued in Helsinki to find suitable locations for the testbed-on-wheels. Two criteria excluded many of the schools from consideration: the testbed-on-wheels required a three-phase power socket at a convenient location on the school property. When the rescue authorities reviewed the container's safety plan, it was also determined that the container's location must situate at a safety distance of eight meters from the school or other buildings, and this made it more challenging to identify schools where the container could be safely and effectively placed (D in Figure 6). After visiting the first 50 schools, only seven suitable schools had been found. Fourth, in September 2023 an open call inviting EdTech companies to apply for testbed services provided by the container for the upcoming spring semester was issued. After a selection process until the end of the year, six companies were chosen to participate in the piloting phase of the movable learning environment.

Fifth, after selecting the container supplier and interior design and implementation team, the container was converted into a testbed-on-wheels. This conversion involved two main levels: customizing the container itself by adding the features needed and converting the interior into an exciting technology-enriched learning environment. The container rental company customized the office container according to the needs, including a door with a protective barrier, windows with bars, lightning, a ventilation unit, a heat pump, two electric radiators, 10 electrical outlets at designated locations, a 32A socket on the exterior, which was requested to have covered with a lockable safety hatch, mounting points on the wall for securing drawers, and frameworks for coat racks and a display screen. This was done in October–November 2023 in the rental company's warehouse. The interior services were provided by an external supplier chosen through an earlier tender process. The interior design was based on predefined requirements, including technological and user group needs as well as a request to create an attractive visual identity, and was refined through workshop to further clarify interior needs. This transformation included the creation of an exciting interior design and logo, wall decals, coat racks, and shoe compartments, and the painting of recycled furniture along with presentation technology and mood lighting, resulting in fully equipped, technology-enriched testbed-on-wheels. This was done in a rented warehouse in December 2023–January 2024. Building the container's interior in the middle of winter required renting a hall space (E in Figure 6).

Sixth, the testbed-on-wheels required service processes to deliver and maintain its value. These processes had two main goals: facilitating learning in the testbed-on-wheels and supporting education in the main school building. To support education in the main school building, a lending service was organized. This allowed teachers to use educational technology in regular classrooms and included procuring educational technologies and their charging cabinets, creating clear instructions for teachers on how to borrow the technology, and arranging separate transportation due to the weight limitations of the containers.

To facilitate learning in the testbed-on-wheels, the project specialist served as the container host. The service processes for the EdTech companies were developed and EdTech startups were provided with guidelines for conducting workshops in the container. Additionally,

https://doi.org/10.52202/080684-0013

various workshop concepts were developed for learners, including an engaging workshop titled 'Collision with a Meteorite.' This workshop (F in Figure 6), collaboratively designed by the host and teachers, features a simulation video to prepare participants for the challenge. The video was produced in collaboration with the media team in January 2024. This customizable workshop is tailored to different age groups, making it accessible and engaging for students aged 5-16 years.

Furthermore, security measures were arranged for the testbed-on-wheels in early January 2024. Based on the tenders, the security service was selected. To prevent vandalism, the containers received a final artistic touch from young graffiti artists aged 13-21 years from the City of Helsinki's Culture and Leisure Division's Graffiti group (Picture 1). After the graffiti artwork was completed in the final weekend of January, the testbed-on-wheels was ready for its first school in the pilot phase.

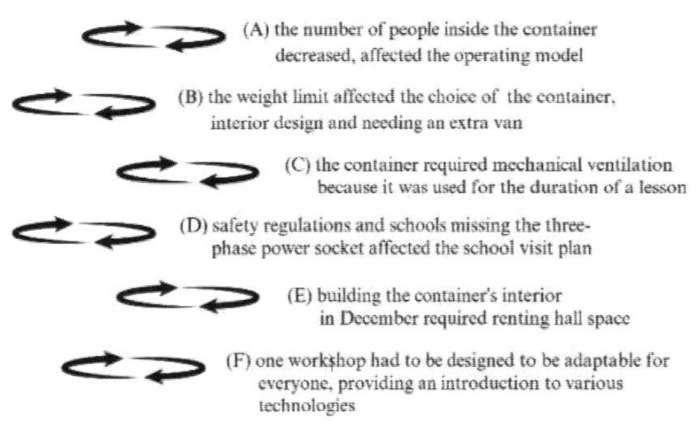

Figure 6: Key results in action taking.

Installation of the smart wall was done afterward in February 2024 during the winter break.

Picture 1: Mobile Testbed Tekla. Picture City of Helsinki.

4.4 EVALUATION: PILOT PHASE

The pilot phase consisted of two tasks (Figure 7), implemented from February to May 2023. These tasks were to evaluate and further develop the operating model of the testbed-on-wheels and to refine the service processes for both the testbed-on-wheels and the regular classroom. During these 4 months, development work was based on 220 workshops with 2,344 participants, including learners and teachers, led by the six EdTech companies and the container host. The host of the testbed-on-wheels had a central role in developing the development of the operating models.

EVALUATING

(1) evaluate and further develop the operating model
(2) refine the service processes
  (2.1) for the classroom-on-wheels
  (2.2) for the regular classroom

Figure 7: Tasks in evaluating.

The operating model was based on the Testbed activity organized by the City of Helsinki. The aim was to develop the container into an easy and accessible way for teachers and learners to test and learn about new technologies. In this project, the testbed-on-wheels serves also as a testbed for EdTech companies and as a technology-enriched learning environment for learners, combining these two perspectives. The delivery of the services required the design and implementation of multiple types of content (Figure 8). Firstly, the EdTech companies planned their workshops for children aged 5-16 and planned feedback collection methods to further develop their products. Using this information, the EdTech companies completed a workshop card based on a standard template developed by the project team. This card was used to promote their EdTech solutions to teachers. The catalogue of these cards provided teachers with a service with a variety of EdTech solutions and related workshops to choose from, allowing them to select the best fit for their group and study topic. To facilitate easy booking for teachers, a tailored workshop calendar was created, showing available slots for each EdTech technology based on the school timetable. Teachers could then directly reserve a suitable time slot.

Before the testbed-on-wheels arrived at a school, the information to the schools to encourage teachers to utilize the testbed-on-wheels and borrow the learning technologies was provided. The project team attended a teachers' meeting introducing them to the workshop alternatives, the workshop and the loanable equipment calendars, and the technologies available for lending. Experience at the second school, where a visit was not allowed, highlighted the importance of these introductory visits: the utilization of the testbed-on-wheels and loanable technologies was significantly lower without it. These visits became a crucial aspect of collaboration and vital for the school's participation in the experiment.
Furthermore, a pre-written message service for the teachers to send to the guardians of the learners was provided. In addition, an information package to the

principal was sent. This package included a list of tasks and responsibilities for the host of the testbed-on-wheels and the school staff.

The operating model was co-developed iteratively during the pilot phase. Twice-weekly meetings were also held to discuss experiences and feedback, and to support ongoing development. Based on feedback, several improvements were made (Figure 8), for example: the workshop calendar was made more user-friendly, different types of booking calendars for loanable equipment were implemented, a maximum number of workshops per school day was established, cleaning was rescheduled to be done during the lunch break, and the EdTech companies quickly learned which aspects of their workshops were effective and engaging and which were not.

Additionally, the service to loan equipment for trying out new EdTech solutions in regular classrooms was further developed during the piloting phase. Initially, the technologies were available in the teachers' office, ready to be loaned out with instructions. However, the utilization of these loaned technologies was low. For instance, during a two-week period in March 2024 when there was no dedicated host for this equipment, the technologies were not used at all. Based on this, it was decided to allocate an additional host to help teachers use the loaned equipment during lessons. It was observed during the pilot phase that when teachers participated in facilitated workshops in the container and received assistance during lessons, they were more willing to try modern technologies with their learners.

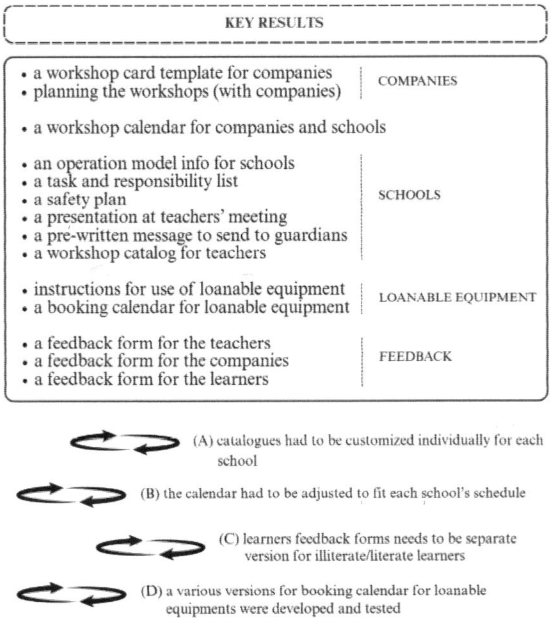

Figure 8: Key results in evaluation.

4.5 SPECIFYING LEARNINGS

Action research cycle with the iterative cycles provides learning points related physical, digital, and social structure of the testbed-on-wheels (Figure 9), aligning with the principle of bits, bricks, and interaction (Früchter, 2005).

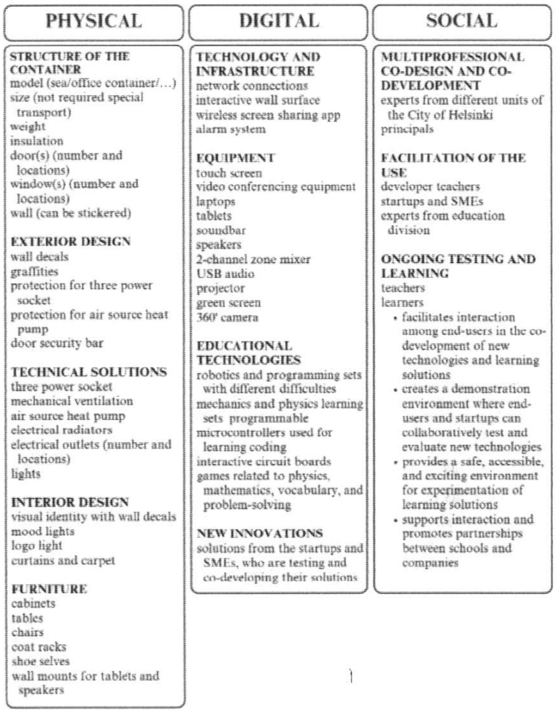

Figure 9: Physical, digital, and social structure of the testbed-on-wheels.

The physical perspectives include structural and technical issues and issues connected to indoor environment. The interior design, furniture and visual outlook were also part of the physical solution. The positive outcome of the concept is connected to mobility and functionality, which is appropriately supporting the pedagogical goals. However, despite the mobility potential not all the locations are suitable for the container. The possibility of suitable electrical solutions determines the successful use of the testbed-on-wheels. Additionally, the container cannot be used by the typical group size for one classroom.

The digital perspective includes the network connections, larger and smaller equipment, and the safe and functional ways to store and charge them. The design and procurement of these elements were important parts of the process to enhance technology-enriched pedagogy conveniently and effectively. The testbed-on-wheels provides low-threshold access to new technology, and a shared test environment is a way to focus on relevant digital solutions to the classroom. The possibility to invest in the solutions which will be used is better after small scale testing. The support and facilitation as a part on

https://doi.org/10.52202/080684-0013

making technological solutions familiar for the learners and teachers was important. The service providers, EdTech companies, could also use test environment for their service development based on the immediate feedback from users. The critical thing is to find time and resources for an introduction visit of the teachers at the school before the actual learning actions for smaller groups take place.

Social perspectives and learning points include the observation of the multi-professional team. It is significant in the design phase, use phase, and continuous development of the testbed-on-wheels. The collaboration within the different units in the city organisation is a valuable source for a successful outcome. The commitment of stakeholders and the information flow between them need to be strengthened in various ways. One can claim that sharing the EdTech requires, first, the right information for the right people at the right time – the fine-toned social environment for the diverse actors. Secondly, it requires aligned and integrated physical and digital testbed-on-wheels, which are easy to access. Thirdly it requires facilitation - user-friendly service processes and contact person, host taking care of the experience of the users: learners, teachers, and the service providers. The ecosystem of school, city, and companies was easy and purposeful to create around the common platform, testbed-on-wheels.

5 CONCLUSIONS

This study involved five steps of action research, from problem identification to planning, action, evaluation and identification of general findings. Action planning, action taking, and evaluating took a total of 14 months (with a one-month summer break in between). Within these phases, several smaller cycles of action inquiry occurred requiring rapid reaction and several adjustments to the plans along the way.

The results show that it is possible to build and deploy mobile testbed-on-wheels for temporary and shared use. This encompasses structural, technical, and functional elements, as well as logistical and outdoor environment issues at the school. Modular building solutions allow for adaptability in meeting changing pedagogical requirements (Blazy et al., 2024), and testbed-on-wheels served as a practical example of such adaptability in action: the Tekla project in Helsinki shows that a shared use, mobile physical learning environment offers learning experiments and tests new technologies in schools and kindergartens. Its main advantages are the quick and low-cost construction of the learning environment, its portability to the yards of different schools and daycare centers every two to four weeks, easy accessibility for teachers and learners from their schools, and the sharing of expensive learning technology tools with several users. Tekla supports key principles of the circularity through its reuse and sharable features: the furniture are recycled, the container had been previously used and will be looped for future reuse after the service life of Tekla. Additionally. several of its components such as the air source heat pump, furniture, presentation equipment, and loanable technologies are transferable to other settings after Tekla's operations. It integrates products and services by offering both an additional learning space for the school and shared use technologies that are actively used by various groups on a daily basis. Furthermore, the reuse and circularity of the container are enabled through renting rather than transferring the ownership. In addition, it is a particularly efficient and globally unique way for EdTech companies to test their new and developing technologies together with end users.

Modular solutions can address space needs by utilizing relocatable classrooms (Blazy, 2024); however, this project demonstrated that, at least in most schools in Helsinki, there is limited readiness to accommodate such units unless the issue of power supply is resolved through alternative means. There are examples of such solutions as well, such as the Energy Positive Portable Classroom, which produces several times more energy annually than it consumes, thanks to its extensive photovoltaic surface and energy-efficient design (Energy Positive Portable Classroom, 2014). The notion is that portable classrooms tend to remain in place after their initial setup, despite their intended mobility (Ander et al., 2004) but in this case the space was actively relocated at least once a month, which posed additional challenges both during the development of the space and the implementation of activities.

At the same time, the development process showed how important it is to commit the right people to the project already at the planning stage. The design of learning spaces calls for a diverse and multidisciplinary team, whose composition may vary at different stages of the design process (Oblinger, 2004), as was also the case in this project. The construction of a container requires technical expertise at least in electrical, network, ventilation, and logistics matters, as well as, for example, expertise in presentation technology and EdTech tools. Support from pedagogical experts was highly valued in the development of workshop content. Educational technology is often introduced without adequate technical support, leading to tensions among teachers and, in some cases, negatively affecting well-being (Fernández-Batanero et al., 2021). The pilot phase showed that simply bringing a learning environment to the school or daycare yard or bringing educational technologies inside the school does not inspire teachers to embrace the opportunities they offer to enrich teaching. Teachers also need to be committed to Tekla in advance and help and guidance are still needed in using educational technologies.

Because low-threshold services for schools were developed for Tekla's physical learning environment, the workshops it offered were well received. The core idea behind creating the service package was that Tekla's coming to schools and participation in its activities should not be burdensome for the teachers. Pre-prepared information materials pre-made visits to teachers' meetings per school, workshop catalogues and booking calendars and, above all, facilitated workshops by EdTech companies and the container host ensured an easy "excursion destination" for teachers and learners in their own school or kindergarten yard. Using modern technologies in teaching can be daunting for some teachers - teachers have reported that their limited competence in using EdTech tools contributed to feelings of stress and frustration, and diminished their sense of control over their work (Huhtasalo et al, 2021) - a mobile, full-service, technology-enriched temporary learning environment removes barriers to trying them.

The research contributes to the theoretical discussion about sharing economy in the educational action environment. Additionally, the research results are interlinked with ecosystem development and management. Technical perspective provides input to modular, temporary and shared use facilities. In practice the benefit is for the stakeholders in the educational field: schools, educational technology providers, and public sector to identify different solutions for learning environments.

Action research method provides insights to the process, role of actors and actions needed in managing multiprofessional experts and stakeholders. The identified projects have used methods like participatory design methodology (e.g. Pedro et al., 2017) and a case study research approach (e.g. Nguyen et al., 2023) and the common conclusion is that the varied ways to collect the data and use methods in research design are needed to solve transdisciplinary problems. Additionally, action research is used in analyzing the change like workplace transformation, in which it was also understood as a sociometrical whole, encompassing not only the organizational dimension but also the physical, virtual, and social dimensions (Andrade-Asikainen, 2022).

Although action research has its advantages, such as solving practical challenges with practical outcomes via positive change (Susman and Evered, 1978; De Oliveira, 2023), it also has limitations. As typically criticized, for example by De Oliveira (2023), the first author of this paper is an actor within the community. This can be interpreted to mean that the results may be biased, and that the external validity is low (De Oliveira, 2023). However, as Ward (2021) explains, actions can be separated between motivating reasons and justifying reasons. The first are reasons for which the person has a motivation for the action and are thus tied to the person's "desires, beliefs, and emotions" (Ward, 2021). The latter refers to "reasons for or against" actions that are not tied to the person but to the world beyond. In this action research, the actions are based on reasons for and against, and these learnings have set a new direction for the testbed-on-wheels project.

Another typical limitation in action research is that the study is situational and challenging to replicate, as is the case here as well. In action research, it is assumed that the relationships between people, events, and things are not fixed, but are defined by the current actors, often based on the context (Susman and Evered, 1978). In this research, it is seen as a strength because the action is based on justifying reasons within a context where the actors understand the relationships between people, events, and things.

The future studies could be conducted about how testbed-on-wheels has been perceived by learners, teachers, education providers, start-ups, and SMEs: whether it has delivered value to all members of the ecosystem as intended, and what kind of value it has delivered. In addition, the study could compare other testbed models offered by the City of Helsinki and explore the needs for which the mobile testbed is best suited. It would also be interesting to have research data on whether the container space itself adds value to the support offered to schools and startups in this approach. Would it be sufficient and and further reduce the use of construction resources to simply have a technology host with technology equipment and edtech tools visit a school for a few weeks at a time and offer guided workshops with startups in the school's own classrooms?

ACKNOWLEDGEMENTS

This research is part of the Resourceful Sharing project partly funded by the EU and authorized by the Helsinki-Uusimaa Regional Council. The authors wish to thank The City of Helsinki providing access to the study environment.

Declaration of Generative AI and AI-assisted technologies in the writing process: During the preparation of this work the authors used DeepL and ChatGPT4.o to check the spelling and grammar or for translation.

REFERENCES

Ander, G. D., Pierce, R. A., Magar, C., & Jones, E. (2004) High Performance Relocatable Classrooms.

Andrade-Asikainen, N. (2022). Supporting well-being in workplace transformations-Critical sociomaterial components and their evaluation in a knowledge-intensive organization.

Baas, A. M. (1973). Relocatable Classrooms. Educational Facilities Review Series Number 10.

Batty, R., Wong, A., Florescu, A., & Sharples, M. (2019). Edtech testbeds: Models for improving evidence. NESTA Report.

https://doi.org/10.52202/080684-0013

Blazy, R., Bodziony, M., Baziak, B., Hrehorowicz-Gaber, H., Błachut, J., Ciepiela, A., Łysień, M., Dudek, J., Hrehorowicz-Nowak, A., & Synowiec, A. (2024). The Example of the "Green Class" Project in Krakow: A New, Mobile System of Educational Facilities That Takes Rainwater Retention into Account. *Sustainability*, *16*(14), 6157. https://doi.org/10.3390/su16146157

Circular Economy (2017), "Making sense of the circular economy – the 7 key elements", available at: www.circle-economy.com/the-7-key-elements-of-the-circular-economy

Energy Positive Relocatable Classroom / Anderson Anderson Architecture" 02 Oct 2014. ArchDaily. Accessed 23 Mar 2025. <https://www.archdaily.com/550780/energy-positive-relocatable-classroom-anderson-anderson-architecture> ISSN 0719-8884

Erickson, K. A. (1971). An Analysis of Enrollment, Facilities, and Grade Organization. Phoenix-Talent School District# 4, Jackson County, Oregon

Fernández-Batanero, J. M., Román-Graván, P., Reyes-Rebollo, M. M., & Montenegro-Rueda, M. (2021). Impact of educational technology on teacher stress and anxiety: A literature review. *International journal of environmental research and public health*, *18*(2), 548.

FNAE. (2014). National Core Curriculum for primary and lower secondary (basic) education. The Finnish National Agency for Education. Available at: https://www.oph.fi/sites/default/files/documents/perusopetuksen_opetussuunnitelman_perusteet_2014.pdf. (Accessed 27 August 2024).

Fruchter, R. (2005). Degrees of engagement in interactive workspaces. *AI & SOCIETY*, *19*, 8-21.

Galal El-Deen Mohamed, E. (2017). Recycled Shipping Containers as a Tool to Practice Modularity in Architecture Studio. *JES. Journal of Engineering Sciences*, 45(No 6), 791-803. doi: 10.21608/jesaun.2017.116887 Globaledtech.org. (Accessed 31.1.2025)

Globaledtech.org. (Accessed 31.1.2025) About - Global Edtech

Huhtasalo, J., Blomberg, A., & Kallio, K.-M. (2021). Digitaalisen opetusteknologian vaikutukset opettajien asiantuntijuuteen ja ammatti-identiteettiin yliopistoissa. *Yliopistopedagogiikka*, *28*(2). https://lehti.yliopistopedagogiikka.fi/2021/12/14/digitaalisen-opetusteknologian-vaikutukset-opettajien-asiantuntijuuteen-ja-ammatti-identiteettiin-yliopistoissa/

Kenttälä, M. (2020). Easy Access Co-Development – yhteiskehittämisen palvelumuotoilua. Master's Thesis. Metropolia University of Applied Sciences. Helsinki. Available at: https://www.theseus.fi/bitstream/handle/10024/352388/Kentta%cc%88la%cc%88_Marjo.pdf?sequence=2&isAllowed=y. (Accessed 29 August 2024).

Koshy, V. (2005). *Action research for improving practice: A practical guide*. Sage.

Kyrö, R., Jylhä, T., & Peltokorpi, A. (2019). Embodying circularity through usable relocatable modular buildings. *Facilities*, *37*(1/2), 75-90.

McNiff, Jean, and Jack Whitehead. You and Your Action Research Project, Taylor & Francis Group, 2009. ProQuest Ebook Central, http://ebookcentral.proquest.com/lib/aalto-ebooks/detail.action?docID=446724. Created from aalto-ebooks on 2025-01-03 20:18:25.

Ministry of Education and Culture, Finland (2023). Policies for the Digitalization of Education and Training 2027. Available at: https://julkaisut.valtioneuvosto.fi/bitstream/handle/10024/165248/OKM_2023_48.pdf?sequence=1&isAllowed=y. (Accessed 27 August 2024).

Nguyen, T.D.H.N., Jeong, J., Ahn, Y. and Shin, H. (2023) An innovative approach to temporary educational facilities: A case study of relocatable modular school in South Korea, Journal of Building Engineering, Volume 76,2023,107097,ISSN 2352-7102,https://doi.org/10.1016/j.jobe.2023.107097.

Nordic EdTech Group (2024). Nordic EdTech/digitalisation: generating evidence through piloting testbeds. International Centre for EdTech Impact. WiKIT. Available at: https://www.foreduimpact.org/reports/nordic-edtech. (Accessed 27 August 2024).

Oblinger, D. (2005). Leading the transition from classrooms to learning spaces. *Educause quarterly*, *28*(1), 14–18.

Oinas, S., Vainikainen, M.-P., Asikainen, M., Gustavson, N., Halinen, J., Hienonen, N., Kiili, C., Kilpi, N., Koivuhovi, S., Kortesoja, L., Kupiainen, R., Lintuvuori, M., Mergianian, C., Merikanto, I., Mäkihonko, M., Nazeri, F., Nyman, L., Polso, K.-M., Schöning, O., ... Hotulainen, R. (2023). *Digitalisaation vaikutus oppimistilanteisiin, oppimiseen ja oppimistuloksiin yläkouluissa: Kansallisen tutkimushankkeen ensituloksia suosituksineen*. Tampereen yliopisto. http://www.urn.fi/URN:ISBN:978-952-03-2780-4 (Accessed 27 August 2024).

OKKA Foundation. Finnish national Sustainable Development Certification of Educational Establishments. 10) Kestävän tulevaisuuden oppimisympäristöt. Available at: 10) Kestävän tulevaisuuden oppimisympäristöt - Oppilaitosten kestävän kehityksen sertifiointi. (Accessed 6 January 2025).

De Oliveira, B. (2023). Participatory action research as a research approach: Advantages, limitations and criticisms. Qualitative Research Journal, 23(3), 287-297.

Pedro, N,, Baeta, P., Paio, A., Pedro, A., Matos, J.F., (2017) Redesigning classrooms for the Future: Gathering Inputs from Students, Teachers and Designers. INTED2017 Proceedings, pp. 7908-7917.

Rogers, E. M. (2003). Diffusion of Innovations. (5th ed.). New York: Free Press.

Silva, H. (1985). *The Children of Mariel from Shock to Integration: Cuban Refugee Children in South Florida Schools* (Vol. 13). Cuban American National Foundation.

Susman, G.I and Evered, R.D (1978) An assessment of the scientific merits of action research, *Administrative Science Quarterly*, 23 (4), pp. 582-603.

Sylvester, T. S. (1988). Relocatable and Modular Classrooms: Booming Business. *School Business Affairs*, *54*(1), 22-23.

Tripp, D. (2005). Action research: a methodological introduction. *Educacao e pesquisa*, *31*, 443-466.

UNESCO (2023). Finland, Technology: Country Profile. Available at: https://education-profiles.org/europe-and-northern-america/finland/~technology. (Accessed 23 August 2024).

Vanbecelaere, S., Adam, T., Sieber, C., Clark-Wilson, A., Boody Adorno, K., & Haßler, B. (2023). Towards Systemic EdTech Testbeds: A Global Perspective. https://doi.org/10.53832/opendeved.0285. Available from https://globaledtech.org/.

Ward, Z. B. (2021). On value-laden science. *Studies in History and Philosophy of Science Part A*, *85*, 54-62.

Wilson, S. E., & Schneider, J. R. (1989). Facility Design for Transient Populations and Reuse. *Educational Facility Planner*, *27*(6), 25-27.

 https://doi.org/10.52202/080684-0013

BEYOND EXPERIMENTATION: TEMPORARY USE AS A SOCIAL CIRCULAR STRATEGY

Chiara Mazzarella[1*], Angela Greco[1], and Hilde Remøy[1]

[1]Department of Management in the Built Environment, Faculty of Architecture and the Built Environment, TU Delft, NL

*Corresponding Author: c.mazzarella-1@tudelft.nl

ABSTRACT
Background and aim. Temporary use of vacant spaces—the short-term activation of properties awaiting transformation—has gained recognition for its potential to foster urban revitalization. While such uses provide a platform for experimentation, accessibility, and social inclusion through participatory and cultural activities, they often remain precarious and underutilized as strategic tools for circular economy. This study aims to explore how specific hybrid approaches to temporary real estate management can transform temporary use into a social circular economy strategy, balancing social values with market logics.

Methods and Data. This research employs a qualitative analysis, first defining a framework from literature and then analysing specific temporary use projects through a retrospective case analysis of three cases by Plateau Urbain (France), communa (Belgium), and Stad in de Maak (Netherlands). Data collection included interviews, project documentation, and field observations, allowing an in-depth exploration of the enabling conditions for successful hybrid approaches in creating social value.

Findings. This study makes three key contributions. First, it conceptualizes collaborative temporary use as a social circular strategy, clearly defining the evolution of the concept and its potential in temporary real estate adaptive reuse. Second, by drawing on the literature on organizational hybridity and case study analysis, it identifies key enabling conditions, such as tweaking the balance between social value and market logic over time to recalibrate impact—that underpin temporary use projects as social circular economy strategies. Third, it offers a framework to determine whether a temporary real estate reuse initiative can function as a social circular economy strategy.

Theoretical / Practical / Societal implications. This study offers theoretical insights into hybrid organizing for urban development and practical recommendations for integrating temporary reuse of real estate into social circular economy frameworks. Societally, it underscores the potential for collaborative temporary use to foster circular urban transformation by balancing economic goals with community-driven social value creation.

KEYWORDS: temporary use, adaptive reuse, social circular economy, real estate management, value creation.

1. INTRODUCTION

Temporary adaptive reuse – the temporary uses of existing real estate– naturally stems from circular practices by reusing vacant real estate and recovering, reusing, or recycling components such as furniture and construction materials. From the first independent urban pioneers of temporary use in Berlin (Oswalt et al., 2012; Senatsverwaltung für Stadtentwicklung Berlin, 2007) to contemporary European research projects (Galdini, 2022; *gE.CO Toolbox | gE.CO Toolbox*, n.d.; *Resources | MESOC*, n.d.), temporary use is based on building reuse for diverse purposes that meet social needs, employing essential modifications to enable functionality while prioritizing material recovery and community engagement.

The past decades have seen growing attention to the temporary use of buildings and public spaces, driven by the diverse benefits and advantages these short-term uses provide to a wide range of urban stakeholders. Beyond pop-up stores or short-term rentals, temporary use that actively engage communities and foster socio-cultural processes have the potential to start placemaking and create diverse kinds of value, namely economic, social, environmental, and cultural (Bragaglia & Caruso, 2022; Karachalis, 2021; Mangialardo & Micelli, 2017; Martin et al., 2019).

Experimental temporary uses include socio-cultural oriented initiatives, that test communal approach to place management and are frequently dedicated to placemaking and non-profit activities. These unconventional practices can generate significant benefits for both people and stakeholders in the real estate sector by enhancing place attractiveness, improving neighbourhood amenities, and contributing to the vibrancy and functionality of urban environments. These practices often rely on hybrid organizing (Mitzinneck & Greco, 2021) involving collaboration among initiating organizations, property owners, and public authorities, to realize their potential benefits. These organizations enable diverse stakeholders to utilize available spaces, incorporating them into place governance and, in some cases, involving users directly in decision-making processes. Temporary uses managed under this model test forms of real estate management, services, and forms of collaboration and sharing.

The intangible benefits of building reuse, social inclusion, and cultural initiatives seem evident, and are replicated even if they have not been clearly evaluated (Munzner & Shaw, 2015). At the same time, these practices face challenges due to economic constraints and social challenges due to their short-term nature (Ferreri, 2020). In traditional economic terms, profit motives dominate investment decisions, whereas in the realm of non-profit, hybrid organizations, resources from the sharing economy, such as time, trust, and availability, become tools to produce social value (Greco, 2024). In turn, intangible assets become instrumental in community-led real estate management and collective practices. Social strategies demand a reimagining of investment incentives, merging financial objectives with ESG principles to create value-driven and impactful decision-making frameworks.

In the built environment, material circularity and use value underpin circular practices in adaptive reuse (Hamida et al., 2025), temporary building uses (Talamo et al., 2020), flexible temporary shelters to reactivate public space (Ginelli et al., 2020) and the broader spectrum of community-oriented management (Greco et al., 2024). Social topic appear in a recent study to conceptualize circular economy (Kirchherr et al., 2023), with concepts of 'social equity', such as human health, well-being, and just transition. In fact, from a doughnut economy perspective (Raworth, 2018), the reuse and recycling of goods and services meet the need for social equity and resource distribution. Hence, experimental temporary uses in buildings can be considered a spontaneous form of social circular economy strategy, for their collective and inclusive approach to building reuse.

However, despite the growing attention to temporary real estate reuse on one hand, and to socially driven circular economy on the other, there is still a lack of studies that specifically address temporary real estate reuse as a strategy for the social circular economy.

Thus, the purposeful embedment of these approaches in temporary real estate reuse planning, combined with the adoption of social circular strategies raise two fundamental questions:

- *What are the factors that make temporary reuse of real estate a social circular strategy?*
- *What are the enabling conditions for hybrid organizations to implement social circular strategies through temporary use?*

Exploring these questions through the lens of the value proposition in sustainable business models (Baldassarre et al., 2017; Greco, 2024) provides a first step towards understanding their potential impact. Drawing from the literature and qualitative analysis of three cases of temporary use, this paper explores which factors can make us consider temporary uses as social circular economy strategies and what are the enabling conditions for temporary uses to serve as social circular strategies in real estate management.

2. THEORETICAL BACKGROUND

The experimental temporality in cities is shaped by the creative potential of temporary urbanism (Bishop, 2015; Madanipour, 2017). It is constituted by ecosystems of temporal events, that take place in public or private real estate. Because private developers are increasingly interested in the integration of informal uses for temporary real estate (Matoga, 2019; Vivant, 2022) private and public organizations are testing forms of collaboration and management.

In the public context, Patti & Polyak, (2015) did an inventory of policies for temporary use. In the relationship between practice and policies for temporary use, they stated that value is created by fostering transparency in real estate management, incentivizing the reuse of vacant spaces, and easing regulatory and financial barriers. However, true innovation in municipal policies depends on coordinated support from various public departments and bodies. Central to this process is trust: without mutual understanding of motivations, objectives, and working methods among actors, such as civic organizations, design studios, developers, and municipalities, effective cooperation in regeneration projects becomes challenging.

In this section, we frame the theory of hybrid organizations as tools, and the social dimension of the circular economy as objectives. This theoretical framework will then be applied to structure the methodology and the analysis.

2.1 HYBRID ORGANIZATIONS

Organizational hybridity refers to the blending of diverse organizational goals that would not typically align within a single organization, enabling the simultaneous pursuit of social, environmental, and economic objectives (Mitzinneck & Greco, 2021). Hybrid organizations blend public, private, and community-driven models, to create flexible frameworks that prioritize social value and outcomes over monetary revenues. They are positioned to contribute to civic wealth creation by addressing complex

https://doi.org/10.52202/080684-0014

societal challenges that traditional organizations or purely market-based solutions often overlook. By integrating public, private, and civil sector approaches, hybrids foster inclusivity, innovation, and value creation across multiple dimensions (Greco, Long, & de Jong, 2021). The benefits of hybridity are manifold. Hybrid organizations excel at leveraging diverse resources, forging cross-sectoral partnerships, and aligning stakeholders around shared goals. This adaptability makes them effective in tackling systemic issues such as inequality, environmental degradation, and access to essential services (Doherty, Haugh, & Lyon, 2014). Moreover, their capacity for sustainable business model innovation enables them to remain responsive to changing societal needs (Greco, 2024). However, hybridity also presents significant limitations. Balancing competing logic can create tensions that strain internal identity coherence (Ebrahim et al., 2014), decision-making processes, and stakeholder relationships (Greco et al., 2021). Additionally, sustaining hybrid organizations requires navigating financial constraints and maintaining legitimacy across diverse audiences, which may hinder their scalability and long-term impact (Doherty et al., 2014).

Despite these challenges, hybrid organizations play an essential role in fostering societal resilience and innovation. Their ability to experiment with unconventional strategies makes them instrumental in advancing novel solutions to pressing societal challenges (Mitzinneck & Greco, 2021). This makes them highly relevant to the theme of temporary use of buildings as a social circular strategy. By leveraging their ability to align diverse stakeholders and opposing goals while fostering collaborative spaces, hybrid organizations are uniquely positioned to create enabling conditions for temporary use projects and maintain their impact so to contribute to social inclusion, cultural vibrancy, and urban sustainability.

2.2 THE SOCIAL DIMENSION OF CIRCULAR ECONOMY

The social dimension has been mostly overlooked in research on circular economy, as highlighted in some literature review on the topic. A systematic review by Padilla-Rivera et al. (2020) identified thematic areas such as *labour practices, human rights, societal impacts, and product responsibility*. These aspects feature the importance of equitable labour practices, diversity, community inclusion, and participatory governance in the circular supply chains. In a different review, Mies & Gold (2021) mapped the social dimension of the circular economy, identifying employment opportunities, education and awareness, health and safety, and government involvement as the most discussed social issues. Their study assessed social aspects across various stakeholder groups, including workers, organizations, consumers, local communities, and society at large. To capture the complexities of sustainability, they emphasized the need for a more diverse consideration of the social dimension in the circular economy, integrating multiple social aspects that extend beyond easily measurable factors directly tied to economic or ecological sustainability: A shift that requires changes in organizational and societal mindsets, supported by education, awareness-raising efforts, and active engagement of diverse stakeholders. In this review, the social circular economy began to go beyond the quality of the labour of the *circular* product.

In a more radical perspective, Savini (2023) explained that the socio-ecological value of waste lies not only in material reuse but in fostering circuits of care and mutual support. This shift aligns with degrowth theory, integrating circular economy principles to challenge conventional paradigms. Instead of prioritizing monetary value, the focus moves toward recognizing the socio-ecological value of waste.

A more recent literature review on the social contribution of circular economy has been based on capability approach variables. It showed inconsistencies in the literature regarding the assessment of the circular economy as a development strategy (Valencia et al., 2023), highlighting contrasting perspectives in its contribution to the socioeconomic system, namely development focused and degrowth. Beyond job creation opportunities, topics like decision-making, collaboration, equity, liveability in cities and quality of life expand the understanding of circular economy as part of a socio-economic system. Valencia et al. (2023) highlight that the *built environment with the growing sharing economies* is a priority for the social dimension of the circular economy.

Thus, based on existing research, we can consider the social dimension of circular economy in the real estate sector at the intersection of social value creation from labour, management, and new sharing economies. Building on this, we define social circular strategy as an approach to circular economy practices that integrates material reuse with social value creation by fostering community participation, equitable governance, and adaptive economic models. Unlike traditional circular economy approaches that focus solely on resource efficiency, a social circular strategy transforms temporary use and hybrid organizational forms into mechanisms for recalibrating the balance between social and economic value over time.

3. MATERIALS AND METHODS

This study employs qualitative analysis methods in two phases:

- Theoretical analysis: A review of articles on the social dimension of the circular economy in temporary real estate reuse was conducted using the Scopus database. Publications on the social aspects of the circular economy were combined with more specific studies focusing on circular economy practices in temporary use contexts. The result is a theoretical framework that identifies and outlines the key factors involved. It is a tool to determine whether temporary uses of vacant real estate can represent a social circular strategy, or to what extent they can.

- Case studies comparison: A retrospective analysis has been done guided by the theoretical framework and structured according to a process-tracing method (Beach, 2020; Beach & Pedersen, 2012; Collier, 2011). Process tracing is a detailed, within-case study approach used to examine causal mechanisms and their effects in a specific case. It helps develop and evaluate theories that connect causes to outcomes within a set of causally similar cases.

It has been employed to disclose from practice what capabilities and enabling factors enable social circular economy in temporary use projects in vacant real estate.

Observing the relationship between the temporary use of vacant properties, hybrid organizations, and the social circular economy, a whole system in some cases (Figure 1). Hybrid organizations, which bring together the public, private, and civil sectors, can form temporary collaborations or other forms of partnership for the reuse of spaces. These collaborations serve both economic and non-economic purposes, such as housing. Key factors of the social circular economy—related to labour, human rights, product responsibility, care, and sharing—are increasingly integrated into real estate projects with a social focus.

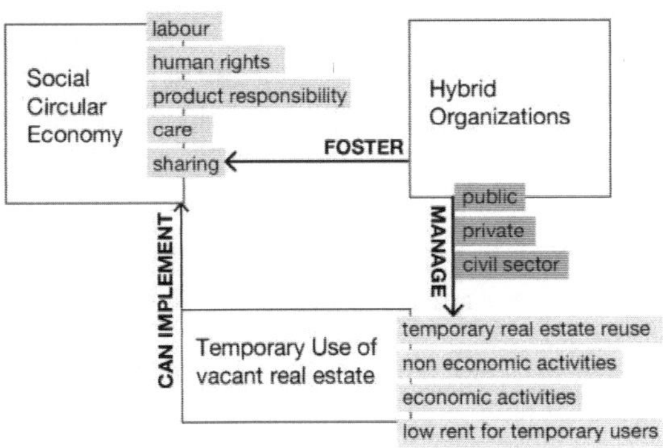

Figure 1. Analysis context.

The factors influencing the development of temporary real estate use projects originate from several sources: the availability of space, such as vacant properties; the demand for use, which may be specific or general and expressed by either property owners or potential users; public policies mandating social or cultural services in neighbourhoods or in emergency situations, such as migrant housing; and real estate rehabilitation projects that initiate reactivation while awaiting permits and final preparations. These factors drive the initiation of a project.

Data have been collected within the context of the NOMAD research project on 15 cases from the Netherlands, Belgium and France (Mazzarella, 2023) from:

- Semi-structured interviews with project initiators, managers, and participants to understand their roles, experience, motivations, and practices.
- Project documentation and archival data, public communications on social networks, and project reports, to analyse operational models.
- Field observations, and participation in community activities.

The analysis has followed a thematic coding process.

4. RESULTS

In this section, we present an analytical framework to investigate the social dimension of circular economy in temporary real estate uses. It has been outlined by the literature that has studied or considered the social value of temporary uses in real estate. The results of the case study analysis are presented based on the framework and the retrospective analysis using the process-tracing method.

4.1 THE SOCIAL DIMENSION OF CIRCULAR ECONOMY IN TEMPORARY REAL ESTATE USES

Adding social value has not been explored as a direct objective of circular economy strategies, but rather as an additional condition within strategies focused on the environmental sustainability. Literature on real estate temporary use recognizes its social value, though it seldomly connects to the circular economy, with only a few exceptions. The case of the real estate sector of the French national railway company (SNCF Immobilier) has been promoting *transient urbanism* strategies together with Plateau Urbain, implementing temporary projects stemming from discourses on the importance of the frugal city, the reuse of existing buildings, and the circular economy (Pinard, 2020). In the Italian context, Roversi et al. (2021) recognized the functional reuse of cultural heritage (Cerreta et al., 2020; Gravagnuolo et al., 2024) as a prerequisite for the circular city, understood as a spatial/territorial manifestation of the circular economy. In the same territorial context, Fatigato & Capaldo (2024) incorporated circular economy actions related to food in their research, integrating them into the incremental temporal phases of a real estate reuse design.

From a non-institutional perspective, (Calzati et al., 2022) analyse the temporary urban commons of two no-profit organizations (also considered in this paper as case studies, i.e. communa and Stad in de Maak), where circular economy is declared to be part of the communa's mission.

In a circular economy perspective, Meslec & Haase (2024) analysed the application of nature-based solutions (NBSs) as a circular strategy and multi-scalar business models to invest in vacant sites. From a material flow perspective, Kawa, Schoor, et al., (2024) examined the material-based design of nine pioneering projects in Brussels and developed a framework of guidelines to support materialization, design, and stakeholder engagement in temporary use projects. Further analysis of

https://doi.org/10.52202/080684-0014

stakeholder ecosystems within temporary makerspaces highlighted their role in fostering community building, exchange, and knowledge transfer in the context of circular practices (Kawa, Galle, et al., 2024).
Thus, the social dimension of the circular economy in real estate temporary use is primarily conceived in relation to the reuse of properties and the implementation of circular economy actions. By integrating the perspective offered by recent literature on the Social Circular Economy (see Section 2.1), we can also consider aspects of social well-being linked to both productive and non-productive activities associated with the different phases of temporary property reuse (Table 1).

Table 1. Group and indicators of social circular economy in temporary real estate reuse.

Group	Indicators
Circular economy actions	Real estate reuse Nature-based solutions (such as gardening, or related to food) Furniture and component reuse Material recycle
Productive activity	Labour conditions Work well-being Start-up of new companies
Social activity	Community building Mutual support Sharing goods and services Knowledge transfer Social cohesion Start-up of new associations

The assessment of social value indicators for the use categories (non-productive activities, productive activities, and circular economy solution) can let us consider a temporary use as a social circular strategy.

4.2 ENABLING CONDITIONS IN TEMPORARY USE

In this section, we focus on an in-depth analysis of three case studies. Temporary occupation, as defined by the Urban Catalyst project (Oswalt et al., 2012), can follow different patterns: Displacement, Subversion, Pioneer, Parasite, Coexistence, Consolidation, Impulse, Free Flow and Stand-In. In any case, temporary activation involves the cooperation of landowners and hybrid organizations managing the temporary use to prepare the site, activate it, ensure its functioning, and eventually vacate it.
To disentangle these mechanisms, we apply a qualitative retrospective analysis using the process-tracing method, focusing on the three innovative cases selected (Table 2). As mentioned, these have been initiated and coordinated by three non-profit organizations committed to temporary uses and real estate management: Plateau Urbain, communa, and Stad in de Maak.
Plateau Urbain is a cooperative specializing in solidarity-based real estate and transitional urbanism. It offers affordable workspaces and, where possible, emergency housing solutions in creative, and socially driven third places across Île-de-France and several major cities, including Lyon, Bordeaux, and Marseille. Additionally, the cooperative provides consulting and support services throughout France (*Plateau Urbain*, 2025b).
Communa is a non-profit organization dedicated to fostering a more affordable, democratic, resilient, and creative city. While temporary occupation is their main approach, they also develop other practical solutions to address the commodification of urban spaces (communa ASBL, 2025).
Stad in de Maak is a no-profit association that explores new, socially inclusive housing models in the city. The foundation oversees buildings that enable collective living for diverse target groups, with 30% of these spaces dedicated to 'commoning', sharing and managing facilities for the neighbourhood and social organizations (Stad in de Maak, 2025).

Table 2. Temporary use projects by no-profit organizations: LAC, Minima, and DGB.

	Les Arches Citoyannes	Maxima/ Minima	De Grote Beer
Organization	Plateau Urbain	communa	Stad in de Maak
City, Country	Paris, France	Brussels, Belgium	Zwijndrecht, Netherlands
Neighbourhood	4th arrondiss.	Forest	Planetenbuurt
Building type	Heritage architecture	Office building	Social housing
Building Owner	BNP Paribas Real Estate, RATP Solutions Ville and Apsys	Municipality of Forest	Trivire Housing Association
Area (m^2)	30000	6000	1500
Objectives	Activating the building and testing uses	Experimenting uses and promoting social initiatives	An autonomous neighbours' house
Functions	Work and leisure	Social activities for neighbours	Social activities for neighbours
Temporary Users (n.)	450 organizations (1000 daily users)	70 organizations	25 neighbours
Duration (years)	3 years (2021-2024)	5 years (2020-2025)	5years (2024-2029)
Kind of Use (UC)	Stand-In, Impulse, Consolidation	Stand-In, Consolidation	Consolidation

Three collaborative temporary use projects managed these no-profit organizations were selected to provide examples and insights on the intersection of social value and economic conditions in temporary reuse and are: Les Arches Citoyennes by Plateau Urbain in France, Maxima/Minima by communa in Belgium, and De Grote Beer by Stad in de Maak in the Netherlands. Their temporary uses have been started and are managed by hybrid organizations with property owners, users, and associations.
The three organizations play a role of intermediaries in the temporary real estate usage. In the case of communa and SidM, Calzati et al. (2022) discussed how these

organizations also work to consolidate their socio-cultural practice through cooperative ownership.

4.2.1 The social circular economy in the real estate temporary use

The retrospective analysis of three of their projects provides a qualitative lens to identify the key factors that enable them as social circular economy strategies. Applying the process-tracing method, we have identified recurrent causes and outcomes in the development of temporary use projects (*Figure 2*).

Factors that define a temporary real estate use as a social circular economy can be identified at different stages of the process.

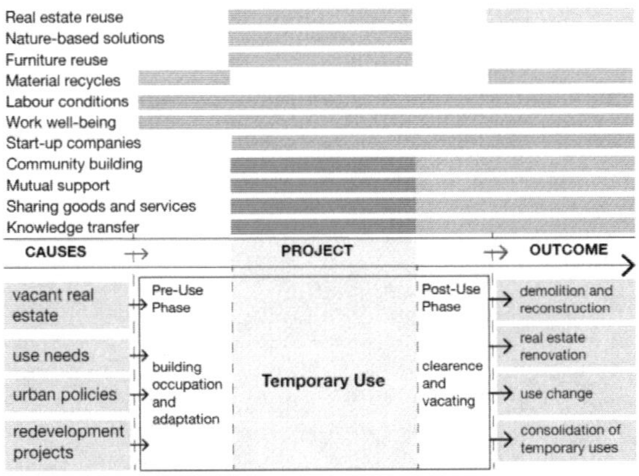

Figure 2. Process-tracing of Social Circular Economy factors and Temporary Use conditions.

During the temporary use phase, the site can host either non-productive or productive activities, both of which can include actions related to the circular economy. In cases where projects are social and inclusive, community building is the central factor in the success of the temporary project. Temporary inhabitants who share living, working, or recreational spaces, when guided by mediators or associations focused on creating social value, are enabled to collaborate in managing collective use decisions.

4.2.2 Case studies analysis

Les Arches Citoyennes is a temporary project of co-working and community spaces run by the non-profit cooperative Plateau Urbain in Paris (France). It is a private investment project aimed at start testing future uses of "Citizen Hospitality" in response to the "Reinventing Paris 3" call for projects in a historical Haussmanian building during the few years before the beginning of its renovation and redevelopment for the permanent project.

The **Les Arches Citoyennes** project has been initiated by Plateau Urbain in the centre of Paris (Plateau Urbain, 2025a) in team with Base Commune, Vraiment Vraiment, Association Aurore in setting up the transitional phase of transforming two historical Haussmann buildings used as offices into housing, shops and services. The temporary use is a prefiguration phase that included the team in the BNP Paribas Real Estate and RATP Solutions in response to a call for the *Réinventer Paris 3* call for projects (*Réinventer Paris 3 : La Reconversion de l'ancien Siège de l'AP-HP – Le Sens de La Ville*, n.d.) (Figure 3). The AP-HP (Assistance Publique–Hôpitaux de Paris) launched a consultation in June 2021, shortly before relocating to the Saint-Antoine Hospital site for the transformation of its former headquarters located in the centre of Paris.

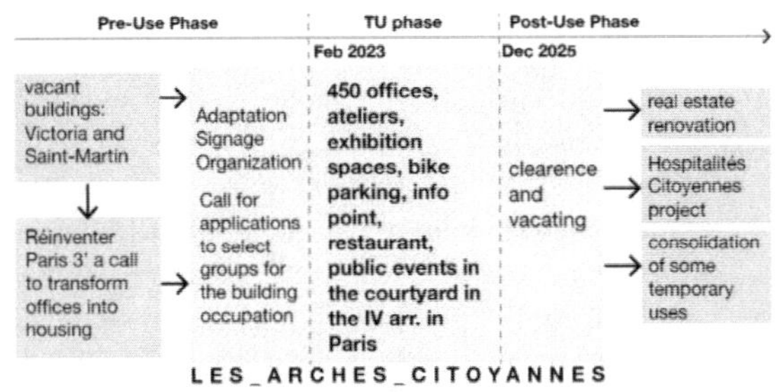

Figure 3. Process tracing of Les Arches Cytoiannes

Private investment enabled the necessary renovations and activation of the space, creating a foundation for its diverse uses. The space was purposefully reorganized to host 450 organizations and activities that open the place to young people and artists, and test future uses, creating economic, social and cultural values for both private and public stakeholders.

Figure 4. Les Arches Cytoiannes, Paris. Open living room at the ground floor and the courtyard (Photos: Chiara Mazzarella, May 2024).

https://doi.org/10.52202/080684-0014

A significant strength of Les Arches Citoyennes is its inclusive management model, which prioritizes affordability and accessibility for diverse users through flexible business model schemes, i.e. ateliers are rented according to the organization income, and the La Cantine restaurant has an agreement based on a fair economy model.

The project provides coworking spaces for social enterprises, artists, start-ups, and has opened the ground floor to public events. The prefiguration of a new urban public space in the courtyard envisions the reuse of the patio as a semi-public area, fostering social interaction and community engagement. The heritage architectural design of the ground floor, or building plinth, remains closed off to the sidewalks, creating a sense of enclosure while maintaining an atmosphere within (Figure 4). Openly accessible furniture encourages flexible and inclusive use of the space, while the presence of the restaurant *La Cantine* serves as an anchor for activity, drawing people in and enhancing the vibrancy of the courtyard as a shared urban space.

Plateau Urbain's expertise in managing temporary use projects emerges in their ability to coordinate with private investors, public authorities, place users and other diverse local organizations. The goal of testing future uses to respond to partners (investors) is a tool to make accessible the 30000m^2 of the buildings to creatives, young people and passing visitor, that thousands of people per day.

Maxima/**Minima** is a temporary project managed by the non-profit association communa in the Region of Brussels (Belgium). It is a temporary use project in a public property that has been made available by the Municipality of Forest for social services to the neighbourhood.

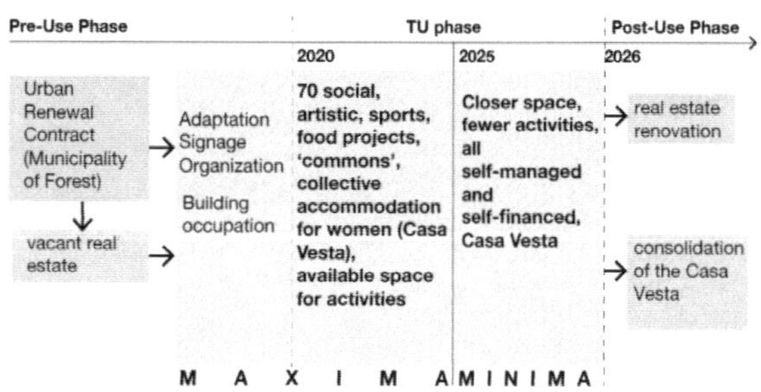

Figure 5. Process tracing of Maxima/Minima.

After five years, and at the time of writing, the project is currently facing challenges of financial self-sustainability without public fundings.

The evolution of Maxima into Minima in Brussels offers insights into the opportunities and challenges of public funded temporary use projects. This case highlights how a large, multifunctional space could be adapted and sustained in the context of urban renovation, by community engagement, and facing financial constraints.

The Municipality of Forest (Brussels) provided the initial access to a vacant 6,000 m^2 property previously used as a private headquarters, enabling the project to take root.

Financial support through the *Contrat de Rénovation Urbaine* enabled the transformation of the space into an accessible and functional place (Figure 5).

Communa has been experimenting with several uses of the site, including local associations, and giving accessible space to neighbours in the courtyard (Figure 6), aligning with the Municipality's vision of creating a permanent public facility by 2026 within the Saint-Antoine neighbourhood in Forest, where a strong associative culture already existed.

The space was configured to support 70 diverse projects, from artistic and cultural initiatives to social and culinary activities, enabling the site to become a hub for community-driven initiatives.

The co-creation of *Casa Vesta*, a collective housing project for women in precarious conditions, in partnership with *Samu Social* association, demonstrated the stakeholders' capability to address urgent social needs building partnerships.

Figure 6. MAXIMA, Municipality of Forest, Region of Brussels. Open space at the ground floor (Photos: Chiara Mazzarella, Nov. 2023).

The Municipality's ownership of the property and its commitment to supporting interim use provided a stable foundation for the project. The financial backing of the *Contrat de Rénovation Urbaine* enabled the transformation of the site and the initiation of community-oriented activities (Figure 5).

https://doi.org/10.52202/080684-0014

The dense associative culture of Saint-Antoine and the neighbourhood's need for space to host activities were key factors in the project's relevance and acceptance. Flexible and participatory approaches allowed the space's functions to evolve based on ongoing dialogue with local actors.
The willingness to experiment with new forms of shared management and multi-purpose uses allowed communa to test and refine models of co-management, social circularity and shared governance, laying the groundwork for future consolidation of the place.
Currently, *Minima is Maxima, but smaller*. Facing the end of public funding at the end of 2024 (Communa ASBL, 2024a), communa had to restructure the project into *Minima*, a shrieked and self-managed version of Maxima with some associations for food distribution, psychosocial support, and activities for young people, but closing the courtyard (Communa ASBL, 2024b).
This transition reflects their ability to adapt autonomous operational models in response to financial limitations, and not renouncing to the place opportunity in name of the organization aims and goals.

De Grote Beer is a temporary project by the non-profit association Stad in de Maak (SidM) in Zwijndrecht (Netherlands). The housing association Trivire asked to SidM to create a community and a commons space at the beginning of a long redevelopment phase of five blocks.

Figure 7. De Grote Beer, Zwijndrecht. Meeting room and the garden yard (Photos: Chiara Mazzarella, Oct. 2024).

During the redevelopment SidM has been asked to support the social transformation of the area. According to the 'Programma veerkrachtige buurten Zwijndrecht 2024-2040' (Coalition Resilient Neighborhoods, 2024) (Figure 8) the residents need more social support and facilities. In this context, the SidM's *commons community center* aims to establish a lasting social canter for the neighbours (Stad in de Maak, 2024b).
The project was commissioned by the housing association Trivire to Stad in de Maak for the 2022-2027 period. Trivire owns and manages residential housing complexes in Zwijndrecht (Figure 8). Over the coming years, some of the buildings will be renovated, while two will be demolished and rebuilt. During this process, many residents will be relocated, and new ones will move into the neighbourhood. Amidst this dynamic transformation, SidM has been revitalizing a ground-floor apartment and an open garden since 2024, creating an accessible social space for the neighbourhood.
SidM's inclusive and collaborative approach began with directly involving residents, organizing convivial gatherings and informal meetings to build connections (Stad in de Maak, 2024a). They don't make open call for submission to fill the space but are looking for human resources in the area that are available to get involved in the community building. Thus, identifying and attracting residents of the neighbourhood has been the organization's first step in this project.

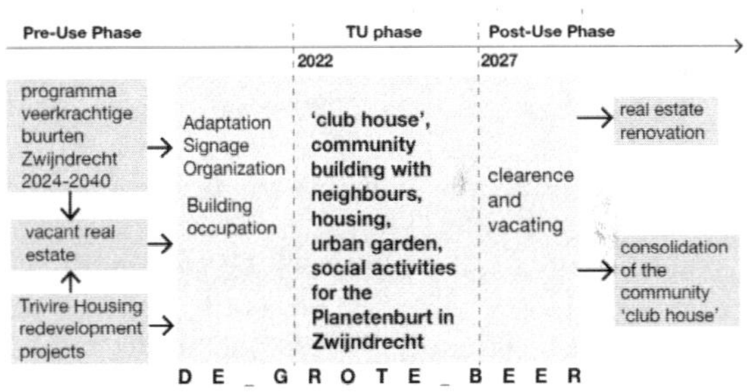

Figure 8. Process tracing of De Grote Beer.

SidM is managing an apartment, a building and a yard with the goal of transforming it into a self-managed social space for the neighbourhood's residents. The objective of the temporary use is to test activities and forms of collaboration by enabling local neighbours to self-organize projects and self-sustainable activities. To get people involved, a SidM member *rings people's doorbells to ask what they need* (Stad in de Maak, 2024a).
This direct community engagement process is progressively forming a constellation of actors, local associations, and new groups of people potentially interested in getting involved into the De Grote Beer social club. In December 2024, a group of residents started the *Tuintje Planetennbuurt* ("Little Garden Planetenbuurt") to make a vegetable garden in the yard. The yard is a garden that is also being used for meetings

https://doi.org/10.52202/080684-0014

and outdoor lunches on temporary wooden structures built by SidM. The organization has funds to support the purchases and expenses of these volunteer-led initiatives. During a conversation, a resident remarked that he had "never seen anything like this in the neighbourhood."

In the same residential complex, temporarily vacant apartments are being managed by *Ad Hoc*, one of the anti-squat companies in the Netherlands that manage empty properties for short periods to prevent them from being left unused.

While the benefit given by anti-squat companies is only in renting properties to lower price, SidM manages spaces as *commons*, adopting an open, inclusive, and unconventional approach that share decision-making power to the users, allowing freedom in temporary adaptation based on their needs and shared use of resources.

At De Grote Beer, the temporary wooden structures and other equipment were reused from a previous project (Vlaardingen Meent).

De Grote Beer project is still in its early stages, but the approach clearly reflects principles of the social circular economy and shared governance. The mechanisms for maintaining these activities remain to be observed as the project develops further.

5. DISCUSSIONS

Temporary use of vacant real estate show to have the potential to be embedded within urban planning as a strategic tool for implementing a social circular economy under specific conditions.

Our research shows that the social circular economy is evolving beyond its initial focus on labour well-being to social value creation through new forms of real estate management by hybrid organizations. We argue that for real estate reuse to qualify as a social circular strategy, it must not only incorporate circular economy activities but also prioritize labour well-being and foster social interaction. This study contributes to the literature on temporary use, by identifying three set of practices that can be conceptualized as social circular strategies, namely: 1) Circular economy actions, such as the real estate functional reuse, the presence of nature-based solutions (gardening, food recycling), reuse of furniture and component reuse, adaptations with material recycling; 2) Productive activities, such as favourable labour conditions, work well-being, entrepreneurial activities, i.e., the creation of start-up of new ventures, and 3) Social activities, such as community building, mutual support, sharing goods and services, knowledge transfer, social cohesion, and the creation of joint new associations. These factors of social circular economy have been verified in the temporary use phases of a property.

Building reuse is a fundamental prerequisite for any temporary use, making it a consistently relevant condition. The presence of nature-based solutions depends on the presence of a greenery, such as in De Grote Beer, where gardening has been an activator for neighbours' engagement and community building. All the cases considered have furnished second hand furniture and reused temporary structures: in the case of De Grote Beer the domo in the garden is moved from the previous project in Vlaardingen.

This study does not explore indicators of the social dimension in productive activities, which would require further in-depth research through interviews. Being the three associations no profit organization with high commitment in social value creation, these indicators could be assessed exploring the work quality of their employees.

Figure 2 highlights that social activities can take place during the use phase and can last if the temporary use is a prefiguration of future uses, as in the three case studies.

Social activities vary greatly across the three cases. LAC operates on the scale of a city, where the users of the workshops and offices know their neighbours and some of the regular visitors. Many independently proposed internal activities have not been successful and have faded over time. The director of LAC reflected that sometimes, all it takes for a social moment is a break and a place to relax, such as the restaurant in the courtyard. The space hosts numerous events and cultural activities, attracting many visitors to the courtyard. It is not possible to identify a single community in LAC but rather a collection of groups and individuals who share social moments centred around art and culture.

At Minima in Brussels, funding cuts have led communa to reduce the number of activities of the building, limiting access to only those associations that have managed to establish a lasting presence in the neighbourhood. Creating communities in places awaiting transformation may seem paradoxical, as once the temporary use ends, each temporary inhabitant will be forced to find another place to live or work. However, in the three cases we have examined, while the temporary inhabitants (i.e., the 450 structures of LAC, the 70 associations of Maxima, and the temporary residents of the buildings in Planetenbuurt) will have to leave, the residents will continue to benefit from the cultural and social services provided to the neighbourhood. Start-ups and initiatives that manage to establish themselves still have the opportunity to carry forward the work developed during the temporary use.

Moreover, some residents of LAC have pointed out that bonds and connections do not necessarily end when a project concludes, and in many cases, the association Plateau Urbain helps its residents find new workspaces within its properties.

Thus, the circular capacity of temporary uses can be assessed based on the expected impacts of each initiative: In the case of LAC, the prefiguration of uses to be consolidated; in the case of Maxima, the activation of social services for residents; and in the case of LGB, the creation of a local resident community. Temporary development requires investments that, within a perspective of social circular economy, should be contextualized according to the interests of investors (whether public or private property owners) and the broader framework of public programs and policies.

One aspect that requires further investigation is temporality in relation to placemaking in their urban contexts (Zhang, 2018), or the ability to sustain a shared economy and foster new bonds of trust in temporary communities in real estate contexts when places in constant transformation.

6. CONCLUSION

This article explored the connection between the social circular economy and temporary uses, highlighting the role of hybrid organizations in the innovative management of experimental temporary use. While the social dimension of circular economy is increasingly explored, a review of the literature reveals that there is limited research on circular dimension of temporary reuse in the real estate context. Hence, this paper conceptualizes the social circular economy within the context of building reuse and presents a framework detailing its practices, drawing from a comparative retrospective multiple case study analysis.

The framework proposed represents an initial identification of three key groups of indicators that can guide the development of social value within the context of temporary property reuse.

The three case studies presented demonstrate that, even under extremely different conditions, temporary uses managed through inclusive and collaborative approaches have the potential to serve as strategies for social circular economy. In particular, the success of temporary real estate reuse projects within the framework of a social circular economy depends on multiple enabling conditions. These conditions are shaped by the characteristics of the property itself, the objectives of the owner, the management approach adopted by temporary use organizers, and the long-term vision for the space. The key factors include:

1. Resources: The availability of real estate, financial, and human resources, is necessary to sustain temporary use operations.
2. Management Approach: The governance model and operational strategies that shape the social value of the temporary use activities.
3. Property Value: The real estate market value and the potential of the space to attract funding or support for renovation projects that can host temporary uses as a prefiguration phase.
4. Urban Policies**:** Municipalities that call for social services or amenities in neighbourhood encourage developers to implement those social circular strategies in temporarily vacant buildings.
5. Short- and Long-Term Objectives: The alignment between immediate use and the broader vision for the area or real estate development define the temporary use management and objective.

These factors determine whether a temporary real estate reuse initiative can effectively function as a social circular economy strategy.

Additionally, the case study highlights how enabling factors such as institutional support, community engagement, and strategic partnerships can drive such initiatives toward long-term impact.

The limitations of this study include the scope of the literature considered and the number of case studies. A systematic review on the topic of social value within the circular economy, applied to adaptive reuse, could identify additional criteria and indicators as tools for developing sustainable property management plans. Further case studies could provide additional insights into the role of hybrid organizations in the context of temporary uses.

REFERENCES

Beach, D. (2020). Process Tracing Methods. In C. Wagemann, A. Goerres, & M. B. Siewert (Eds.), *Handbuch Methoden der Politikwissenschaft* (pp. 699–719). Springer Fachmedien Wiesbaden. https://doi.org/10.1007/978-3-658-16936-7_43

Beach, D., & Pedersen, R. (2012). *Process-Tracing Methods: Foundations and Guidelines*. University of Michigan Press. https://doi.org/10.3998/mpub.10072208

Bishop, P. (2015). From the Subversive to the Serious: Temporary Urbanism as a Positive Force. *Architectural Design*, *85*(3), 136–141. https://doi.org/10.1002/ad.1913

Bragaglia, F., & Caruso, N. (2022). Temporary uses: A new form of inclusive urban regeneration or a tool for neoliberal policy? *Urban Research & Practice*, *15*(2), 194–214. https://doi.org/10.1080/17535069.2020.1775284

Calzati, S., Santos, F., & Casarola, G. (2022). On the (Non)Institutional Disclosure of Urban Commons: Evidence, Practices and Challenges From the Netherlands and Belgium. *Frontiers in Sustainable Cities*, *4*, 934604. https://doi.org/10.3389/frsc.2022.934604

Cerreta, M., Elefante, A., & La Rocca, L. (2020). A Creative Living Lab for the Adaptive Reuse of the Morticelli Church: The SSMOLL Project. *Sustainability*, *12*(24), 10561. https://doi.org/10.3390/su122410561

Coalition Resilient Neighborhoods. (2024). *Programma Veerkrachtige Buurten Zwijndrecht 2024—2040.*

Collier, D. (2011). Understanding Process Tracing. *PS: Political Science & Politics*, *44*(4), 823–830. https://doi.org/10.1017/S1049096511001429

Communa ASBL. (2024a, September 5). *C'est peut-être la fin de Maxima !* Instagram. https://www.instagram.com/p/C_ijJkut2q8/

https://doi.org/10.52202/080684-0014

Communa ASBL. (2024b, December 16). *MAXIMA SE TRANSFORME ET DEVIENT MINIMA*. Communa ASBL. https://communa.be/maxima-se-transforme-et-devient-minima/

Communa ASBL. (2025). *Communa mission*. Communa ASBL. https://communa.be/en/home/

Doherty, B., Haugh, H., & Lyon, F. (2014). Social Enterprises as Hybrid Organizations: A Review and Research Agenda. *International Journal of Management Reviews*, *16*(4), 417–436. https://doi.org/10.1111/ijmr.12028

Ebrahim, A., Battilana, J., & Mair, J. (2014). The governance of social enterprises: Mission drift and accountability challenges in hybrid organizations. *Research in Organizational Behavior*, *34*, 81–100. https://doi.org/10.1016/j.riob.2014.09.001

Fatigato, O., & Capaldo, S. (2024). Third Places and Transitory Uses: The Experimental Project "Hall - Hemicycle Agrifood Living Lab" in the East of Naples. In O. Gervasi, B. Murgante, C. Garau, D. Taniar, A. M. A. C. Rocha, & M. N. Faginas Lago (Eds.), *Computational Science and Its Applications – ICCSA 2024 Workshops* (Vol. 14824, pp. 253–270). Springer Nature Switzerland. https://doi.org/10.1007/978-3-031-65332-2_17

Ferreri, M. (2020). Learning from temporary use and the making of on-demand communities in London's Olympic "fringes". *Urban Geography*, *41*(3), 409–427. https://doi.org/10.1080/02723638.2019.1679527

Galdini, R. (2022). Interpreting contemporary urban complexity: The fluid city paradigm. *Sociologia Urbana e Rurale*, *128*, 14–26. Scopus. https://doi.org/10.3280/SUR2022-128002

gE.CO toolbox | gE.CO toolbox. (n.d.). Retrieved 31 January 2025, from https://www.gecotoolbox.io/en/

Ginelli, E., Pozzi, G., Lazzati, G., Pirillo, D., & Vignati, G. (2020). Regenerative urban space: A box for public space use. In *Research for Development* (pp. 137–147). Scopus. https://doi.org/10.1007/978-3-030-33256-3_14

Gravagnuolo, A., Angrisano, M., Bosone, M., Buglione, F., De Toro, P., & Fusco Girard, L. (2024). Participatory evaluation of cultural heritage adaptive reuse interventions in the circular economy perspective: A case study of historic buildings in Salerno (Italy). *Journal of Urban Management*, *13*(1), 107–139. https://doi.org/10.1016/j.jum.2023.12.002

Greco, A. (2024). Changing Constellations of Hybridity Through the Phases of Sustainable Business Model Innovation. *Academy of Management Proceedings*. https://doi.org/10.5465/AMPROC.2024.54bp

Greco, A., Long, T., & De Jong, G. (2021). Identity reflexivity: A framework of heuristics for strategy change in hybrid organizations. *Management Decision*, *59*(7), 1684–1705. https://doi.org/10.1108/MD-10-2019-1369

Greco, A., Van Laar, B., Remøy, H., & Gruis, V. (2024). Accelerating circularity systemically: Three directions for impactful research. *Npj Urban Sustainability*, *4*(1), 45. https://doi.org/10.1038/s42949-024-00183-8

Hamida, M. B., Remøy, H., Gruis, V., & Jylhä, T. (2025). Circular building adaptability in adaptive reuse: Multiple case studies in the Netherlands. *Journal of Engineering, Design and Technology*, *23*(1), 161–183. https://doi.org/10.1108/JEDT-08-2022-0428

Karachalis, N. (2021). Temporary Use as a Participatory Placemaking Tool to Support Cultural Initiatives and Its Connection to City Marketing Strategies—The Case of Athens. *Sustainability*, *13*(4), 1652. https://doi.org/10.3390/su13041652

Kawa, G., Galle, W., & De Temmerman, N. (2024). Temporary makerspaces: The transformative potential of temporary projects and productive activities in circular cities. *IOP Conference Series: Earth and Environmental Science*, *1363*(1), 012047. https://doi.org/10.1088/1755-1315/1363/1/012047

Kawa, G., Schoor, X. V., Galle, W., & Temmerman, N. D. (2024). *Designing Temporary Use: Prototyping a Framework towards Material-Wise Projects*. *14*(1888). https://doi.org/10.3390/ buildings14071888

Kirchherr, J., Yang, N.-H. N., Schulze-Spüntrup, F., Heerink, M. J., & Hartley, K. (2023). Conceptualizing the Circular Economy (Revisited): An Analysis of 221 Definitions. *Resources, Conservation and Recycling*, *194*, 107001. https://doi.org/10.1016/j.resconrec.2023.107001

Madanipour, A. (2017). *Cities in Time*. Bloomsbury Academic.

Mangialardo, A., & Micelli, E. (2017). New Bottom-Up Approaches to Enhance Public Real/Estate Property. In S. Stanghellini, P. Morano, M. Bottero, & A. Oppio (Eds.), *Appraisal: From Theory to Practice* (pp. 53–62). Springer International Publishing. https://doi.org/10.1007/978-3-319-49676-4_5

Martin, M., Deas, I., & Hincks, S. (2019). The Role of Temporary Use in Urban Regeneration: Ordinary and Extraordinary Approaches in Bristol and Liverpool. *Planning Practice & Research*, *34*(5), 537–557. https://doi.org/10.1080/02697459.2019.1679429

Matoga, A. (2019). Governance of temporary use. *Proceedings of the Institution of Civil Engineers*

- *Urban Design and Planning*, *172*(4), 159–168. https://doi.org/10.1680/jurdp.18.00052

Mazzarella, C. (2023, 2025). *NOMAD. Nomad MAnagement of Urban Development. The complex value of temporary communities*. CORDIS - EU Research Results. https://cordis.europa.eu/project/id/101109623

Meslec, M., & Haase, M. (2024). *Multi-scalar Business Models for Advancing Circular Economy in Real Estate Development*. *489 LNCE*, 471–481. Scopus. https://doi.org/10.1007/978-3-031-57800-7_43

Mies, A., & Gold, S. (2021). Mapping the social dimension of the circular economy. *Journal of Cleaner Production*, *321*, 128960. https://doi.org/10.1016/j.jclepro.2021.128960

Mitzinneck, B., & Greco, A. (2021). Organizational Hybridity. In R. W. Griffin (Ed.), *Oxford Bibliographies in Management*. Oxford University Press. https://doi.org/10.1093/OBO/9780199846740-0204

Munzner, K., & Shaw, K. (2015). Renew Who? Benefits and Beneficiaries of Renew Newcastle. *Urban Policy and Research*, *33*(1), 17–36. Scopus. https://doi.org/10.1080/08111146.2014.967391

Oswalt, P., Overmeyer, K., & Misselwitz, P. (2012). *Urban Catalyst. The Power of Temporary Use | Klaus Overmeyer, Philipp*. DOM publishers. https://www.naibooksellers.nl/urban-catalyst-the-power-of-temporary-use-philipp-oswalt-klaus-overmeyer-philipp-misselwitz.html?___store=english&___from_store=default

Patti, D., & Polyak, L. (2015). From practice to policy: Frameworks for temporary use. *Urban Research & Practice*, *8*(1), 122–134. https://doi.org/10.1080/17535069.2015.1011422

Pinard, J. (2020). Developing 'Transient Urbanism' as a New Urban and Real Estate Strategy: The Case of the French National Railway Company (SNCF). In L. Andres & A. Y. Zhang (Eds.), *Transforming Cities Through Temporary Urbanism* (pp. 141–154). Springer International Publishing. https://doi.org/10.1007/978-3-030-61753-0_10

Plateau Urbain. (2025a). *Les arches citoyennes homepage*. Les Arches Citoyennes. https://lesarchescitoyennes.fr/

Plateau Urbain. (2025b). Plateau Urbain. https://www.plateau-urbain.com/

Réinventer Paris 3: La reconversion de l'ancien siège de l'AP-HP – Le Sens de la Ville. (n.d.). Lesensdelaville.Com. Retrieved 30 January 2025, from https://lesensdelaville.com/mission/reinventer-paris-3--la-reconversion-de-l-ancien-siege-de-l-ap-hp

Resources | MESOC. (n.d.). Retrieved 31 January 2025, from https://www.mesoc-project.eu/resources

Roversi, R., Longo, D., Massari, M., Orlandi, S., & Turillazzi, B. (2021). Cultural heritage as activator of circular urban dynamics. *TECHNE*, *22*, 218–226. Scopus. https://doi.org/10.36253/techne-10607

Savini, F. (2023). Futures of the social metabolism: Degrowth, circular economy and the value of waste. *Futures*, *150*, 103180. https://doi.org/10.1016/j.futures.2023.103180

Senatsverwaltung für Stadtentwicklung Berlin. (2007). *Urban Pioneers. Temporary Use and Urban Development in Berlin*. jovis. https://www.naibooksellers.nl/urban-pioneers-temporary-use-and-urban-development-in-berlin.html?___store=english&___from_store=default

Stad in de Maak. (2024a, November 8). *A new landing site in Zwijndrecht – City in the Making*. https://www.stadindemaak.nl/een-nieuwe-landingsplaats-in-zwijndrecht/

Stad in de Maak. (2024b, November 8). *Een nieuwe landingsplaats in Zwijndrecht*. https://www.stadindemaak.nl/een-nieuwe-landingsplaats-in-zwijndrecht/

Stad in de Maak. (2025). *Who we are – City in the Making*. https://www.stadindemaak.nl/who-we-are/

Talamo, C., Lavagna, M., Monticelli, C., Atta, N., Giorgi, S., & Viscuso, S. (2020). Re-NetTA. Re-Manufacturing Networks for Tertiary Architectures. In S. Della Torre, S. Cattaneo, C. Lenzi, & A. Zanelli (Eds.), *Regeneration of the Built Environment from a Circular Economy Perspective* (pp. 303–314). Springer International Publishing. https://doi.org/10.1007/978-3-030-33256-3_28

Valencia, M., Bocken, N., Loaiza, C., & De Jaeger, S. (2023). The social contribution of the circular economy. *Journal of Cleaner Production*, *408*, 137082. https://doi.org/10.1016/j.jclepro.2023.137082

Vivant, E. (2022). From margins to capital: The integration of spaces of artistic critique within capitalist urbanism. *Journal of Urban Affairs*, *44*(4–5), 490–503. Scopus. https://doi.org/10.1080/07352166.2020.1811115

Zhang, A. Y. (2018). Thinking temporally when thinking relationally: Temporality in relational place-making. *Geoforum*, *90*, 91–99. https://doi.org/10.1016/j.geoforum.2018.02.007

 https://doi.org/10.52202/080684-0014

EXPERIENCE MEETS NEW IDEAS AND CHALLENGES - A STRATEGY-MODEL FOR CONSIDERATE BUILDING IN RURAL AREAS

Ingrid Svetoft[1*], Lars-Åke Mikaelsson[1], and Karin Johansson[1]

[1]Mid Sweden University

*Corresponding Author: ingrid.svetoft@mau.se

ABSTRACT

Background and aim. A challenge when building in rural areas is to minimize the negative effects on climate, environment and to avoid conflicts between local and national interests. In the Interreg project SOURCE (Sustainable and nature pOsitive development of housing for ReCreational usE) the aim is to learn more about building recreational homes with a minimal negative impact. The research project supports business growth by collaborating and transfer knowledge between university and schools, local companies and organizations. The aim is to develop a circular building process based on local conditions in rural areas, by mutual learning.

Methods and Data. A model with *"five strategies for considerate recreational-houses"* will be further explored and developed: **1. Share**, existing buildings. **2. Convert**, use and update existing buildings **3. Condense**, add new buildings. **4. Simplify,** identify local materials that can be used **5. Active maintenance,** an agile process.

Findings. Knowledge from local conditions can be of importance when planning a house without connection to municipal supply systems. Persons that are familiar to local traditions, materials, resources and conditions can have "tacit knowledge" to be transferred in well planned meetings with students.

Practical/Societal Implications. Challenges for building in rural areas will be identified and solved: Transports to the construction site. The design of the foundation. Heating, electricity, water-supply and waste planned for an off-grid solution. Respect for regulations and national interests. Local cultural and social values. The whole process must therefore be accurate planned in dialogue, taking care of different perspectives, and experiences.

KEYWORDS: Considerate building, local knowledge, mutual learning, off-grid solutions

1 INTRODUCTION

The construction and property sector has a fundamental problem, namely being fragmented. The gaps between the involved actors in a traditional building process, are caused by different factors. A separation between design and construction and lack of trust are examples on what causes fragmentation. This phenomenon causes difficulties to keep the wholeness and to foresee the consequences of all choices that must be made in a planning and building process. The negative impact of the building process on the environment and climate is also a challenge to deal with. In Sweden between 6 and 40 % of the total of factors disturbing the environment can be related to the building and property sector (Boverket 2025). This makes it even more important to use all available knowledge and to co-operate to achieve the best solutions.

In the countryside, especially in areas with high natural values, it is of great importance to minimize the negative effects when building houses, roads and other infrastructure. This is important to maintain the unused and original conditions. In more exploited areas there are also possibilities to re-construct natural habitats and areas. There are several good examples from Norway to study and learn from.

As a perspective on circularity this paper will present a wide spectrum of factors and models that can support the aim to reach a better circularity.

The SOURCE project will continue for some time, so this paper can be called a status report that reflects the current research material. Models and ideas will be developed and tested further on.

https://doi.org/10.52202/080684-0015

1.1 CHALLENGES

The National Board of Housing, Building and Planning's environmental indicators are based on data from the Statistics Authority, Statistics Sweden. Overall, the sector accounts for 6 to 40 percent of negative environmental impact in Sweden in the areas that are followed up with the Housing Authority's environmental indicators. It should also be added that the sector contributes to additional emissions in other countries through the import and transport of construction products.
Environmental indicators shows an updated picture of the environmental impact from the construction and property sector:

- greenhouse gases 22%
- nitrogen dioxides 19%
- particles 20%
- energy use 34%
- hazardous chemical products 9%
- environmentally hazardous chemical products 4%
- waste 39%

Another challenge is the well-known problem with a fragmented sector where knowledge spread to different actors and with a process that is longitudinal. This is like a relay race that creates several gaps between actors and usable knowledge. These gaps can be bridged and supported by circular models. The challenge is to create a more circular way of using existing knowledge from all members in the project team. Knowledge-circularity can develop and contains routines for how to loop the feed-back and can therefore support a better and more sustainable process. The complexity with different stakeholders needs a tight collaboration with a joint ambition. The learning process to work more sustainable, starts early in an education, and can be supported through connection to external contacts, that combines theory and practise. The models presented in this paper can support knowledge management and better communication and hopefully minimize fragmentation. During further research and collaboration with students and companies the models will be implemented and developed. The authors long and aggregated experience from both theory and practise from different parts of the construction sector, are forming the basis of models presented in this paper.
Research, pedagogical development in education and collaboration with external actors, are three important tasks for employees at universities. The aim is to show how these different tasks can be used and executed in close collaboration with other actors, for everyone's benefit.

1.2 PROBLEM AREA

Fragmentation causes gaps between the actors and their knowledge and their usable experiences. The challenge is to manage a process with better collaboration where this can be better used in a building process.

The gaps in construction sector can be described as an effect of a separation of design and construction (Nawi et al, 2014). Fragmentation caused by a lack of feedback loops or co-ordination between the design and construction process can also widen the gap. Another cause is lack of communication in the supply chain due to actors' different languages and communication culture. A process without focusing on the clients and their involvement through both design and construction process, hinders knowledge integration. One-off projects with unique conditions and temporary relationships, cause an adversarial culture with a lack of trust and mismatches between actors in the project team.
Traditional fragmented processes can be changed with a common ambition to work towards sustainable goals.
In case studies on big scale projects, good examples are presented (Svetoft, 2009) The key factors are: time to build trust and collaboration between the actors involved. Using everyone's' knowledge early in and through the whole process, as well as working together with focus on the end-users' requirements, give clearly good results. Both practical and theoretical knowledge must be transferred in a mutual learning process by the actors involved, when you want to achieve a beneficial effect in an environmentally and climate-friendly construction process. Experiences from local conditions, cultures and traditions require mutual learning and respect. Actors with different roles in a building project, can get inspiration from local companies and from students and pupils involved.

1.3 AIM AND RESEARCH QUESTIONS

Circularity in the building and property sector have different angles to study furthermore. In rural areas there are challenges both to identify and then to handle. The aim is to minimize the negative impact caused by building projects in rural areas, that have sensitive environmental conditions and high values. SOURCE's main goal is to guide small and medium-sized businesses (SMB) in mountain municipalities in Trøndelag and Jämtland-Härjedalen into a more nature-positive recreational housing development. Theory will be put into practice through workshops in pilot studies and a toolbox.
The questions are:
How can we build in rural areas with minimal negative effects on the environment and climate?
How can we support knowledge transfer and combine practice and theory?

2 PERSPECTIVES ON CIRCULARITY

This paper will discuss the wideness of Circularity and reflect on the research project SOURCE that can give new perspectives when dealing with the challenge to build recreational houses in rural areas with minimal negative effects. Building material is one aspect as well as circularity in economy and business models. When adding the perspective on how to manage and take decisions in circularity, both practical and theoretical knowledge must

https://doi.org/10.52202/080684-0015

be transmitted and used. Especially when local conditions, culture and traditions are to be considered.
The overarching research question for this study is what constitutes examples of sustainable and circular construction. To answer this, it is necessary to specify the concepts and what they stand for. Sustainable construction goes back to the Rio Declaration in 1992 and Agenda 21 with the mission to the countries of the world to combine technology, economy and ecological sustainability into a new lifestyle based on solidarity, which means ecological, economic and social sustainability. This was followed up at the UN conference in 1996 in Istanbul with an action program for sustainable construction and housing, which means economy with physical resources and consideration of biological, economic, organizational, social, historical, cultural and aesthetic resources (Atlestam et. al, 2015).
The concepts of sustainable construction and sustainable community construction also include many other aspects, for example energy, land planning and land use, consideration of existing environments, to design with nature and not against it, to design so that places are safe and accessible to people of all ages and with different functional variations. It also includes respecting the natural cycle of water by avoiding, for example, hardening surfaces, draining natural wetlands or emptying groundwater reserves, which can cause floods, landslides and sinkholes.
The principle of circularity built on the four laws of ecology was already launched in 1971 by the biologist and ecologist Barry Commoner in the book "The Closing Circle". Commoner argues that the entire business community globally must be subject to a program of ecological reconstruction. Gösta Ehrensvärd, professor of biochemistry, presented similar thoughts in the book "Före – Efter"("Before – After") by (Ehrensvärd, 1971), where the author predicts that if industrial society is not quickly converted to cycle-adapted production, it will lead to global ecological collapse as early as around 2050. In the Club of Rome's report "Limits to Growth" (Meadows et. Al, 1972) the researchers present a similar scenario.
Ellen MacArthur foundation has suggested definitions and three design principles of Circular Economy:

"Circular economy: A systems solution framework that tackles global challenges like climate change, biodiversity loss, waste, and pollution. It is based on three principles, driven by design: eliminate waste and pollution, circulate products and materials (at their highest value), and regenerate nature."

or in short:

"Circular economy – an economy designed to keep materials in use, eliminate waste and regenerate natural systems." (www.ellenmacarthurfoundation.org)

2.1 ENVIRONMENTAL GOALS AND CLIMATE REGULATIONS

Business models and environmentally friendly work must be combined. By using the circularity as a vision, the long-term thinking is supported instead of short-term strategies and longitudinal processes. Both sixteen national and seventeen global goals for a better environment and climate helps to change the way of taking decisions and collaborate with new strategies.
It is crucial that good ambitions are supported by regulations and law. In Swedish law, steps are taken by the government to support the building and property sector towards a more environmentally friendly way of working. The law to declare the energy use was introduced in 2006: *"The purpose of the law is to promote efficient energy use and a good indoor environment in buildings."* (Regeringen, SFS 2006:985)
The law about climate declaration is a new way to regulate the whole building process and encourages dialogue between actors involved to find new and better solutions for the climate. Boverket declares the aim with the law:
"By calculating the climate impact, the builder's knowledge increases, which in turn makes it easier to take measures in the construction process that reduce the climate impact." (Regeringen, SFS2021:787)

2.2 KNOWLEDGE AND EXPERIENCE MANAGEMENT

2.2.1 Organizational and mutual learning

Construction processes are generally complex, with many actors involved at various stages. The information about what is to be built must be transferred from one actor to another is often done in a linear process during a limited period. Sustainability adds even more complexity to a project, not least because the meaning of the word sustainability varies between different actors. New building techniques or materials may be needed, as well as specialized expertise from other areas. (Jonasson et.al, 2020)
Practice from companies and theoretical knowledge from academy meets through pupils and students and create development. Local culture and conditions are important as well as influences from other places with similar conditions. Single-loop learning mean that you can learn from mistakes to do better next time: share, learn and do better and double-loop learning (Argyris & Schön, 1978). It occurs when error is detected and corrected in ways that involve the modification of an organization's underlying norms, policies and objectives. Reflection is important to achieve this change. Similarities can be found in theories in design and in organizational learning (Senge, 1990). The study of group learning and building a shared vision is essential. To have joint goals and keep up a creative atmosphere also includes a good leadership in the process. To use knowledge from different actors and learn together can be used from the very start in a planning and building process. Consultants, builders, craftsmen, clients and

public actors, companies can meet students and pupils and be able to learn more together.

At the sustainable building engineering program at Mid Sweden University there is a more than thirty years' experience to work integrated with research, education and development of construction projects together with the building companies.

An interesting example is the project Storsjö Strand, a new township in Östersund, using a strong interactivity and a triple helix process with the municipality, developers, and the university. The role of the university was, by an action research approach, to create involvement in the process and to document and evaluate it. (Jonasson et.al, 2014)

2.2.2 A pedagogical management model by Penta-Helix

Rapid changes, innovations and social development require collaboration and co-operation. Collaboration and mutual learning create joint benefit and local implementation power. The model Penta-Helix (figure 1) can support each actor to see its role in mutual learning. This model was developed in a project when a non-profit organization. The home village association at Bjäre is planning a building in the living history museum in Båstad. Local companies, networks and public actors form a reference group that can give feed- back and have a role as external supervisors and contacts.

The project involves students and pupils who are invited to create a meeting place, a "Food culture house", to meet and learn about local food, based on a historical and cultural perspective. The collaboration with all five actors has given very positive results and ideas in a process that will continue further on.

Fundings from the county's "Fund for environment" supports the costs for the small but important things like coffee, lunch, travel costs and communication. Students and pupils' results have been presented and discussed at workshops and meetings with representatives from all five parts in the model Penta-Helix.

Pupils from practical courses and programmes takes care of several of the chores that are connected to cultivation. Thesis by students was presented at posters and were available in the main building in the living history museum. Every year new students and pupils can participate and contribute with new ideas.

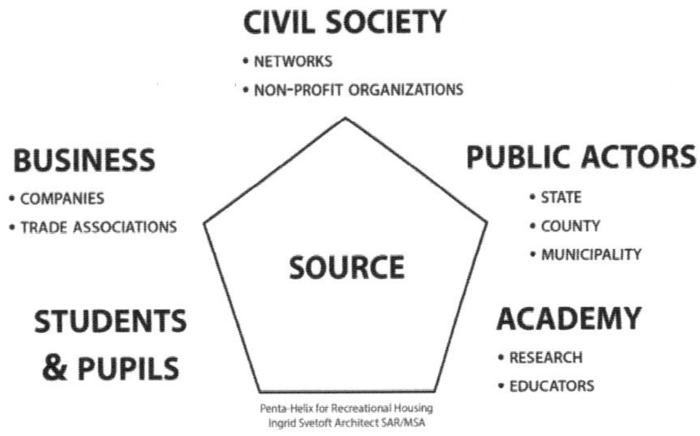

Figure 1: Organizational learning by Penta-Helix model (Authors, 2025)

2.2.3 Civil society

The energy and commitment from civil society, networks and non-profit organizations is always an important resource to involve. The definition of this group is usually of a local form due to local conditions and traditions. There can be formal and informal networks based on a group of citizens, special interests and hobbies for example. Some of the actors can be part of a regional, national or international organization.

2.2.4 Public actors

Public actors as municipal officials and politicians can prepare and take decisions that are important implementation force to a project. Representatives from state, county and municipality can all reflect on their decisions on planning. By sharing their knowledge about regulations and law a better understanding can be achieved among all actors involved.

2.2.5 Academy

Within academy there are knowledge, research and findings with both national and international perspective. In the mission for a university there are three main things to work with: research, pedagogy and collaboration. Teachers, researchers and lecturers are the link to students and pupils in courses and when supervising thesis.

2.2.6 Students & Pupils

Students and pupils from different schools can give a fresh perspective on important societal and environmental issues. Their studies, thesis and examinations can be combined with real cases and give a win-win situation. This group often gives a dynamo-effect to a project. We can also talk about reversed mentorship when companies for example receives new ideas from students. Or questions that make the companies reflect on their own business.

2.2.7 Business

Local companies and trade associations can get new perspectives and share their practical experience to the other actors involved. Their role can also be to transfer local culture, conditions and tradition.

https://doi.org/10.52202/080684-0015

2.3 INTEGRATED PLANNING

Integrated Planning (IP) is a construction site management tool. IP integrates the different planning skills used by site managers, construction workers and craftsperson's into an interactive group which manages a production planning process from the earliest stages to the end of a building project (Mikaelsson, 2017).

The studies which provided the basis for this tool, were performed over three decades, tested, longitudinally evaluated and refined the IP model for use in modern sustainable building sites.

The refined model (figure 2), Integrated Planning for Sustainable Building Production (SBP), includes the factors: leadership, health and safety, quality management and environmental management (Mikaelsson, 2017).

Figure 2: SBP, model for Integrated Planning for Sustainable Building Production (Mikaelsson, 2017)

2.4 GOING FORWARD BY LOOKING BACK-RECREATIONAL HOUSING IN A HISTORICAL PERSPECTIVE

There are many good examples to learn from that has a durable progression over time. We often have an ambition to create new ideas and plans but forget to look at previous achievements.

It is hardly possible, nor desirable, to formulate any unambiguous definition of the concept of sustainable and circular construction. The concepts slide into each other together with other concepts such as eco-building, green building and environmentally friendly construction. What can be perceived as conceptual confusion can, on the other hand, be an expression of something positive, such as that dear child has many names.

Similar reasoning can be applied to the definition of recreational houses. After all, the concept is quite new and stems from the time when the concept of "leisure time" arose as opposed to working time. It was only during early industrialism with wage labour that a clear distinction was made between working time and leisure time. The early labour movement fought for eight hours of work with eight hours of rest and eight hours of leisure. A broad definition of a recreational house can thus be the house where you live in your spare time. The architect Anders Nyqvist, who designed Sweden's first sustainable recreational house village, Rumpans Ekoby, started from this definition in the book "Rumpans Ekoby - From vision to realization:

"The architect's view of the concept of holiday home can be summed up in the following sentence: A recreational house is a house where you live in your spare time." (Nyqvist, 2019)

The houses in the Rumpan village were designed in accordance with this so that they could be used all year round. The architect himself moved with his family after a few years and settled permanently in Rumpan's eco-village.

The best summary of the vision for Rumpans Ekoby and the vision of sustainable construction can probably be found in Anders Nyquist's introduction to the book:

"Rumpan's eco-village is a housing vision that has been able to be implemented without compromise. When the ideas were launched, the words ecovillage, long-term sustainable construction, life cycle analysis or cycle-adapted construction were not in the vocabulary. What we strive for in today's construction was already 50 years ago in the description of Rumpan's eco-village but expressed in other words.

The idea of the village has been easy to understand, and the loyalty of the settlers has been great. The village is a social experiment based on community. Everyone involved has been able to contribute with their knowledge. The village is completed by 25 families who had limited knowledge of how to build a cycle-adapted village in balance with nature. Learning from each other and helping each other have been our guiding stars. The purpose of this book is to document our journey from an overgrown agricultural landscape to a living village". (Nyqvist, 2017)

3 THE SOURCE PROJECT

In the Interreg project SOURCE (Sustainable and nature pOsitive development of housing for ReCreational usE) the aim is to learn more about building recreational homes with a minimal negative impact.

The SOURCE project consists of 7 Work Packages with research questions ranging from restoring damaged nature to developing regenerative tourism.

Theory will be put into practice through workshops in pilot studies and results in a Toolbox.

3.1 PROJECT DESCRIPTION

The development of simple cycle-adapted concepts for sustainable recreational houses is needed. A model for this will be developed via the Industry Council for Sustainable Community Building at Mid Sweden University. It means, in short, that teachers and researchers in dialogue with companies will develop construction solutions that can be built in collaboration

https://doi.org/10.52202/080684-0015

with the education for construction workers at the regional high school educations.
The goal is that establishment and operation of the houses will contribute to less utilization of resources and that the resources are kept longer in the cycle. That means developing prototypes and concepts that generate and produce, both for people, climate and biodiversity, rather than consuming. It could be, for example, outlining new ways to use solutions, to present a typical house or construction methods that can replace conventional methods and thus reduce the impact of future construction projects. Old methods and materials can also be more sustainable and circular as they are cleaner and easier to reuse.

3.2 METHOD

The work package WP4 aims to apply a research approach called 'Change Agent' (Kørnøv et al., 2010, Kørnøv et al., 2011). Change agent in this context means that the Researchers' do not only show good examples of sustainable houses, but instead jointly develop a prototype together with the industry and academia. That means that the researchers act as change agents. The researchers are not just observers but contribute to the development of a field.
A model for circular concepts for sustainable houses will be developed via the Industry Council for Sustainable Community Building at Mid Sweden University. Teachers and researchers in dialogue with companies will develop construction solutions that can be built into small modules in collaboration with the education for construction workers at the regional upper secondary education.
The models developed from earlier research and studies presented in this paper, will be implemented and tested in projects with a natural continuation. New sustainable projects and collaborations and restart of former projects gives conditions for collaboration in a Penta-Helix formation. Existing contacts with students' external actors in ongoing education will provide opportunities to use the succession model for student collaboration. And the model for integrated planning can support actual planning and construction processes for sustainability.

3.3 A MODEL FOR CONSIDERATE RECRETATIONAL HOUSES

Work Package WP4 Circular Economy, revolves around two research questions:
WP4.1 How can recreational housing development help keep resources in the cycle longer?
WP4.2 What measures can reduce the total footprint of recreational housing?

Our aim are answers that contributes to a more circular economy. It's about where we build, how we build: how big, with what materials and technical solutions, how people travel to the site, what they do there, and how often the houses and resources are used.

Two conditions are specific to holiday homes:

- They are often empty for a large part of the year.
- It is a growing problem if you build on untouched natural land.

In many rural areas there are already existing buildings that can be used significantly more than today. Increased use of these buildings helps to preserve the value of investments made and can provide increased income, reduced costs, new local service offerings, new jobs and, in the long term, even occupancy.
Instead of focusing on just building new houses, we propose a process that identifies and evaluates local conditions in the first step; to continue to build upon, and do more with, resources that already exists.
A model (figure 3) with "five strategies for considerate recreational housing" will be further explored and developed in the SOURCE project. These five strategies are share more, convert, condense (densify), simplify and maintain. We have chosen to give maintain, or active management, a special position as a unifying strategy (the glass), while the other four (ingredients) depend on the conditions at the site.

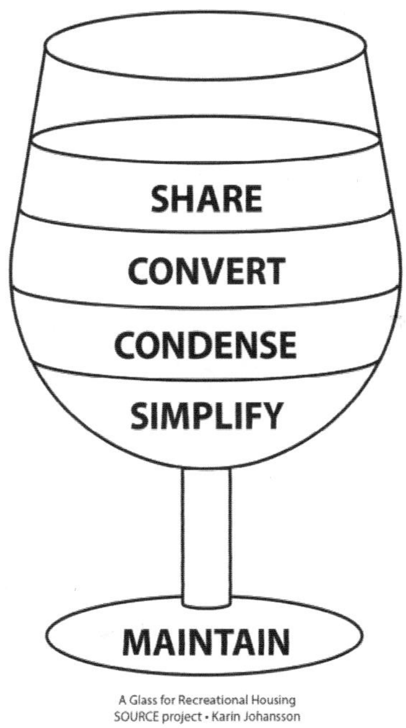

Figure 3: A Glass for Considerate Recreational Housing (Authors, 2025)

3.3.1 Share more

Both buildings and equipment like boats, fishing rods and bikes can be shared. There are however several thresholds for increased sharing. One is practical, regarding keys, cleaning and supervision, and another is emotional, it is one's second home and there may be objects that are personal, fragile or have great sentimental value. And of

https://doi.org/10.52202/080684-0015

course, the desire to be alone in a place, a longing for solitude.
Increased sharing can be achieved in several ways: Swap a home through home exchange services. Rent out homes that you already have. Your permanent home can be someone else's holiday home. Furnish and remodel to make room for more people, together or separately. Buy a co-op instead of your own holiday home. Create a room or cupboard with things to share. Develop local services that facilitate sharing.

3.3.2 Convert

Update and use existing buildings that currently have a low utilization rate. Reuse entire houses instead of just parts.
Convert buildings that are not used or have a low level of use, in rural areas and small towns, and get a historical and practical context around which you can continue to build concepts. Create lifestyle homes and themed tourism.
These buildings can be agricultural buildings, barns, summer houses, cottages, public buildings like schools or community center's, closed shops or empty office or industrial premises.

3.3.3 Condence

Create more beds in existing houses. Connect more houses to existing infrastructure. Extend and develop new seasons on already developed land, especially around infrastructure that has already involved major interventions, such as ski resorts, golf courses, trail systems and houses with a high year-round standard.

3.3.4 Simplify

Identify methods and local materials that can be used and re-used. Start from the site. Build small-scale and square-smart with simple technical solutions and locally produced materials. Make it reversible and easy to move. An allotment within cycling or public transport distance is still a good solution.

3.3.5 Maintain- active maintenance and management

Active management is an agile process. It means continuously maintaining and adapting buildings based on changing conditions and needs. A prerequisite is choosing good materials that can be maintained. Fix problems as soon as symptoms appear, before major damage has developed. Avoid replacing functioning parts. Use things in their place. Define needs and see what can be done with what already exists. Supplement with materials and techniques that are adapted to the building's construction system.

In summary:

• Use what already exists

• Limit new production, overconsumption and dormant resources.

3.3.6 Business potentials

Increase revenue opportunities:

• ROT, renovation, remodelling and extension offers with sustainable materials and methods

• Local services that facilitate sharing, for example reception, property management, cleaning, storage, service packages: adapt the house to the person coming. Destination development - new and longer seasons with themes, activities, courses, packages, food

• Cooperative housing. Further research should be done to investigate the conditions, practical and economic advantages and disadvantages, and possible consequences for local communities of converting holiday homes to cooperative housing with multiple co-owners.

3.4 SUCCESSION MODEL FOR STUDENT/PUPIL COLLABORATION

The Penta-Helix model can be used in combination with a Succession model to support knowledge development. Supported by companies, researchers, trade associations, schools and public actors.
When actors with different experiences exchange their ideas, new perspectives can be unfolded. That is one of the positive effects when theory and practice meets in research projects. In the SOURCE project possibilities to include students in their university studies has been developed and gives the framework for a model (figure 4). In contacts with Jämtlands gymnasium and the program for building the idea is to create a long-term co-operation. The Succession model includes four steps:

1. Course project
2. Thesis.
3. Internship and workplace introduction to profession.
4. Profession

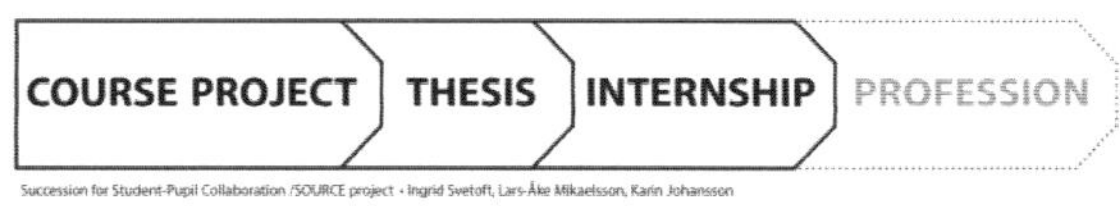

Figure 4: Succession model for student/pupil collaboration (Authors 2025)

3.4.1 Course project

In universities and schools there are many possibilities to use cases and projects that are performed by private and public actors. In course projects networks and persons can be connected that bridges between academy, civil society and municipalities and local companies. These projects are crucial to plan in detail to match the schedule for all parts. The time invested by the external actors can be paid back from the students/pupils by a short presentation or documentation of their results.

3.4.2 Thesis

Different kind of thesis can balance between a formal template and a case study with external contacts. For the supervisor it is of great importance to remind the student/pupil to have critical perspective on the topic and question to study as well as work independently. The

https://doi.org/10.52202/080684-0015

external contact can get a new perspective on their own organization and work. For the supervisor there is often a positive effect when the student/pupil is highly motivated in their writing process. Being part of a research project or a real project gives new knowledge and several contacts that can be useful in the future.

3.4.3 Internship

When contacts are established when collaborating in course projects or thesis a relation for further collaboration can be achieved. When working as an external supervisor there are possibilities to present the company or public organization for a forthcoming employee. The student/pupil can also learn more about a prospective employer. In this model an internship gives both parties a possibility to see if there are interest to continue in a work relationship.

3.4.4 Profession

If the collaboration in project, thesis and internship is of mutual interest, a long-term employment relationship can be formed. Companies can more easily work with the employment process if contacts and collaboration have been performed. It takes a lot of time and effort to handle an add and then go through all applications. Companies do not only employ their staff but also keep their employees. Trust and loyalty towards all parts of the work relationship can be grounded during education.

4 CONCLUSIONS

Challenges for building in rural areas will be identified and solved: Transports to the construction site. The design of the foundation. Heating, electricity, water-supply and waste planned for an off-grid solution. Respect for regulations and national interests. Local cultural and social values. The whole process must therefore be accurate planned in dialogue, taking care of different perspectives, and experiences and follow regulations. The actors, methods and models presented in this paper can support a planning and building process to minimize the negative climate impact.

Circularity is a fundamental principle for all sustainable construction, thus also for the construction of recreational houses. IVL The Swedish Environmental Institute wrote in 2018: "In practice, circular construction is often about minimizing waste in various ways or increasing material recycling." (www.ivl.se)

To be able to circulate, the building materials must meet certain requirements, for example, be non-toxic, and able to be dismantled, reused or recycled, and not create waste. Many building materials are composites (consisting of several materials) that are impossible to take apart and contain toxic additives. Other materials are non-toxic (eg brick) and can be used but we do not have systems to handle them. Industrially produced building materials and details can be delivered with large quantities of packaging. Sometimes it is therefore also meant that the materials should be local, fossil-free, low-processed, natural materials that are not taken from the earth's crust, for example unburnt minerals and straw instead of wood. This can sometimes be problematic in relation to legislation where building materials must have declarations and meet standards. In a broader sense, the term "circular" can also refer to how the residents of the house close local loops through how they have arranged water supply, water purification, latrine and kitchen waste management, for example through rainwater collection, sand filters, bio-purification plants, root zone (productive purification), composting or biogas production.

The models, methods and good examples presented in this paper can all support possibilities to co-operate and combine practical experience and theoretical knowledge. In the SOURCE project these models will be presented, tested and hopefully implemented with the ambition to minimize negative climate effects. For local companies' collaboration with other actors and students/pupils, can give positive effects for growth. When learning together and discussing challenges new ideas and new knowledge occurs. The network and collaboration in project and thesis also provide contact with future employees. To face climate and societal challenges together, collaboration and circularity in a broad perspective can give positive effects. For individuals, companies as well as to society.

ACKNOWLEDGEMENTS

Thanks to all actors in the research project SOURCE, colleagues and students.

REFERENCES

Argyris C & Schön D (1978), Organizational learning, A theory of action perspective. Reading, MA: Addison-Wesley.

Atlestam G., et. al, (2015), Ekobyboken, VOTUM

Boverket(250131),Miljöindikatorer (https://www.boverket.se/sv/byggande/hallbart-byggande-och-forvaltning/miljoindikatorer---aktuell-status/)

Commoner B., (1971) The Closing Circle, (Cirkeln sluter sig), Aldus/Bonniers

Ehrensvärd G., (1971), Före – Efter,Aldus/Bonniers

Ellen Macarthur Foundation (2025) (https://www.ellenmacarthurfoundation.org/about-us/ellens-story) (https://www.ellenmacarthurfoundation.org/topics/circular-economy-introduction/overview)

IVL Svenska miljöinstitutet (2025) (https://www.ivl.se/)

Jonasson J., Danielski I., Mikaelsson L-Å., Fröling M., (2014), Approach for Sustainable Processes for Built Environment in Triple Helix Cooperation – The case of Storsjö Strand in Östersund. Linnaeus ECO-TECH ´14, Kalmar, Sweden, November 24-26, 2014

Jonasson J., Mikaelsson L-Å., Persson U., (2020), Knowledge management of sustainable construction processes. Beyond 2020 (full paper) 20200108

Kørnøv, L., Lyhne, I., Vannem-Larsen, S. & Merrild-Hansen, A. (2011), Change Agents in the Field of

https://doi.org/10.52202/080684-0015

Strategic Environmental Assessment: What Does it Involve and What Potentials Does it Have for Research and Practice? Journal of Environmental Assessment Policy and Management, 13, 203-228.
Nawi M, Baluchi N & Bahauddin A (2014), Impact of Fragmentation Issue in Construction Industry: An Overwiew. Building Surveying, Facilities Management and Engineering Conference (BSFMEC 2014)
Kørnøv, L., Vammen-Larsen, S., Merrild-Hansen, A. & Lyhne, I. (2010) Change Agents & Impact Assessment. Nordic Research Day on Impact Assessment. Aalborg University.
Nyquist A., (2019), Rumpans Ekoby, Från vision till förverkligande, Anders Nyquist Arkitektkontor AB
Meadows et. al, (1972), Limits to Growth, A report for the Club of Rome's project on the predicament of mankind
Mikaelsson, L-Å., Larsson, J. (2017), Integrated Planning for Sustainable Building Production – An evolution over three decades, Journal of Civil Engineering and Management.
Regeringen (2006), Energy declarations, SFS 2006:985
Regeringen (2021), Climate declarations, SFS 2021:787
Senge P (1990) The Fifth Discipline: The Art and Practice of the Learning Organization, Currency USA
Svetoft I (2008), Arkitektens roll och brukarmedverkan, PhD thesis, Dept of Construction Sciences Lunds University, Lund

Proceedings of the Circular Building Sector Conference 2025, Lund, Sweden, 1-3 June 2025

REAL ESTATE COMPANIES AND CIRCULAR BUSINESS MODELS

Per Anker Jensen[1*], Henrik Juul Nielsen[2]

[1]DTU Engineering Technology

[2]Niras

*Corresponding Author: pank@dtu.dk

ABSTRACT

Background and aim. Many Real Estate Companies (RECs) develop, rent out and manage their buildings as landlords. However, as the buildings are used by their tenants, landlords do not have much opportunity or incentive to optimize the use and thereby reduce the need for space. On the contrary, to secure their investment, they have strong incentives for long-term rental contracts and expand their building portfolio, while commercial tenants have incentives for flexibility in their lease contracts. Besides RECs' commercial interests, they also meet increasing expectations from the public and potential tenants to behave in an ethical and sustainable way. This paper investigates how RECs balance these dilemmas and develop their business models.
Methods and Data. The paper is based on interviews with six RECs in Denmark.
Findings. The business models vary a lot and so does the focus on sustainability. The purely profit driven RECs lack incentives to become more sustainable. Legislation is the main drivers for these companies. State owned and ethical driven investors and administrator companies owned by pension funds have clear strategies towards being more sustainable. A new administrator company had recently been established by a pension fund particularly to meet the increased sustainability challenges with a strong focus on circularity.
Theoretical / Practical / Societal implications. The paper adds to the growing literature on circular buildings with particular focus on management of the use and operational phase. It can give RECs inspiration to develop circular business models.

KEYWORDS: Real estate, investors, administrators, circular, business models

1. INTRODUCTION

It has become a mantra that "*the most sustainable buildings are those that are not built*". However, it is just as important to take care of existing buildings and extend their lifespan and relevance to the purpose they serve; preferably keeping the use value and embodied CO_2 infinitely through repurposing and sufficient maintenance.

The Real Estate (RE) sector has a major role in changing the Built Environment (BE) to become more sustainable. The sector consists of many different types of both private and public organisations. In this paper we apply the roles and responsibilities described by Jeppesen, (2024) for Property Management: Portfolio Management, Asset Management, Property Administration, Renting-out, Building Client, Operation Management, and Caretakers. Real Estate Companies (RECs) can cover one or more of these roles with many interdependencies between companies, which often changes due to out- and insourcing. This makes the sector very complex.

The most important circular actions for existing buildings are to prolong the lifetime and increase the intensity of use as shown in section 2. Renting-out secures a steady income to the RECs and the building owners, which creates clear incentives to develop and maintain buildings to ensure a long lifetime. On the other hand, RECs, who are renting-out do not have much opportunity or incentive to optimize the use and thereby reduce the need for space of their tenants. Contrarily, to secure their investment, they have strong incentives for long-term rental contracts and to expand their building portfolio, while commercial tenants have incentives for flexibility in their lease contracts.

https://doi.org/10.52202/080684-0016

Besides RECs' commercial interests, they also meet increasing expectations from the public and potential tenants to behave in an ethical and sustainable way. This paper investigates how RECs balance these dilemmas and potentially develop circular business models within Real Estate and Facilities Management (REFM). The paper is based on an interview survey among RECs in Denmark. The paper starts with a literature review of Circular Economy (CE) in general and in relation to the BE and REFM, and sustainable business models. After the literature review follows sections on methods and data, findings, and discussion and conclusions.

2. LITERATURE

2.1 ACCELERATING THE CIRCULAR ECONOMY IN EUROPE

The European Environment Agency – EEA - has recently published a report on how the circular economy can generally be accelerated in Europe (EEA, 2023). The report has the following main messages:

- 100% circularity is impossible, so it is crucial to prioritize the reduction of resource consumption and move towards a less material-intensive economy
- Maximizing the usability of existing products requires a significantly greater intensity of use per product and much longer life
- Success of a circular economy depends heavily on returning significant quantities of high-quality secondary raw materials to productive use

REFM has a significant role to play, especially in relation to the first two points considering the fact, that construction and buildings account for the largest share of environmental burdens. Figure 1 shows a model for circularity from the report, and here the use phase is the most important as indicated by the sizes of the circles.

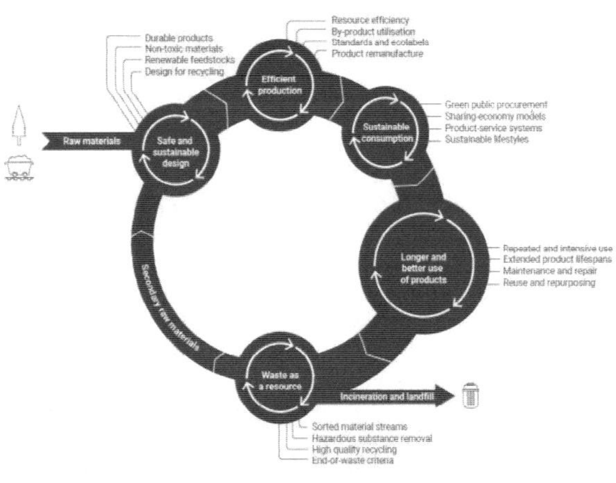

Figure 1: Model of circularity (EEA, 2023)

The second bullet point in the EEA report stresses the two most important circular actions for existing buildings of intensification the use of buildings and prolonging lifetime. Both actions lead to a reduced demand for new buildings. Demand management in general leading to a reduced need for space is similarly important.

In relation to waste, many R models have been developed, e.g. 3R with reduce, reuse and recycle. The report contains a 9R model as shown in Figure 2. This is divided into Before use. During use and After use, and as many as four of the nine measures relate to the use phase.

In a REFM context, they can be termed: Long-life use and maintenance, Recycle and share, Repair, and Renovate and transform. Like much other literature on CE, the starting point is manufactured products. Therefore, intensification of use through space optimization is missing in Figure 2 even though it is included in Figure 1.

The EEA report highlights that the circular economy deals with more than waste management. It's more about keeping the value of materials high and making sure they last longer and designing unnecessary material out of the economy. This requires new business models and a transition from ownership models to service-based solutions.

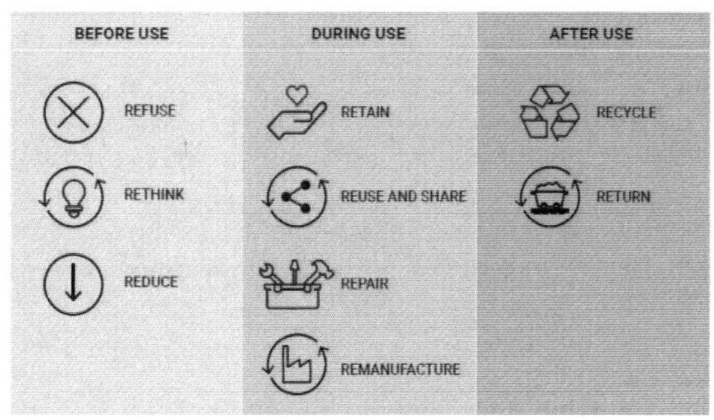

Figure 2: Actions for increased circularity (EEA, 2023)

Although the EEA report does not specifically address REFM, it highlights the importance of the use phase. FM can contribute to a more sustainable future by integrating both environmental and social factors (Jensen and Nielsen, 2024).

2.2 CE DEFINITIONS AND PRINCIPLES

Bocken et al. (2021a) claim that Circularity is the new normal and states: *"Circular solutions are **not inherently sustainable** – unless designed to be so"*. And many previous definitions of Circularity have the same relative relationship between circularity and sustainability. But with the release of the new ISO 59000 family this relative relationship has changed. According to ISO 59004 (2024): *"Circular economy is an economic system that uses a systemic approach to maintain a circular flow of resources, by recovering, retaining or adding to their value, while contributing to sustainable development".*

Thus, a CE system must at least contribute to a sustainable development to be compliant with the ISO 59004 definition.

Previous definitions for CE have described principles for CE. Apart from a slight variation in semantics, the most dominant definitions have evolved around the four main principles *Narrow*, *Slow*, *Close* and *Regenerate*, see Figure 3.

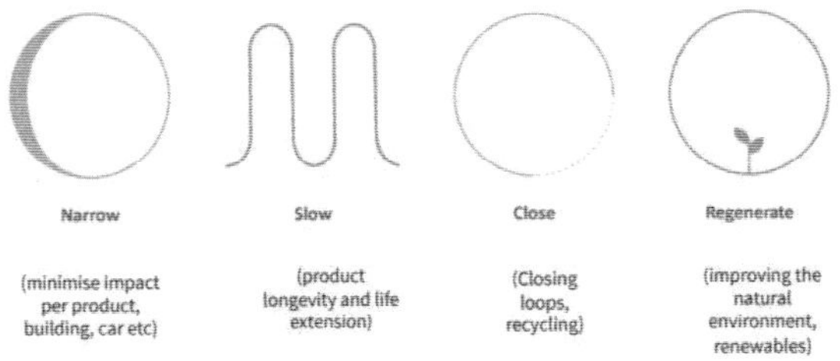

Figure 3: Four common CE Principles (Bocken et al., 2016; 2021b)

ISO 59004 (2024) introduces 6 principles of CE:

- **Systems Thinking:** Adopts a long-term perspective and considers impacts on environmental, social, and economic systems
- **Value Creation:** Restores, preserves, or adds value through effective solutions
- **Value Sharing:** Distributes value fairly among stakeholders to promote collaboration and social equity
- **Resource Management:** Sustainable management of resources to ensure their availability for future generations
- **Resource Traceability:** Maintains data to track resources through value chains
- **Ecosystem Resilience:** Protects and regenerates ecosystems and biodiversity

The four ISO 59004 principles, *Value Creation*, *Value sharing*, *Resource Management* and *Ecosystem Resilience* have many similarities to the previously described principles *Narrow, Slow*, *Close* and *Regenerate*. But the two remaining ISO 59004 principles, *Systems Thinking and Resource Traceability* address the need for a strategic approach and data generation to enable a business model for CE that both include a strategic and systemic approach to CE and documents improvement. *System Thinking* and *Resource traceability* can also be described as principles that are general to any circular action, to ensure that such actions are compliant with a defined circular strategy and to ensure that any action taken to elevate circular value creation can be documented.

2.3 BE, REFM AND CIRCULARITY

Where circularity in the built environment has traditionally been defined as reduction of waste and reuse of building materials, circularity in FM and building operations spans over several of the circular principles. This includes not only life extension through maintenance, renovation and transformation and more intensive use through space optimization but also sharing economy and service-based solutions, which can optimize resource utilization and reduce environmental impacts. By integrating these broader principles, REFM can contribute significantly to a more sustainable and circular future.

Kyrö (2020) investigates, how the real estate sector can transition to circular economy. The main result is the identification of four different approaches to RE management shown in the two-by-two matrix in Figure 4.

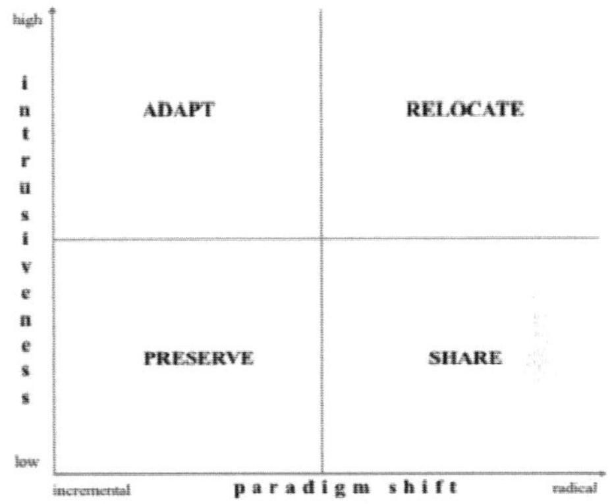

Figure 4: RE approaches to CE (Kyrö, 2020)

A paper by Jensen and Nielsen (2024) develops the topic of Circular FM from the perspective of an in-house FM-organization and presents a case study of how FM can create added value by optimization of a property portfolio, existing buildings and workplaces. It partly builds on The Value Building, see Figure 5.

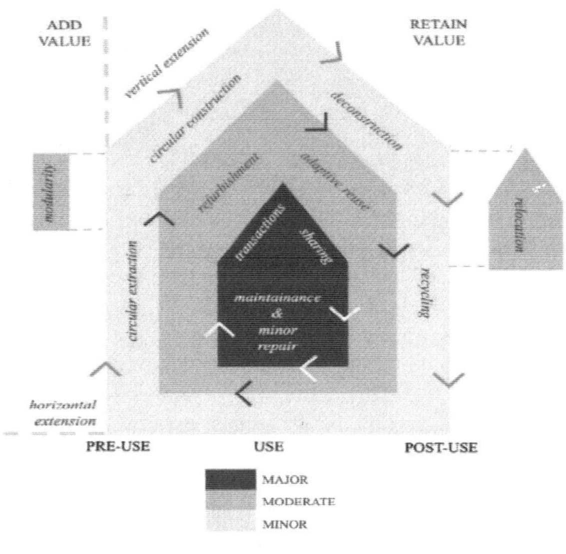

Figure 5: The Value Building (Kyrö a2nd Lundgren, 2023)

https://doi.org/10.52202/080684-0016

The Value Building is like the 9R model shown in Figure 2 divided in: Before use, During use, and After use with During use being the most important. It includes what Kyrö and Lundgren (2023), call 12 "Circular Building Categories in the Built Environment". We will for simplicity call them actions like Jensen and Nielsen (2024), who also added the action 'space optimization' during use. This is a subdiscipline in FM called Space Management and includes more than the action *sharing* in the Value Building. There is a clear difference between an RE owner and landlord perspective and the user organization perspective of FM. When a company itself owns and uses a building, there is a difference between a long-term owner perspective and a short-term user perspective. But this is reinforced, when the building is owned and rented out by an external part. This is what creates the Landlord-tenant dilemma, which has been widely discussed in connection with energy renovation (e.g. Åstmarsson et al., 2013). From the above it becomes obvious that the same dilemma applies to Space Management.

Space Management (SM) is essential from a circularity perspective of intensifying use and supports the circular principle "Narrow" as described in Konietzko et al. (2020) and Bocken et al. (2016). It is also compliant with the ISO 59004 principle of *Value Creation* by ensuring that built space provides as much value to the organization as possible. By optimizing the use of space, SM is essential in "narrowing" and rightsizing the amount of space to support the user needs. Thereby not only the resources spent on the producing physical buildings are reduced, but also the recuring resources spent on energy, maintenance and services related to that space. This is often overlooked in various models of CE - as mentioned also in relation to Figure 2, while it is included in Figure 1 from the same EEA report. RE owners and lessors have opposite economic interests and incentives to increase the amount of space to be rented out as mentioned in the introduction section. This is one of the dilemmas that form the background for this study.

The Value Building includes *transaction*. Kyrö and Lundgren (2023) writes on this: "variations that would be considered circular in this context include more flexibility in lease agreements and shorter-term, even pay-per-use access. The servitization of spaces and space-as-a-service business models have also gained foothold in recent years." Thus, *transaction* and renting-out space is not a circular action in itself, but it can include circular actions.

The Danish Facilities Management association (DFM) made a study in 2022 on the sustainable practice among RECs on the private rental housing market in Denmark (Rasmussen & Bøytler, 2022; Rasmussen et al., 2023). Among the conclusions were that sustainability by many owners is regarded as a necessity for the long-term profitability of property investments.

This is supported by a recent research paper, which concludes, that *"ESG achievements of the real estate industry has significant impact on investor decision-making. As the concept of global sustainability continues to grow, real estate investors will prioritize evaluating how buildings perform regarding environmental stewardship, societal contribution, and corporate governance when choosing investment projects."* (Liu, 2024).

2.4 CIRCULAR BUSINESS MODELS

A business model (BM) describes what value a business provides to its customers, how it creates the value, and how it generates income. There are many definitions and models of BMs. The four block model (FBBM) by Christensen et al. (2016) is one of the simplest models and proven to be useful for research purposes (Berg et al., 2021). The FBBM framework includes Priorities divided in Value proposition and Profit formula, and Capabilities divided in Resources and Processes as shown in Figure 6. The authors behind the model stress that the interrelationships and balance between the four blocks are essential. The value proposition is the most important element.

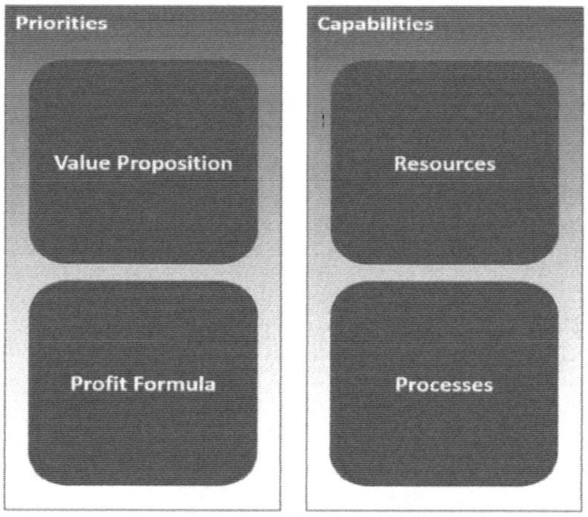

Figure 6: A simplified version of the FBBM framework (Berg et al., 2021; based on Christensen et al., 2016).

Estarrona et al., (2019) investigate new business models being developed in asset provision and management through servitization of space and a social push towards CE. Jensen and Nielsen (2024) review the literature on CE and the BE and shows that it often refers to business models (Baniya, 2023; Kyrö, 2020; Lundgren et al., 2023; Murano et al., 2020).

The paper by Baniya (2023) has CE and FM in the title and uses the term Circular FM. It is a literature review and focuses on environmental sustainability for facility service providers on the three scopes: Procurement, building use, and end of life, which also resembles the division in Before, During and After use in Figure 2 and 4. The paper finds that subtle changes in the core facility function, such as in products' purchase approach, delivery of ongoing maintenance and refurbishment of building assets, and end-of-life management, possess the potential to enable circularity. The paper primarily has an operational environmental focus, while Jensen and Nielsen (2024) have a more strategic focus on buildings and space use from the perspective of an in-house FM-organisation and impacts on holistic sustainability.

3. METHODS AND DATA

This paper presents an explorative study. The overall research question for the paper is: *How can Real Estate Companies (RECs) balance different dilemmas and become more sustainable by adapting circular strategies and business models?*

The empirically study is based on interviews with six RECs in Denmark selected to present maximum variation and the interviewees being managers on senior levels. The selection of interviewees took as starting point the same as used by Rasmussen and Bøytler (2002), which included six carefully selected interviewees from different types of RECs in Denmark. Two of the people were still in the same company and willing to be interviewed, while the other four had either changed company or did not respond. Instead, other interviewees were recruited to cover different types of companies.

The interviews were conducted in late 2024 and early 2025. They lasted approx. one hour. Some interviews were done online and others by physical meetings. The interviewees received a short project description beforehand. Interviews were recorded with consent and written minutes summarizing the interviews were sent to the interviewees for comments and acceptance. The list of the interviewees is shown in Table 1.

Table 1: Interviews

Inter-viewee	Function	Company	Date
1	Technical Manager	Østervold (EØ)	02-12-2024
2	Technical Director	Newsec (NS)	10-12-2024
3	Senior Strategic Procurer	Heimstaden (HS)	10-12-2024
4	Head of Sustainability	Freja Ejendomme (FE)	10-12-2024
5	Head of ESG	ATP Ejendomme (AE)	09-01-2025
6 and 7	CEO and Head of ESG	PKA Ejendomme (PE)	09-01-2025

Both private, institutional and public RECs are represented, and they cover all the main business types among RECs with a variation in types of facilities as well. The last interview had two interviewees.

The development of a professional market for property investments is a relative recent phenomenon in Denmark. According to the chief economist in the Danish Property Association it did not evolve before the 1970s (Larsen, 2010). Property development is defined it as *"Transformation of a site from one condition to another in such a way that is creates added value in itself or another form of return"* (Leikvam and Olsson, 2022 – translated from Norwegian).

Property Management can be divided in: Portfolio Management (PM), Asset Management (AM), Property Administration (PA), Renting-out (RO), Building Client (BC), Operation Management (OM), and Caretakers (CT) (Jeppesen, 2024). Table 2 shows a categorisation of the six companies according to this classification.

Table 2: The companies' property management activities

	FE	HS	EØ	PE	AE	NS
PM	x	x	x			
AM		x	x	x	x	x
PA		x	x	x	x	x
RO		x	x	x	x	x
BC		x	x	x	x	x
OM		x	x	x	x	x
CT		x	x	x	x	x

4. FINDINGS

The BMs of the RECs are analysed using the FBBM framework in section 4.1. For processes the actions regarded as circular are identified. The positioning of the companies in relation to being Investors/developers or Managers and have an active or passive customer relationship is analysed in section 4.2.

4.1 BUSINESS MODELS

The BMs of the companies, as interpreted by the authors, are shown as basic BMs in Table 3 and with recent developments in the BMs in Table 4. The basic BMs in Table 3 shows how the RECs traditionally have operated to be economic and social sustainable (profitable or viable), while Table 4 shows how the BMs have developed in recent years partly with the aim to become more environmentally sustainable. Table 5 shows how the processes in the last row of Table 4 can be interpreted in relation to sustainability and circularity. In the following each company is described with ownership, business

https://doi.org/10.52202/080684-0016

types and activities related to sustainability and circularity.

Newsec (NS) is an investor-owned Sweden-based limited company (A/S) and a large international property administrator for different investors. Their approach to office buildings is to rent them out fitted to the need of their new tenants. They have two inhouse architects to help new tenants to develop the space plans and interior design. When a contract has been signed and the tenants move in, it is up to them, how they use the space.

The circular actions of NS include the basic maintenance and occasionally renovation and modernization to improve energy performance, market value and keep up building condition. Earlier the pension funds were not interested in investing in energy improvements, because the reduction in energy consumption would only benefit the tenants (Landlord-tenant dilemma), but this is changing, because sustainability has become a competitive parameter. NS also aim at increasing reuse of building materials and have in their project departments standards descriptions including for instance reuse of kitchens and floors.

It has become a trend, that one must find new ways to utilize buildings and accommodate new user needs. Earlier, the landlord could determine the conditions, but now they need to be much more user focussed. Thus, NS is involved in developing new offerings in collaboration with investors, for instance the office hotel concept Union developed together with the pension fund PFA. Office hotels increase flexibility for customers to adapt to changes and optimize their space utilization. On multi-tenants building NS have quarterly customer meetings.

NS experiences that some customers demand sustainability certifications like DGNB and BREEAM in Use. Some international investors set higher requirements on sustainability reporting than Danish investors. NS established an ESG department 10-12 years ago, which now has approx. 12 staff in Denmark, mostly engineers, and it is still expanding and does ESG reporting.

PKA Ejendomme (PE) is a limited company (A/S) and was established in 2023 as a subsidiary of the pension fund PKA and is fully owned by the mother company. They are like NS a specialized property administration company, but only with property directly owned by the mother company in Denmark. Before establishing the new subsidiary PKA had outsourced almost all of their property management except portfolio management to NS and another administration company. The reason for establishing PE was that the increasing requirement for integrated sustainability necessitates a change process, and an outsourcing strategy is not suitable in this situation.

Thus, PE focus on managing sustainable and up-to-date buildings. Their strategy is to own and hold with the aim to secure the PKA members' pensions for the next 40 years. PE has developed a business model, which to a much higher degree incorporates sustainability in their everyday practice in contrast to the more traditional approach with sustainability as an add-on in various projects. The philosophy is to have all value-adding activities in-house, and only rent administration is still fully outsourced.

PE's headquarters is established in an existing building near Copenhagen. They had LCA calculations made when they planned the refurbishment and the move to this new location and were surprised how much the embedded CO_2 in furniture counts. Thus, they planned the refurbishment, so that they removed as few building components like wall surfaces and reused as much of building components and furniture as possible. They have worked a lot with colours to trick the eyes. They use similar principles in the property they manage. Building materials should stay on the cadastre as far as possible. PE also collects building components like doors, separating walls and light fittings and have them stored at their building contractors.

PE market office spaces fitted out generically and with information about number of workplaces, so they can be taken into use without rebuilding. If customers want changes in interior layout and design, they will have to pay for that themselves. PE aim to get close relationships with their customers to know their needs for space and facilities and their development plans to retain them as long as possible; particularly when a lease contract is close to termination. If a customer needs more space, they will not have to pay for moving out of their existing facilities, if they have a larger lease in another building. PE do not have office hotels, but they have several multi-user office buildings.

PKA was the first in Denmark to have a new building for renting-out housing certified by the Nordic Swan label in 2016. They also work with the building certification labels DGNB and LEED. PE plan to have their whole building portfolio certified with the new Nordic Swan label for building operation. Sustainability reporting is done by the individual pension funds in PKA.

ATP Ejendomme (AE) is a limited company (A/S) and is a fully owned subsidiary of the state-owned pension fund ATP. They are like PE specialized in property administration with property directly owned by the mother company in Denmark. AE has a sustainability strategy that focuses on reducing the CO_2 footprint in their buildings and construction projects. Their buildings must comply with the applicable CO_2 limit values and be flexible and adaptable to future needs. Circularity is part of their strategy, especially in relation to reducing the CO_2 footprint of materials for new construction. They do CO_2

accounting and have an overview of consumption data. However, they do not yet do specific reporting on how large a proportion of sustainable materials they use.

AE try to create transparency and demand sustainability from their tenants, but the demand for data is primarily driven by their own needs. They make clear demands on their business partners to comply with sustainability and ethical standards. Their partners also expect transparency and sustainability in their projects, and society has an increasing expectation that they operate sustainably and ethically. The financial sector is also very aware of sustainability standards and expects them to live up to them.

One of AE's biggest challenges is to balance the need for flexibility in buildings with sustainability requirements and financial considerations. They work with flexibility and sustainability in both existing buildings and new projects and try to keep buildings relevant to users over time. New construction must comply with high sustainability standards and be flexible enough to be adapted to changing needs. AE are considering conversion options from office to housing, especially based on the experiences from the corona pandemic, where housing unlike commercial building was not subject to lockdown. They plan to increase the proportion of residential properties in their portfolio and reduce the number of offices and shopping centres. This is part of their risk management strategy.

AE report in accordance with EU Taxonomy, and from 2025 they are subject to CSRD as part of the ATP Group. Specifically for properties, they follow the new industry standard Real ESG. They use and recognize the international certification standards DGNB, LEED and BREEAM.

Østervold (EØ - Ejendomsselskabet Østervold) is a small family-owned limited company (A/S), which invests in and manage existing buildings in Denmark. The company has no defined sustainability policy and is not yet covered by EU's ESG reporting, but they are preparing to report. EØ invest in buildings to hold and manage them to prolong building lifetime. They have established office hotels in provincial towns to increase their offerings, reduce redundancy and meet market demand.

Heimstaden (HS) is a large investor-owned Sweden-based limited company (A/S) and is an investor and administrator specialized in housing, HS maintain, renovate and modernize buildings to raise the energy labels and rent, and they have changed their policy from replacing faulty fridges to repair if possible. ESG-reporting has started in a Danish environmental unit with two people, and they have an ESG department at the headquarters in Sweden.

Freja Ejendomme (FE) is a state-owned limited company (A/S) with the purpose of developing redundant public property to be sold for new use and thereby create a maximum surplus for the public. They develop the property to have an approved local plan before selling. FE priorities architectural value and has a strict sustainability policy, which they are measured on. FE are at the moment working on including sustainability requirements in sales conditions. For instance, they rejected a bidder, when they became aware that the potential buyer wanted to demolish the buildings to build new.

4.2 POSITIONING OF THE COMPANIES

The case companies vary according to company type and customer relationships. Another differentiation of the companies is whether they are specialized in specific facilities or are more diverse. The case companies span:

- The investor and developing company Freja Ejendomme, which take over all kinds of redundant state property to sell off to private investors for new use, and has a passive relationship with customers,
- Investor and managing companies Østervold, who only invest in existing buildings to hold and with a passive relationship to customers except for office hotels, and Heimstaden, who both invest in new and existing buildings with a specialization in housing with relationships to representatives of tenants and private housing associations,
- Property administration companies Newsec, who administrates for different investors, and ATP Ejendomme and PKA Ejendomme, who only administrate property owned by their mother companies - a state and private pension funds, respectively, and with an increasingly closer relationships with their tenants.

https://doi.org/10.52202/080684-0016

Table 3: Basic Business Models of the Companies

Company / BM block	NS: Newsec	PE: PKA Ejendomme	AE: ATP Ejendomme	EØ: Østervold	HS: Heimstaden	FE: Freja Ejendomme
Value proportion	Rent-out housing estates and commercial prop. for various private investors and for private housing associations in Northern Europe.	Rent-out prop. owned by the mother company and manage the prop. strategically and develop close relations to their tenants in DK	Rent-out prop. owned by the mother company and manage prop. in Denmark.	Rent-out carefully selected housing estates and commercial buildings in central locations based on investments in existing buildings to hold in DK.	Rent-out housing owned by the company in Northern Europe.	Develop and sell redundant public facilities.
Profit formula	Income from prop. adm. and mgmt. Expands by taking on more properties from investors and associations.	Income from prop. mgmt. Develops according to the mother company's portfolio.	Income from prop. mgmt. Develops according to the mother company's portfolio.	Income from prop. mgmt. Develops according to the company's own portfolio.	Income from prop. mgmt. Develops by building new and procuring existing property portfolios.	Create max. surplus for the public.
Resources	In-house staff for prop, mgmt. and technical experts for mgmt. of M&O services. Marketing staff to increase market share.	In-house staff with technical experts for mgmt. of M&O services. Rental adm. is outsourced.	In-house mgmt. staff with technical experts for mgmt. of M&O services.	In-house staff to procure new and manage existing prop. and technical experts for mgmt. of M&O services.	In-house mgmt. staff to administrate existing prop. and technical experts for mgmt. of M&O services.	In-house staff to develop projects. Building consultants assist in developing projects.
Processes	Adm., mgmt. and M&O.	Adm., mgmt. and M&O.	Adm., mgmt. and M&O.	Adm., mgmt. and M&O.	Adm., mgmt. and M&O.	Project development.

Table 4: Developments in Business Models of the Companies

Company / BM block	NS: Newsec	PE: PKA Ejendomme	AE: ATP Ejendomme	EØ: Østervold	HS: Heimstaden	FE: Freja Ejendomme
Value proportion	Modernize investor-owned prop. Change from seeing investors as their customers to stronger focus on the tenants. New developments with investors, for instance office hotel concept.	Daughter company started in 2023 based on insourcing major parts of mgmt. and services with particular focus on value-creating services. Nordic Swan label, DGNB and BREEAM certifications.	DGNB, LEED and BREEAM certifications.	Establish office hotels to develop their offerings to reduce redundancy and meet market demand.	Normally own to hold but the need to free capital has forced them to sell flats to be individual owner-occupied.	
Profit formula	Income from projects to increase rent from modernization and changing	Occasionally modernization projects to increase rent income.	Consider conversion from offices to housing. Part of risk strategy to increase the	Occasionally expanding portfolio and increasing rent from modernization	Occasionally modernization projects to increase rent income.	Five assessments criteria: Economic added value, Environmental

	offices to housing and offering additional services in office hotels.		proportion of housing and reduce commercial property.	and offering additional services in office hotels.		sustainability, Culture and city life, Architectural quality, Special societal considerations.
Resources	ESG department with approx. 12 people.	Sustainability reporting done by each pension funds in PKA.		Preparing for ESG.	ESG-reporting by a small Danish environmental unit.	
Processes: Sustain-ability / Circularity	Maintain, renovate and modernize to keep up building condition. Office hotels increase flexibility for customers. Customers demand sustainability certifications.	Sustainable og up-to-date property. Offer standard offices. Re-use, e.g. of kitchens. Modernize buildings to improve energy labels and sustainability certification of buildings	Sustainability strategy that focuses on reducing the CO_2 footprint in buildings and construction projects and circularity of materials for new construction.	Invest to hold. Office hotels increase flexibility for customers and their possibilities to optimize space.	Modernize buildings to improve energy labels. Change from replacing fridges to repair if possible.	Sustainability requirements are being implemented in sales conditions. Climate and environment is one out of five areas in CSR-policy.

Tabel 5: Interpretation of Sustainability / Circularity processes

Company	Sustainability/Circularity processes included in last row of Table 4	Interpretation of actions
NS: Newsec	Maintain, renovate and modernize Office hotels increase flexibility for customers Customers demand sustainability certifications	Prolongs lifetime of buildings Increases intensity of use and reduces the need for space Ensures sustainable quality and potentially prolong lifetime
PE: PKA Ejendomme	Focus on sustainable and up-to-date property Offer standard offices Re-use, e.g. of kitchens Modernize buildings to improve energy labels and sustainability certification of buildings	Ensures sustainable quality and potentially prolongs lifetime Reduces need for bespoke adaptations of buildings and demand for building materials Prolongs lifetime of components Reduces energy and CO_2 as well as potentially prolong lifetime
AE: ATP Ejendomme	Sustainability strategy that focuses on: • Reducing the CO_2 footprint in buildings and construction projects • Circularity of materials for new construction.	Reduces CO_2 Prolongs lifetime of building materials
EØ: Østervold	Invest in buildings to hold Office hotels increase flexibility for customers and their possibilities to optimize space	Potentially prolongs lifetime of buildings Increases intensity of use and reduces the need for space
HS: Heimstaden	Modernize buildings to improve energy labels. Change from replacing fridges to repair if possible	Reduces energy and CO_2 as well as potentially prolong lifetime Potentially prolongs lifetime of equipment
FE: Freja Ejendomme	Sustainability requirements are being implemented in sales condition Climate and environments is one out of five areas in CSR-policy	Reduces energy and CO_2 as well as potentially prolong lifetime Reduces energy and CO_2

https://doi.org/10.52202/080684-0016

5. DISCUSSION

5.1 SUSTAINABILITY AND CIRCULAR PROCESSES

The sustainability and circular processes in Table 5 shows that the following processes are used in typical situations.

For operation of existing buildings:

- Maintain, renovate and modernize
- Focus on up-to date property
- Repair rather than replacing e.g. fridges
- Invest in buildings to hold
- Sustainability certification

For development of existing buildings

- Establish office hotels
- Reuse of e.g. kitchens
- Modernize to improve energy label
- Transform from offices to housing

For transactions:

- Offer standard offices
- Sustainability requirement in sales conditions

For new building projects:

- Circularity of materials
- Sustainability certifications

The effects of the sustainability and circularity actions are:

- Reduce energy consumption and CO_2 emissions
- Prolong the lifetime of buildings
- Increase intensity of use
- Reduce the need for space
- Ensure sustainable quality

5.2 CIRCULAR ACTIONS AND ISO 59004 PRINCIPLES

The circular actions in the use phase included in the Value Building, see Figure 5, and in Jensen & Nielsen (2024) are:

- Transactions
- Sharing
- Space optimization
- Maintenance and minor repairs
- Refurbishment
- Adaptive reuse

These identified circular actions have a close relation to the ISO 59004 defined principles. They do however not correspond 1:1 to the ISO 59004 defined principles but have significant elements of overlapping. Table 6 shows a comparison of the six circular actions and the ISO 59004 principles.

Table 6: Comparison of circular actions and ISO 59004 principles

Circular Actions	ISO 59004 principles
Transactions	System thinking
Space optimization	Value creation
Sharing	Value sharing
Maintenance and minor repairs	Resource management
Refurbishment	Resource traceability
Adaptive reuse	Ecosystem resilience

As an example, *Transaction* can be described as an action that relates to *System thinking*, *value creation* and *resource management*. However, *Transaction* can be described as less related to *Value Sharing*. In the other direction the *System Thinking* and *Resource traceability* principles are relevant to all circular actions whereas the remaining principles have a varying degree of relationships to the Circular actions. It is therefore important to distinguish between the circular actions and the ISO 59004 defined principles.

5.3 CIRCULAR ACTIONS IN RELATION TO FINDINGS

Renting-out is a basic aspect of a **transaction**-based BM applied by RECs. Transactions are not circular as such, but can potentially include circular actions, which both can ensure continuous use and long lifetime due to the economic incentives. The BM gives clear incentives to prolong the economic lifetime of buildings and can secure the necessary funds to keep up the technical lifetime and the functional lifetime.

Renting-out can also be seen as a part of **sequential sharing** of space and thereby secure a basic intensity of use. It also includes **maintenance and minor repairs** as a basic circular action part of professional operation management. The BM of RECs also includes incentives to **refurbishment** and renovation to improve energy labels and modernize to increase rent. These actions improve buildings' technical lifetime. Changing market conditions leading to redundancies also gives incentives to **adaptive reuse**, for instance from office to housing, and providing elements of **simultaneous sharing** in office hotels and co-working spaces. Thereby the functional lifetime of buildings can be prolonged.

The serviced-office model includes sharing as an aspect, which provides more **intensive use of space**.

Whether a customer/user-focused model leads to more sustainability or circularity very much depends on the investors and the tenants. Newsec, who serve various investors, experience that particularly some foreign investors set high sustainability requirements, and they also experience a demand for sustainability certifications from some investors. They do not experience that most potential or actual tenants have high demands for sustainability aspects except for energy consumption and cost.

6. CONCLUSION

The study shows that the focus on sustainability and the business models among RECs varies to a high degree. The public investor and developer company Freja Ejendomme is probably not a typical representative for investors/developers by having a high focus on sustainability and not being purely profit-driven. In general, RECs lack incentives to put strong efforts on sustainability and circularity. Public regulation is the main driver.

State owned and ethical driven investors and administrator companies owned by pension funds have clear strategies towards being more sustainable. A new administrator company PKA Ejendomme has recently been established by a pension fund particularly to meet the increased sustainability challenges with a strong focus on circularity. Among administrators there is a strong trend towards being more customer focused.

Investors can diversify their investments away from growth in building volume towards a circular business model that addresses an increased sustainable market demand. When they do build new, they can develop as sustainably as possible as shown for instance by the public pension fund ATP and their administrator ATP Ejendomme. They can also base their business model on securing a long lifetime of buildings by proper maintenance and by refurbishing buildings they own or buy and adapting them to changing market needs, e.g. by changing offices to housing and creating more shared facilities.

The focus on circularity in the building industry is mostly to reuse building materials. It is important to change the focus towards sustaining whole buildings and preserving the embedded CO_2. The Value Building includes relocation as a circular action in the after-use phase. This involves moving whole buildings to new locations, where they are needed and/or more secure. Rising sea levels and increase in monster-rain, earth-slides and flooding will force some buildings to be either demolished – accidentally or intentionally - or moved. Moving buildings is a well-known practice, for instance of historical buildings by open air museum operators, but it needs to be developed into a more general practice for buildings in use and considered, when designing new buildings.

The paper adds to the growing literature on circular buildings with particular focus on management of the use and operational phase. Most research on circular buildings has until now mostly focused on the design phase. The paper can give RECs inspiration to develop circular business models.

The study focuses on a limited number of Danish RECs but many of them operate in Northern Europe and for international investors, so the results have a certain degree of generality and particular the circular activities and the identified models are regarded as general.

ACKNOWLEDGEMENTS

The study has been made without any external funding. The authors thank the interviewees for participating in the survey.

REFERENCES

Ástmarsson, B., Jensen, P.A. and Maslesa, E. (2013). Sustainable Renovation of Residential Buildings and the Landlord/Tenant Dilemma. *Energy Policy*, Vol. 63, pp. 355-362.

Baniya, B. (2023). Circularity in Facility Management: Conceptualisation and Potential Areas for Circularity-Oriented Actions. *Sustainability*, 15, 8460. https://doi.org/10.3390/su15118460

Berg, J. B., Thuesen, C., Ernstsen, S. K. and Jensen, P. A (2021). Reconfiguring the construction value chain: Analysing key sources of friction in the business model archetypes of AEC companies in strategic partnerships. *Construction Management and Economics*, Volume 39, Issue 6.

Bocken, N. M., De Pauw, I., Bakker, C. and Van Der Grinten, B. (2016). Product design and business model strategies for a circular economy. *Journal of industrial and production engineering,* 33(5), 308-320.

Bocken, N. M. P., Stahel, W., Dobrauz, G., Koumbarakis, A., Obst, M. and Matzdorf, P. (2021a) *Circularity as the new normal: Future fitting Swiss business strategies*. Maastricht Sustainability Institute.

Bocken, N., Kraaijenhagen, C., Konietzko, J. and Schuit, C. (2021b). *Experimenting with new business model strategies for the circular economy.* Chapter in: Experimenting with new business model strategies for the circular economy. Elgar Publishing, DOI: 10.4337/9781800037309

Christensen, C.M., Barman, T. and van Bever, D. (2016), The hard truth about business model innovation", *MIT Sloan Management Review,* Vol. 58 No. 11, pp. 30-30..

EEA (2023): *Accelerating the circular economy in Europe - State and outlook 2024.* EEA Report 13/2023. European Environment Agency. https://www.eea.europa.eu/publications/accelerating-the-circular-economy

Estarrona, U. M. de, Seneviratne, D., Villarejo, R. and Galar, D. (2019). The New Asset Management: Implications of Servitization in Circular Economy. *Journal of Industrial Engineering and Management Science*, 2018(1), 109–120. https://doi.org/10.13052/jiems2446-1822.2018.006

 https://doi.org/10.52202/080684-0016

ISO 59004 (2024). *Circular economy — Vocabulary, principles and guidance for implementation*. International Organization for Standardization.

Jensen, P.A. and Nielsen, H.J. (2024). Exploring Circular Facilities Management: A Case of Portfolio, Building, and Workplace Transformation. *Proceedings of the 23rd EuroFM Research Symposium*, 10-11 June 2024, London, UK.

Jeppesen, H.D. (2024). *Nu går vi tættere på ejendommene...* LinkedIn, 18. April 2024.

Konietzko, J., Bocken, N. and Hultink, E. J. (2020). A Tool to Analyze, Ideate and Develop Circular Innovation Ecosystems. *Sustainability*, 12(1), 417.

Kyrö, R. (2020). How can the real estate sector transition to circular economy? Approaches for Real Estate Management. *The 19th EuroFM Research Symposium, EFMIC 2020*, 3.-4. June 2020 (pp. 15-24).

Kyrö, R. and Lundgren, R. (2023). The value building – a conceptual model of circular business models in the building context. *The 22nd EuroFM Research Symposium Research papers*, 09-11 November 2023, Istanbul, Türkiye.

Larsen, M.M (2010): Den private udlejningssektor nu og i fremtiden. Chapter 10 in: Den private ejendomdsudlejnings betydning for samfundet, Ejendomsforeninger Danmarks 150 års jubilæum.

Leikvam, G. and Olsson, N. (2022): Eiendomsutvikling. 2. edition, SBN: 9788245040098. Fakbokforlaget.

Lundgren, R., Kyrö, R. and Olander, S. (2023). The lifecycle impact and value capture of circular business models in the built environment, *Construction Management and Economics*, DOI: 10.1080/01446193.2023.2279743

Liu, Y. (2024). The Impact of ESG Performance on Investors' Decision-Making in the Real Estate Industry: Based on Green Building Certification and Facility Management. *Highlights in Business, Economics and Management*, Volume 45, pp. 124-129.

Munaro, M.R., Tavares, S.F. and Bragança, L. (2020). Towards circular and more sustainable buildings: A systematic literature review on the circular economy in the built environment. *Journal of cleaner production*, 260, 121134.

Rasmussen, H.L. and Bøytler, T. (2022): *Vedligehold og klimabelastning – en undersøgelse af praksis i boligudlejningsejendomme.* Realdania.

Rasmussen, H.L, Jensen, P.A., Bøytler, T. and Nielsen, S.B. (2023): Maintenance and Climate Impact – A Study of the Practice for Private Rented Out Housing Estates. *CIB W070 Conference on Facilities Management and Maintenance, 08th-12th May 2023, NTNU Trondheim. IOP Conf. Series: Earth and Environmental Science 1176 (2023),* IOP Publishing doi:10.1088/1755-1315/1176/

PRODUCT-SERVICE-SYSTEM (PSS) IN TAIWAN'S PUBLIC HOUSING PROJECTS- DEVELOPMENT, BARRIERS, AND FACILITATORS

Yun-Tsui Chang[1*] and Shang-Hsien Hsieh[1]

[1]Department of Civil Engineering, National Taiwan University, Taiwan

Corresponding Author: d05521009@ntu.edu.tw

ABSTRACT
Background and aim. Since 2018, various public sectors in Taiwan have started introducing the "circular economy" concept and implementing new business models, such as Product-Service-System (PSS), into their new housing projects. After several years of construction and operation, the feasibility of this new model has not yet been explored. This study aims to disclose whether the implemented PSS in these new public housing projects has reached its original goals of enhancing building circularity (e.g., extending products' lifespan, remanufacturing products) and long-term profit.

Methods and Data. In this preliminary study, data related to the original plans and actual performances of the implemented PSS is collected via interviews with project owners of these pilot public housing projects. Discussions on the potential reasons behind its success/ failure and suggestions to other project owners who plan to adopt a similar model are also made in the interviews.

Findings. The study identified several barriers to the success of the PSS model in Taiwan's public housing projects, including contract ambiguities, inappropriate bidding methods, and miscoordination during construction and installation. In the meantime, potential facilitators are also identified, including a more mature PSS ecosystem, supportive governance systems, coordinated management schemes, and increased resident sustainability awareness.

Theoretical / Practical / Societal implications. PSS for building operations in Taiwan's public housing projects is a pioneering experiment. Their experiences provide valuable insights for other Taiwanese projects and guide countries that newly adopt PSS in the building industry, helping them transition towards a more sustainable, circular built environment.

KEYWORDS: Circular Business Model, Product-Service-System (PSS), Product-as-a-Service (PaaS), Public Housing

1 INTRODUCTION

Despite recent advances in building energy efficiency and urban liveability, the built environment remains largely based on a linear model, where materials are extracted, used, and discarded as waste (Arup, 2022). This model generates substantial structural waste and positions the built environment as one of the largest consumers of raw materials, as well as a significant source of waste and carbon emissions. For instance, buildings account for around 50% of resource extraction and consumption in the EU, 30% of its annual waste generation, 40% of energy consumption, and 36% of energy-related greenhouse gas emissions (EU, 2022). Moreover, urbanization will rise from 55% to over 66% of the global population by 2050, doubling the size of the built environment and straining urban systems like water, energy, and waste networks (UN, 2018).

To address the environmental problems stated above, governments, academia, and practitioners worldwide aim to transform the building and construction industries from their linear model into a circular one (Guerra et al., 2021). Many tried to identify suitable strategies and approaches for their building projects (Tseng et al., 2021). However, implementing a circular economy in the building industry is more challenging than others because of the customized nature of a building and the complex compositions and distinct lifespans of different systems within (Pomponi & Moncaster, 2017). Another key research question to be answered is how new business models can foster a circular economy in the building industry (Munaro et al., 2021). Among them, Product-Service-System (PSS) is renowned as one of the most powerful tools for transitioning society to a resource-efficient, circular economy (Tukker, 2015). A more general definition of PSS is "a mix of tangible products and intangible services designed and combined

https://doi.org/10.52202/080684-0017

so that they are jointly capable of fulfilling final customer needs" (Tukker and Tischner, 2006).
Following the international trend, Taiwan's central and local governments started embracing the concepts of circular economy, nurturing several public housings as pilot projects since 2018 (Chang & Hsieh, 2019). Besides adopting new design approaches, building materials, and construction methods, new business models such as PSS are also introduced in these projects. The PSS model applies a vast range of building services, including air-conditioning, lighting, appliances, sanitary fixtures, furniture, and elevators. After operating for several years, how this newly implemented model performs requires further investigation. Meanwhile, existing research on the integration of PSS in circular housing remains fragmented, with a focus on technical challenges and business models, yet lacking empirical studies tailored to specific housing contexts (Ghafoor et al., 2023, 2024). Moreover, there is insufficient understanding of the role of project owners in the adoption and implementation of PSS, highlighting the need for more comprehensive, context-specific insights.

This study aims to disclose whether the implemented PSS in these public housing projects has reached its original goals of enhancing building circularity (e.g., extending products' lifespan, reusing or remanufacturing products) and long-term profit. In this preliminary study, data related to the original plans and actual performances of these PSS is collected via interviews with project owners of these pilot projects. Discussion on the potential reasons behind its success/ failure and suggestions to other project owners who plan to adopt a similar model are also made in the interviews.

2 LITERATURE REVIEW

Housings account for a majority of global building stocks, pushing demand for natural resources, leading to severe environmental impacts (Zhang et al., 2024). Its shift to a Circular Economy (CE) offers a solution, yet most efforts focus on technical challenges related to building lifespan and complexity, neglecting housing-specific issues. Research highlights the need for new business models to facilitate circularity, with the Product-Service System (PSS) emerging as a promising approach. However, its role in circular housing remains poorly understood, with fragmented literature across multiple fields.

Ghafoor et al. (2023) employ an integrative review to explore the relationship between PSS and CE in housing, proposing a conceptual framework that positions PSS as a life cycle strategy to enhance efficiency, longevity, and sufficiency in energy, material, and space use in housing. It also examines the economic and social value of PSS, along with its potential impact on current housing industry structures and the transition towards PSS-based practices. Finally, the paper identifies gaps in existing research and outlines directions for future study, practice, and policy development.

Their proposed framework provides a good knowledge base to identify what types of PSS are adopted in our study cases and what benefits should be examined. Moreover, one of the key directions they have proposed for future study is the systemic analyses mapping the transition from traditional models to PSS, particularly regarding its influence on roles, relationship dynamics, and power associations. For instance, inspecting the impact of regulatory, financing, and contractual mechanisms is suggested to evaluate their attractiveness and identify potential diffusion ways (Britton et al., 2021). Empirical studies tailored to specific housing contexts would help develop contextualized insights, moving beyond generic recommendations to account for the unique characteristics of various housing categories. Our case study of Taiwan's public housing adopting PSS can fill this research gap by delivering valuable, context-specific insights into these mechanisms' practical application and outcomes in a unique housing setting.

To bridge the gap mentioned above, Ghafoor et al. (2024) conducted a multiple-case analysis study, examining five leading practitioners through interviews and document analysis. The research resulted in an empirically grounded framework of 14 guiding principles, offering actionable insights for PSS in circular housing. These principles adopt key business model aspects, including value proposition, value creation & delivery, and value capture. The study also highlights the pivotal role of intermediary-led collaborative value networks in hastening PSS and CE adoption in housing.

While Ghafoor et al. (2024) developed their study based on interviews with practitioners from service-providing companies involved in PSS, our research expands the scope by offering insights directly from project owners. This view is crucial as project owners play a significant role in the decision-making related to adopting and implementing PSS in housing projects. By engaging with project owners, our study provides knowledge on the challenges, priorities, and opportunities they face when integrating circular economy principles. This approach bridges the gap between the service providers and those responsible for executing these projects, ensuring a more comprehensive framework for PSS in the housing sector.

Meanwhile, Azcarate-Aguerre et al. (2022) provided a comprehensive exploration of PSS in the building sector, focusing on facades-as-a-service (FaaS) through multi-stakeholder collaboration and pilot projects in the Netherlands. Their work highlights the role of advanced technical integration, asset tracking, and performance-based contracts in enabling circularity. However, such studies predominantly focus on technologically mature, supplier-driven models. In contrast, this study addresses the adoption of PSS in Taiwan's public housing sector, where governance constraints, procurement rigidity, and resident behaviors emerge as key barriers.

3 METHOD

In this preliminary study, data related to the original plans and actual performances of the implemented PSS is collected via interviews with project owners of these pilot public housing projects. Discussions on the potential reasons behind its success/ failure and suggestions to other project owners who plan to adopt a similar model are also made in the interviews.

3.1 STUDY CASE AND INTERVIEWEE SELECTION

Two public housing projects in Taiwan provide the most diverse services via the PSS model, i.e., Taisugar Circular Village (TCV) in Tainan and Bade No.3 Social Housing (B3SH) in Taoyuan. More information about these cases is shown in Tables 1 & 2. More information about the two interviewees is shown in Table 3, who represent project owners' viewpoints, providing valuable data and insights. It is important to note that the sample size is limited to two projects and two interviewees, which may not fully represent the broader spectrum of experiences across all public housing projects in Taiwan. The influence of this limitation on the results should be considered when interpreting the study's conclusions.

Table 1: Basic information on the selected study cases

info	TCV	B3SH
picture		
location	Tainan	Taoyuan
built year	2021	2023
housing unit	351	524
PSS type	use-oriented (Tukker, 2004)	
PSS service	9 items	3 items
bidding type	the lowest bid	the most advantageous

Table 2: Basic information of the PSS in the selected study cases- their service provider type and service length

service	TCV	B3SH
air-conditioning	PW-10 years	PM-10 years
lighting	PW-10 years	
appliance	PW-10 years	
sanitary fixture	PW-10 years	
water heater		PM-12 years
furniture	PW-6 years	PM-12 years
mattress	PW-6 years	
smart door lock	PM-10 years	
waste disposer	PM-10 years	
elevator	PM-20 years	

*PM stands for product manufacturers, while PW stands for product wholesalers.

Table 3: Basic information of the interviewees

info	TCV	B3SH
position title	engineer in the Office of Land Development	chief engineer in the Office of Housing Development
role in project	composing PSS contracts, bidding, supervising service providers	
experience	10 years	30 years

3.2 INTERVIEW QUESTION DESIGN

Interviews were conducted by following the four main questions as shown below:

(1) What are the original motivations/ expectations for adopting the PSS model in your project?

(2) What are the actual performances/ outcomes of the adopted PSS in your project?

- Was the product lifespan extended after planned maintenance and repair? Was the product reused/ remanufactured after its end-of-life?
- Did your service provider profit? Did you save money by reducing cost of maintenance/ repair?

(3) What are the reasons behind the success/ failure of the adopted PSS in your project?

(4) What are your recommendations to other project owners who want to adopt the PSS model?

4 RESULT

4.1 ORIGINAL MOTIVATIONS/ EXPECTATIONS

The TCV interviewee responded that the influence of national policy was one of the key drives for them to adopt a new circular business model (i.e., PSS). Founded in 1946, Taiwan Sugar Corporation (TSC) is a state-run enterprise in Taiwan. It has gradually transformed from the "Sugar Based Production and Sales Business" into a diversified business entity that covers the agricultural, industrial, and commercial industries. Following the "Five Plus Two Industry Innovation Plan" released by the central government in 2018, TSC has incorporated the circular economy concept for innovative items such as new agriculture, pig farming modernization, and resource recycling (biogas energy and biomass material). TSC also invested in proprietary housing construction since 1986. TCV is the pilot project demonstrating their determination to incorporate circular economy principles into their land development to achieve a sustainable living environment.

In addition, the TCV interviewee answered that reducing future maintenance costs is the key expected benefit of the PSS model in their pilot project. Housing construction and rental are two of TSC's main businesses in the office of land development. Responding to the rising need for more social housing by the central government, TSC will soon construct and operate more public housing. Finding a

https://doi.org/10.52202/080684-0017

cost-saving, robust, and sustainable model is critical to the Office of Land Development.

As for B3SH, the interviewee responded that pressure from their competitors was one of the key drives for them to adopt the PSS model in their public housing projects. Since 2014, the Taipei City Government has started constructing many public housings to fulfill its new mayor's political promise (Chen & Rietdijk, 2025). Inspired by TCV, it also aimed to incorporate circular economy principles in its new public housing projects, including the PSS model (Tseng et al., 2021). Meanwhile, as the fastest-growing city in Taiwan, the Taoyuan City Government also urged the construction of many public housings for their citizens. Inspired by TCV and the Taipei City Government, it started adopting the PSS model in its public housing projects in 2022 to prove itself to be an innovative and green government.

In addition, the B3SH interviewee answered that relieving its financial pressure is the key expected benefit of the PSS model in their pilot project. As mentioned above, city governments in Taiwan started constructing many public housings. Nevertheless, this has become a giant financial burden to them. The PSS model enables them to split the cost of building services into numerous months and years instead of paying the full purchase fee in the construction phase at once.

4.2 ACTUAL PERFORMANCES/ OUTCOMES

4.2.1 Regarding Circularity

Regarding whether the implemented PSS enhances the building circularity (e.g., extending products' lifespan, reusing or remanufacturing products), responses from the two cases vary. The TCV interviewee said their service providers often refuse to repair the provided products, blaming that they were damaged during the construction/ installation, or report that the products were too damaged to repair, charging extra for replacement. The interviewee thinks this might be related to their bidding method being the lowest bid. Many of their service providers may have cut costs to win the bid, be inexperienced in PSS, regard PSS as another payment method, and not recognize the benefits that the circular economy can bring.

On the other hand, the B3SH interviewee said that their service providers maintain and repair their products well based on their contracts. This might be related to the fact that all their service providers are product manufacturers capable of gathering and storing sufficient components of their products for repair. Nevertheless, finding enough rooms for such storage becomes a new challenge for them.

4.2.2 Regarding Profit

According to the TCV interviewee, their service providers profit well since they barely fulfill their responsibility of maintaining and repairing the products they provide based on their contracts. In contrast, the TSC's management team at TCV had to maintain and repair those products, adding extra labor and financial burden to themselves.

On the other hand, according to the B3SH interviewee, their product providers have profited limitedly since many residents in public housings use their products wrongly. Moreover, the air-conditioning provider mainly produces their product overseas, and their profit has shrunk largely because of the large difference in exchange rates. They have been reluctant to extend this PSS model to more social housing projects.

4.3 ENCOUNTERED BARRIERS

4.3.1 Regarding Planning

According to its interviewee, one of the key barriers that TCV encountered was the ambiguity in their PSS contract. For instance, whether maintaining the air-conditioners includes cleaning their filters was poorly defined. The contracts did not include what measures could be taken to resolve the different views on the fixability of products and accountability of damages.

The TCV interviewee shared that this barrier mainly results from their lack of experience and knowledge of PSS. As mentioned above, TCV was the first pilot project in Taiwan to implement PSS for various building services. Very limited information was available for the project owner team to compose a well-defined contract. Moreover, the interviewee added that PSS for building services is a new business model for the building industry in Taiwan. Limited companies knew about it and were willing to adopt it when their project started. Hence, they were concerned that a well-defined contract might reduce the number of companies who want to join their bidding.

4.3.2 Regarding Bidding

Interviewees say unsuitable bidding methods are a key barrier to the PSS model's success. As shown in Table 1, TCV and B3SH have taken different bidding means: the lowest bid vs. the most advantageous. As mentioned earlier, the TCV interviewee thinks that the lowest bid might result in inexperienced service providers or those who simply cut costs to win the bid, providing low-quality services and failing to achieve the goals of enhancing building circularity. The TCV project owner team once considered adopting a different bidding method, but their superior agency did not approve it. In the meantime, the B3SH project owner team used the most advantageous method, which ensures that the selected suppliers have the best qualifications and experience to deliver a successful project. Their PSS performances also turn out to be more satisfactory.

Besides the unfitting bidding methods, another key barrier related to the PSS bidding process is the complexity of setting bidding budgets for different building services. The TCV interviewee elaborated that different building services own very different maintenance and repair ways. For instance, repairing lighting fixtures usually involves merely the replacement of lamps, requiring the least effort and costs. On the other hand, maintaining and repairing air-conditioners are more difficult and involve more building interfaces (e.g., between their piping and interior finishing). Furthermore, like appliances, the new models of air conditioners are released quickly and with higher efficiency, increasing the difficulty of repairing the old-modeled air conditioners due to the limited stock of old components, thus increasing the budget for repair and replacement. The TCV interviewee added that it was very challenging to provide an appropriate bidding budget because they lacked knowledge and experience in providing these building services.

4.3.3 Regarding Construction

Another key barrier that both interviewees pointed out is the miscoordination during the building construction/ service product installation phase. The TCV interviewee shared that the installer of these products was their contractor instead of the service provider. Sometimes, the contractor installs those products according to their custom, which differs from the service providers. Furthermore, miscoordination occurred in their late construction phase, i.e., some air-conditioners were misplaced outdoors under poor conditions, which might have caused damage to these conditioners. All these situations lead to difficulty in clarifying the accountability of the product's damages.

4.3.4 Regarding Operation

Another key barrier that both interviewees mentioned is the misbehaviors of the residents. Many residents in these circular housings do not identify with the importance of sustainability. Many regard these public housings as "a temporary dwelling place" rather than their "home." These narratives result in misbehaviors towards the items provided in their dwelling units, e.g., leaving the windows open while running air conditioning or tampering with thermostats, leading to energy inefficiency and increased wear on mechanical components.

4.4 POTENTIAL FACILITATORS

4.4.1 More Mature PSS Ecosystems

From the TCV interviewee's point of view, they have countered many barriers because the PSS model for building services in Taiwan is at its beginning phase. No companies in Taiwan have ever provided those services via PSS. Very few companies know about this model and are willing to adopt it. This might be related to the limited number of rental housing management firms in Taiwan, leading to the low demand for PSS. As a result, the TCV project owner team has very limited information to refer to for their PSS contracts and bidding documents, and they have a very limited number of bidders to choose from.

Following the rising number of public housings and rental housing management firms in Taiwan, the demand for PSS is likely to increase along with the growth in the supply of better building services via PSS, fastening the maturity of the PSS ecosystem in Taiwan. Once more knowledgeable and experienced service providers appear in the market, they can offer more diverse and comprehensive services for project owners to choose from according to their demands, fostering healthy competition. More available information will also enable project owners to make better contracts and bidding documents.

4.4.2 More Supportive Governance Systems

As mentioned earlier, an inappropriate bidding method is one of the key barriers to the success of the PSS model. The TCV interviewee said they had to adopt the lowest bid instead of the most advantageous due to the limitation of their superior agency and the Government Procurement Act (GPA) in Taiwan. Pure financial leasing procurement lacks sufficient heterogeneity, and if the amount is not large enough, it is difficult to adopt the most advantageous bid under GPA. Furthermore, the value of leased goods is usually greater than the proportion of services, making it difficult to outsource using a service model. Therefore, he suggested that the current regulation and mindset of the superior need to change to align with the new concepts of circular economy.

As mentioned in the previous section, increasing demand for PSS is critical to fasten the maturity of the PSS ecosystem in Taiwan. Both interviewees advised that the government should propose encouragement or incentive mechanisms to increase the willingness of agencies to allocate relevant budgets, thereby expanding the market. For instance, they could require a certain proportion of leasing procurement based on the project cost amount.

4.4.3 More Coordinated Management Schemes

In order to avoid miscoordination during the building construction/ service product installation phase, the TCV interviewee suggested that the product installer should be the service provider instead of the contractor. This can reduce the risk of future disputes about the accountability of products' damages during installation. Nevertheless, this will raise challenges in the project design and construction phase since there are more stakeholders with which to coordinate. The TCV interviewee advised that a Professional Construction Management (PCM) team can facilitate stakeholder communication and ensure better coordination during the construction phase.

4.4.4 More Sustainable Behaviours & Partnerships

Both interviewees agreed that more measures should be taken to help residents identify and adopt a more sustainable lifestyle. Many public housings in Taiwan are certified green and smart buildings. Besides offering their inhabitants a secure place to live, they also provide chances for them to experience a new way of living. These housings have adopted many schemes to enhance community cohesion, e.g., the Youth Innovation Project (YIP) (Yu et al., 2023). The schemes that can be adopted to enhance the residents' awareness of sustainability should be explored.

https://doi.org/10.52202/080684-0017

The B3SH interviewee added that the PSS model that B3SH adopted is business-to-business (B2B) based, i.e., the PSS contracts are signed between the city government and service-provider companies. This model ensures these service providers obtain better loans from banks. However, the frequent and enormous need for product maintenance and repair become a great challenge to these service providers. Some residents have complained that they wish more choices for these services. The B3SH interviewee shared that some of their new public housing projects later adopted a new PSS model, Business-to-Customer (B2C) based, i.e., their contracts are signed between the residents and the service provider. Whether this new model is more successful in terms of its sustainability or financial performance is to be explored.

5 DISCUSSION

This study provides valuable empirical insights into the implementation of PSS within Taiwan's public housing sector, offering a unique perspective from project owners. This contrasts with the broader, more conceptual work of Ghafoor et al. (2023), whose integrative literature review outlines a comprehensive framework for PSS in a circular economy for housing but lacks specific contextual details. While Ghafoor et al. (2024) offer actionable guidance through a multiple-case study approach, our findings complement their work by a localized understanding of the barriers and facilitators at play, specifically within the Taiwanese governance context. Moreover, our focus on project owners addresses a gap between the perspectives of service providers, which are a focus of Ghafoor et al. (2024).

One notable area of convergence across past studies is the identification of barriers to PSS implementation. While this study emphasizes governance-related challenges specific to Taiwan's context, such as contract ambiguities and miscoordination during construction and installation, the research of Ghafoor et al. (2023, 2024) highlights more systemic gaps, including regulatory mechanisms and the need for well-defined intermediary roles to accelerate PSS adoption. Moreover, this study identifies facilitators that resonate with the literature, including supportive governance and collaborative frameworks. However, it distinguishes itself through its emphasis on the maturity of the PSS ecosystem and the heightened sustainability awareness of residents as key enablers. While Azcarate-Aguerre et al. (2022) highlight the significance of technical integration, asset tracking, and performance-based contracts, our research underscores the importance of broader systemic conditions in enabling the successful deployment of PSS in public housing.

6 CONCLUSION

This study explores the implementation of the product-service system (PSS) to foster a circular economy in Taiwan's public housing projects. Data was collected via interviews with project owners of two pilot projects. These interviews have provided valuable insights into the original motivations and expectations for adopting the PSS model, the actual performances and outcomes, the challenges faced during implementation, and the potential facilitators to overcome these barriers.

Our findings suggest that while the PSS model offers significant promise for enhancing building circularity, challenges persist, particularly in contract clarity and the operationalization of maintenance and repair services. While TCV struggled with its service providers, B3SH benefitted from more successful service execution due to a more qualified supplier selection process (see Figure 1).

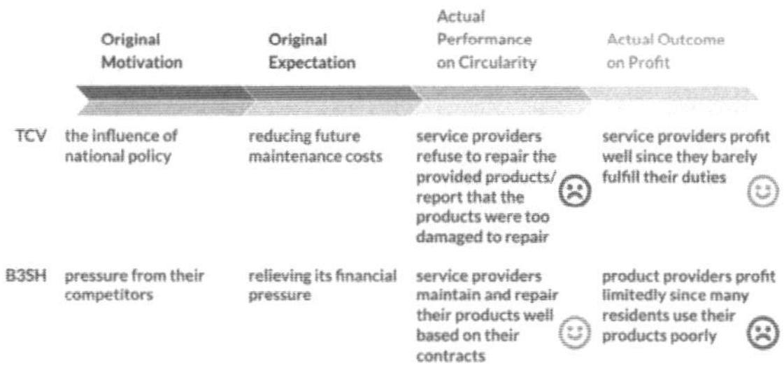

Figure 1: Summary of different motivations/ expectations and actual outcomes on circularity/ profit of two study cases

The study identified several barriers to implementing the PSS model in the two public housing projects (see Figure 2). These included ambiguities in the PSS contracts, particularly regarding the responsibilities for product maintenance and repair, which led to disputes and confusion. Second, the inappropriate bidding methods also hindered the success of the model, with TCV's use of the lowest bid resulting in poor service quality, while B3SH's more advantageous bidding process proved more effective. Additionally, issues such as miscoordination during installation and residents' misbehaviors further complicated the operation of the PSS model.

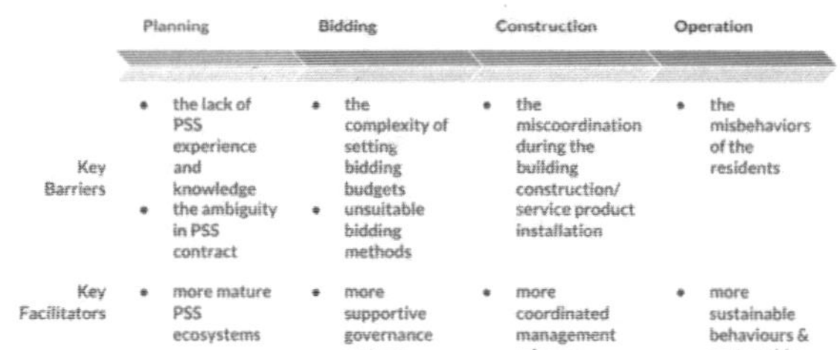

Figure 2: Summary of key barriers and facilitators

The study also identified several potential facilitators that could enhance the effectiveness of the PSS model in future public housing projects (also see Figure 2). Key facilitators include the maturation of the PSS ecosystem in Taiwan, driven by increased demand for public and rental housing and the supply of more experienced service providers, improving service availability, and fostering healthy competition. Additionally, supportive governance systems could facilitate PSS adoption, including policy adjustments to encourage more advantageous bidding methods and financial incentives. Coordinated management schemes involving service providers directly in installation could reduce miscoordination and disputes. Finally, fostering sustainable behaviors among residents and exploring new stakeholder partnership models could improve the PSS model's overall success in achieving its sustainability and financial goals.

This study offers academic implications by filling gaps in existing research on the integration of PSS in circular housing. While much of the current literature focuses on technical challenges and business models, it remains fragmented and lacks empirical studies tailored to specific housing contexts. This paper contributes new, context-specific insights into the adoption and implementation of PSS in Taiwan's public housing sector, particularly on the role of project owners, which has been insufficiently explored in prior research. On the non-academic side, the findings provide actionable recommendations for industry practitioners and policymakers. The study underscores the importance of clearer governance frameworks and informed decision-making in adopting circular business models, offering guidance for scaling the adoption of sustainable practices in public housing and facilitating broader transitions towards a circular economy in the building industry.

Our future research will further expand the understanding of the implemented PSS by incorporating interviews with service providers and occupants to gain a more holistic view of the system's performance. Engaging with service providers will illuminate the challenges of delivering services under PSS contracts. At the same time, feedback from occupants can offer insights into the user experience and sustainable behavior. Additionally, exploring new contract models, such as Business-to-Customer (B2C), and the dynamics between new stakeholders in this model could offer valuable insights into its potential for enhancing sustainability and financial performance. Another key area for further study could also include examining how circularity can be achieved through the actions of product wholesalers, particularly in relation to the value derived from materials within default products. While the connection to circularity is more apparent with product manufacturers, understanding how wholesalers perceive and leverage materials as a source of value is crucial. Finally, developing a maturity assessment framework for project owner organizations will be valuable in evaluating the readiness of organizations to implement and scale PSS models effectively. This framework could guide organizations in assessing their capabilities and aligning resources to ensure the success of circular economy initiatives in housing projects.

ACKNOWLEDGEMENTS

We greatly thank Mr. Su from Taiwan Sugar Corporation and Mr. Chen from Taoyuan City Government for their time and valuable insights, which have greatly contributed to our research.

REFERENCES

Arup (2022). From principles to practices: first steps towards a circular built environment.

Azcarate-Aguerre, J. F., Klein, T., Konstantinou, T., & Veerman, M. (2022). Facades-as-a-Service: the role of technology in the circular servitisation of the building envelope. *Applied Sciences*, *12*(3), 1267.

Britton, J., Minas, A. M., Marques, A. C., & Pourmirza, Z. (2021). Exploring the potential of heat as a service in decarbonization: Evidence needs and research gaps. Energy Sources, Part B: Economics, Planning, and Policy, 16 (11-12), 999-1015.

Chang, Y. T., & Hsieh, S. H. (2020). A review of Building Information Modeling research for green building design through building performance analysis. Journal of Information Technology in Construction, 25.

Chen, Y. L., & Rietdijk, C. E. (2025). FROM GLOBALIZING TAIPEI TO LEARNING AMSTERDAM: Referencing as a Politicizing Strategy for Urban Development in Taiwan. International Journal of Urban and Regional Research.

European Commission (EU) (2022). Green Deal: New Proposals to Make Sustainable Products the Norm and Boost Europe's Resource Independence. Press release.

Ghafoor, S., Hosseini, M. R., Kocaturk, T., Weiss, M., & Barnett, M. (2023). The product-service system approach for housing in a circular economy: An integrative literature review. Journal of Cleaner Production, 403, 136845.

Ghafoor, S., Kocaturk, T., Hosseini, M. R., Weiss, M., & Barnett, M. (2024). How to deploy the PSS towards a circular economy in housing? A multiple-case study. Journal of Cleaner Production, 477, 143821.

Guerra, B. C., Shahi, S., Mollaei, A., Skaf, N., Weber, O., Leite, F., & Haas, C. (2021). Circular economy applications in the construction industry: A global scan of trends and opportunities. Journal of Cleaner Production, 324, 129125.

Munaro, M. R., Freitas, M. D. C. D., Tavares, S. F., & Bragança, L. (2021). Circular business models: Current state and framework to achieve sustainable buildings. Journal of Construction Engineering and Management, 147(12), 04021164.

Pomponi, F., & Moncaster, A. (2017). Circular economy for the built environment: A research framework. Journal of Cleaner Production, 143, 710-718.

Tserng, H. P., Chou, C. M., & Chang, Y. T. (2021). The key strategies to implement circular economy in building projects—A case study of Taiwan. Sustainability, 13(2), 754.

https://doi.org/10.52202/080684-0017

Tukker, A. (2004). Eight types of product–service system: eight ways to sustainability? Experiences from SusProNet. Business strategy and the environment, 13(4), 246-260.

Tukker, A. (2015). Product services for a resource-efficient and circular economy–a review. *Journal of cleaner production*, *97*, 76-91.

Tukker, A., & Tischner, U. (2006). Product-services as a research field: past, present and future. Reflections from a decade of research. *Journal of cleaner production*, *14*(17), 1552-1556.

United Nations (UN) (2018). 2018 revision of world urbanization prospects. Population Division-United Nations.

Yu, H. C., Lin, T. H., & Dąbrowski, M. (2023). Beyond conditionality: Community placemaking in Taiwanese social housing management. Planning Practice & Research, 38(1), 43-61.

Zhang, S., Ma, M., Zhou, N., Yan, J., Feng, W., Yan, R., You, K., Zhang, J. & Ke, J. (2024). Estimation of global building stocks by 2070: Unlocking renovation potential. *Nexus*, *1*(3).

ASSESSING THE ECONOMIC BOUNDARY CONDITIONS FOR REUSING PRECAST CONCRETE ELEMENTS IN CONSTRUCTION

Olli Vigren[1*], Helena Westerlind[2], and Tove Malmqvist[3]

[1]KTH Royal Institute of Technology: Civil and Architectural Engineering; Stanford University: Center for Integrated Facility Engineering & Scandinavian Consortium for Organizational Research

[2]KTH Royal Institute of Technology: Architecture and Technology

[3] KTH Royal Institute of Technology: Sustainable Development, Environmental Science and Engineering

*Corresponding Author: olliky@kth.se

ABSTRACT

Background and aim. The reuse of Precast Concrete Elements (PCEs) offers a promising method to reduce emissions in construction. However, economic feasibility remains a significant barrier to widespread implementation. While technical challenges and value creation within supply chains have been explored, limited research addresses the economic aspects.

Methods and Data. This study constructs a supply chain model to compare standard demolition, PCE reuse, and construction with virgin materials. We investigate economic factors influencing building owners' decisions to donate or sell PCEs, building buyers' choices to use reclaimed materials, and the profitability of individual actors and the overall supply chain. Using 54 data sources, we identify cost and profitability drivers and analyze key decisions through economic theory and cost management perspectives.

Findings. Building owners have strong incentives to donate or sell PCEs for reuse, while buyers' decisions are highly context-dependent. Key costs in PCE reuse include deconstruction, refurbishment, storage, and transportation, while cost reduction drivers stem from savings on landfill fees, material costs and production costs. Long-term profitability depends on economies of scale, new markets, and innovation.

Implications. Investments can already focus on the most promising opportunities, but further research on cost structures, regulatory impacts, technological innovations, and supply chain dynamics is essential to guide decisions. Economies of scale, learning curves, and technological advancements offer significant potential to improve economic feasibility.

KEYWORDS: Circular Economy, Construction costs, Economic feasibility, Finance, Investment, Precast concrete element, Sustainability.

1 INTRODUCTION

Precast Concrete Elements (PCEs) are structural components of a building that are manufactured off-site and then transported to the construction site for assembly. Common PCEs in the building stock include components such as beams, columns, wall panels, and slabs.

Reuse of PCEs involves salvaging concrete elements from buildings condemned for demolition and reassembling them in new construction projects. This process may commonly involve intermediate storage and refurbishing and reconfiguring elements to meet new design requirements.

In recent years, industry, scholarly, and policy interest in reusing PCEs in construction has been on the rise. This interest is largely driven by concrete's significant contribution to global CO_2 emissions, estimated to be approximately 5–8% (Silfwerbrand, 2020), mostly due to

https://doi.org/10.52202/080684-0018

Portland cement manufacturing. Reuse is one alternative strategy for reducing these emissions in construction (Al-Najjar and Malmqvist, 2025), alongside other technologies, such as alternative cement binders (e.g., Gartner & Sui, 2018), carbon capture, and the use of alternative fuels. The reuse of PCEs should also be compared to alternative supply chains and material flows, such as the crushing and recycling of concrete elements for road construction, as well as the current practice of manufacturing conventional, carbon-intensive concrete elements. Such comparisons can promote understanding of the relative advantages of different strategies for reducing CO_2 emissions and other waste in the sector.

The implementation of alternative methods of manufacturing and building with concrete will ultimately rely on both *technical* and *economic* feasibility, which is largely determined by the pace of innovation. As with any complex production system, innovations can involve product innovation (e.g., concrete), process innovation (e.g., manufacturing and construction methods), and business model innovation (e.g., value creation and capture within supply chains).

Today, the pace of innovation is largely influenced by incentives established through legal frameworks. For reuse, the relevant legal frameworks are the European Climate Law and the EU's Fit for 55 package (European Union, n.d.), the EU Emissions Trading System (European Commission, n.d.), and various European Commission initiatives. Examples of these initiatives are the Construction Products Regulation, the Energy Performance in Buildings Directive, and the Transition Pathway for the Construction Ecosystem (Circular Economy Stakeholder Platform, n.d.), along with other legislation such as waste prevention laws. For example, a new emissions trading system, ETS2, has been introduced to cover emissions from buildings, road transport, and additional sectors (European Commission, n.d.). This system, which is set to become fully operational in 2027, complements ETS1 and other European Green Deal policies targeting these sectors. On top of this, there are also national legislation, regulation, and standards. Examples include the climate declaration act in Sweden (Regeringskansliet, 2021) and updates in the Danish building regulations (Social- og Boligministeriet, 2024).

It is expected that these regulations, when enforced, will incentivize innovation to reduce CO_2 emissions in the construction sector. Research should aim to estimate the impact of these regulations on the industry, although the combined effect of this legislative cocktail can be challenging to predict, particularly as emerging regulations are often subject to compromise.

This creates uncertainty, especially when estimating economic feasibility. From the industry's perspective, uncertainty hampers long-term investments, such as those in new PCE reuse technologies and capacity. Nevertheless, this legislation aims to incentivize such investments in the construction industry, making the status quo of manufacturing with conventional carbon-intensive concrete elements less competitive in the market. Therefore, it is reasonable to assume that economic feasibility will increasingly favour low-carbon technologies, such as the reuse of PCEs.

Yet, little is known about the economic feasibility of reusing PCEs. Previous research has primarily addressed the technical challenges of reusing PCEs (e.g., Dervishaj et al., 2023a; Dervishaj et al., 2023b; Räsänen & Lahdensivu, 2023; Suchorzewski et al., 2023), carbon saving potential (Al-Najjar and Malmqvist, 2025), and value creation within supply chains and ecosystems (e.g., Harala et al., 2023; Riuttala et al., 2024; Sairanen et al., 2024; Aarikka-Stenroos et al., 2021), contributing both to theoretical knowledge and the implementation of reuse practices in the sector (ReCreate Project, n.d.; Återhus Project, 2023).

Consequently, there is a research gap in understanding the economic feasibility of reusing PCEs. There is also a practice-driven need within the construction industry to address the uncertainty surrounding the economic feasibility of PCEs reuse, which hampers investments and broader implementation. Therefore, *the purpose of this study is to analyze the economic boundary conditions related to the reuse of PCEs.*

2 METHODOLOGY

2.1 APPROACH TO THE ANALYSIS OF ECONOMIC BOUNDARY CONDITIONS

The discipline of economic feasibility analysis sits at the intersection of engineering studies and business studies, particularly cost management. Economic feasibility generally means that a proposed solution is financially viable and cost-effective, ensuring the benefits outweigh the costs. Cost management as a discipline focuses on cost structures related to products and services, business operations, and supply chains (Kulmala et al., 2002; Paranko, 2012).

Ideally, cost analysis should be conducted for each actor in a supply chain, as well as for the supply chain as a whole. This would highlight the contributions of different actors to the supply chain and identify opportunities for optimizing the entire system (Eriksson et al., 2019; Vigren and Eriksson, 2025). It is generally understood that every company within the supply chain must be profitable in the long term for their business operations to continue.

Firstly, such analysis would require each company in the supply chain to be aware of its own costs (Agndal and Nilsson, 2009; Suomala et al., 2010) and to provide access to this information for analysts. This is rarely the case, as cost analysis demands significant effort within individual companies, and sharing detailed cost data with external parties beyond standard external reporting practices is quite uncommon (Suomala et al., 2010).

Secondly, in nascent supply chains, such as those focused on the reuse of PCEs, cost data may not be well-collected or structured (Vigren, 2022), as initial projects are typically exploratory pilot projects with ad hoc reporting processes.

Thirdly, the reuse of PCEs is a focus of many innovations that may quickly alter the current cost structure of PCEs reuse. While construction processes involving reused PCEs remain unconventional and not business-as-usual, various process innovations — such as pre-deconstruction audit methods, inventory modeling of donor buildings, efficient and smart deconstruction methods, and optimized storage and logistics for elements (Huuhka et al., 2024) — can significantly influence cost analysis.

Fourthly, and finally, an analysis of the cost structures of emerging supply chains, if retrospective in nature, may not account for future business planning needs, as cost structures are likely to change with economies of scale, such as in production and logistics (Besanko et al., 2010). In other words, firms would operate with high unit costs in nascent supply chains, where the number of elements is limited, but significantly lower unit costs in mature businesses, where PCE sourcing, project planning, and construction are business-as-usual.

Another important reason for lower unit costs is the learning curve effect (Besanko et al., 2010), where increased experience and repetition lead to reduced costs and improved efficiency over time. Companies wishing to invest in new capabilities must anticipate future cost levels, for example, when making capital-intensive investments like production facilities and warehouses. Additionally, they need information on the future size of the market, for which no analyses are currently available: Can an investor expect the market of reused PCEs to grow, and if so, when?

Despite these limitations, which are mostly due to the lack of data, much can still be achieved. Focused economic analysis can yield valuable insights that benefit both theory and practice. As Geroski (1997) states: "Strategy decisions often turn on 'how much?' or on 'how big?' a particular effect is, something which can be of some importance when a nifty new strategy idea gets translated into a business plan." This means that even an approximate calculation can be useful for investors and managers if the order of magnitude is correct. In such cases, the analysis can support future-oriented investment decisions.

Accordingly, we will conduct focused exploratory analyses, shifting the focus from directly analyzing the cost structure of reusing PCEs to analyzing the economic boundary conditions associated with their reuse. Economic boundary conditions are the contextual factors that define the constraints for reusing PCEs. These conditions outline the financial and resource-based limits for reusing PCEs and are not limited to a single case study.

2.2 DATA AND ANALYSIS

Figure 1 presents different supply chains related to concrete elements. The first is standard demolition, where concrete elements are deconstructed and sent to concrete recycling plants, backfilling, or landfills, representing the status quo in waste management and recycling. The second is reuse, which is currently being piloted in countries such as Sweden, Finland, Germany, and the Netherlands. The third is construction using virgin materials. Many comparisons between the supply chains could be made. Additionally, these supply chains could be divided into even more specific stages or work tasks (Crowston, 1997).

First, it's reasonable to compare standard demolition and reuse from the perspective of the building donor or seller. Economic principles suggest that, all other things being equal (ceteris paribus), the building donor or seller will choose disposal of material through standard demolition or reuse based on the costs or profits associated with these options. However, other considerations, such as interest in more sustainable alternatives, are of course relevant. Nevertheless, analysis should aim to compare these incentives, and policy should aim to adjust these incentives in favour of reuse.

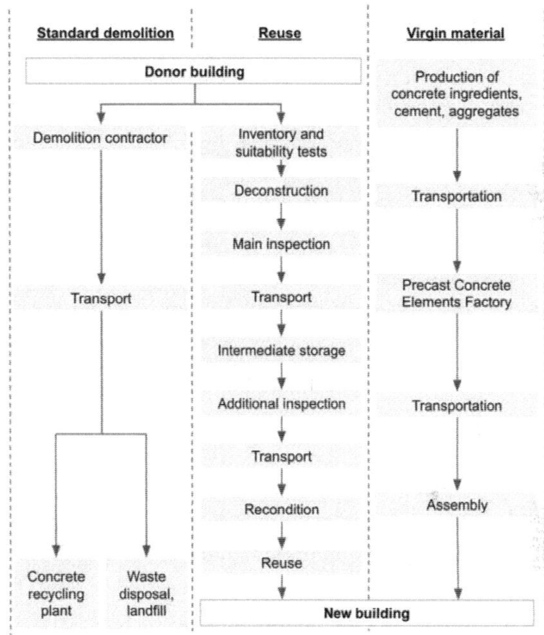

Figure 1: Supply chain of reusing PCEs.

Second, it is reasonable to compare the cost structure of the reuse and virgin material supply chains from the perspective of the buyer of a new building. Economic theory suggests that the buyer would choose the cheaper option if the quality is the same, or the higher quality option if the price is the same, assuming no other factors influence the decision. Therefore, analysis should aim to compare these alternatives, and policy should aim to adjust the incentives in favour of reuse.

Third, and finally, economic principles also suggest that all actors in the supply chain need to remain profitable in the long term in order to stay in business. This means that the analysis could focus on the profitability of each actor, and then on the supply chain as a whole.

We will consider these three cases and ask the following questions:

https://doi.org/10.52202/080684-0018

1. Which economic factors influence building owners' decisions to donate or sell PCEs for reuse?
2. Which economic factors influence building buyers' decisions to choose reuse over virgin materials?
3. Which economic factors influence the profitability of individual actors within supply chains and the supply chain as a whole?

The data cited in this analysis is derived from existing academic literature, industry reports, public databases and websites, internal firm data, and other unpublished documentation (Table 1). To gather this information, we conducted an extensive search and document analysis of 54 data sources focusing on construction costs related to PCE reuse. The academic articles are published studies about PCE reuse and the industry reports and databases/websites cover published insights on PCE reuse or construction costs. Other reports and internal firm documents were obtained through the ReCreate Project (n.d.) and focus on the production cost calculations of three new residential buildings in Sweden from the building owner's perspective, along with industry and stakeholder perspectives on costs. The documents mostly focus on projects from Nordic countries or the Netherlands.

However, the availability of structured datasets is limited, and the data is fragmented. While this supports the analysis of economic boundary conditions, it prevents more in-depth calculations related to economic feasibility.

Table 1: Data sources.

Document type	Document count
Academic article	15
Database/Website	11
Industry report	21
Other report	4
Firm internal document	3
Total	54

3 ANALYSIS OF ECONOMIC BOUNDARY CONDITIONS

This section is organized around the aforementioned questions.

3.1 WHICH ECONOMIC FACTORS INFLUENCE BUILDING OWNERS' DECISIONS TO DONATE OR SELL PCES FOR REUSE?

PCE donors or sellers are real estate owners in the process of decommissioning buildings, and hence they must decide whether to opt for standard demolition or PCE reuse. Typically, they contract a demolition company and other experts to assess the deconstruction, demolition, and waste disposal needs and costs. The building owner is a key decision-maker; as owners, they have legal responsibilities related to decommissioning, and as clients, they control which suppliers are engaged for these tasks (Engström and Hedgren, 2012; Vigren et al., 2022). To fulfil these responsibilities, they enter into contracts with these suppliers. Furthermore, despite their important role, building owners might not necessarily have the expertise to fully understand or control what happens further down the supply chain (Vigren, 2024), especially in PCE reuse, which remain an uncommon practice in the construction sector (Engström and Hedgren, 2012).

We choose to use the term "building donors" because empirical examples from PCE reuse projects in Sweden, Finland, Germany, and the Netherlands show that real estate owners have donated PCEs for reuse (ReCreate Project, n.d.). Decisions about material disposal have considerable economic implications.

First, donating or selling PCEs can reduce waste disposal costs, as elements donated or sold for reuse reduce the amount of material sent to concrete recycling plants, backfilling, or landfills.

Second, there are expectations that the sale of salvaged PCEs could generate additional revenue for building owners and other actors in the reuse value chain (Svedmyr, 2024; Riuttala et al., 2024; Återhus Project, 2023). By establishing partnerships with organizations specializing in material reuse, building owners may monetize components that would otherwise be discarded. This would imply the creation of what could be considered a new market for reused PCEs. Entrepreneurial circularity actors, such as Blocket, CCbuild, Loopfront, and Palats, are driving the growth of digital platforms and marketplaces for recycled materials.

Third, by donating or selling PCEs, building owners can contribute to national recycling goals and achieve other sustainability or circularity targets. Achieving such targets may be a subject of many economic incentives, including tax incentives, grants, subsidies, compliance or avoidance of sanctions related to legislation, or eligibility for government-sponsored sustainability initiatives. Additionally, participation in circular economy practices may open doors to green financing opportunities, lower insurance premiums tied to sustainable operations, or enhanced market competitiveness through positive branding and alignment with corporate social responsibility goals.

These economic advantages, coupled with environmental benefits, can make the practice of donating or selling PCEs a compelling strategy for real estate owners aiming to contribute to sustainable construction and resource efficiency. Therefore, building owners are likely to already have net positive incentives to pursue reuse activities over demolition.

However, these incentives are highly dependent on the country and specific context, influenced by factors such as transportation costs, the availability of suppliers, and the demand for reused PCEs. These local conditions determine which options are available. For example, in rural areas with high transportation costs, low availability of suppliers, and low demand for new buildings, reuse might not be an option. Furthermore, deconstruction is

https://doi.org/10.52202/080684-0018

more expensive than destructive demolition, and it remains uncertain who would bear these costs in a PCE reuse value chain.

We concur with Svedmyr et al. (2024) and Küpfer et al. (2023) that increased availability of data from public and private sources would enable more comprehensive analysis in the future.

3.2 WHICH ECONOMIC FACTORS INFLUENCE BUILDING BUYERS' DECISIONS TO CHOOSE REUSE OVER VIRGIN MATERIALS?

Building buyers are real estate owners considering the construction of a new building or a major renovation of an existing one. From their perspective, many factors related to the new development may have economic consequences. Buildings need to be usable, buildable, operable, and sustainable, and the choice of building materials may have several implications for all these qualities (Fischer, 2017). Furthermore, as with donating or selling PCEs, economic benefits such as tax incentives, grants, subsidies, green financing opportunities, compliance, or branding may become important economic drivers for PCEs (e.g., Riuttala et al., 2024).

Nevertheless, the cost of acquiring a building is, of course, a central concern for buyers. Therefore, it is relevant to compare the cost structures of new buildings based on reused PCEs and virgin materials. The cost of the building's structural frame essentially represents the total "budget" or room for economic flexibility with regard to PCE innovations, such as reuse. To illustrate this, we analyzed the total cost of apartment buildings using data obtained from a Swedish real estate owner.

Let us assume an apartment building costs €15,000,000, and the structural frame and roof system account for 20% of the total costs — €3,000,000. This estimate was considered reasonable by a representative from a precast concrete building systems provider. Now, if we assume a 25% price increase (€750,000) in the cost of the structural frame and roof system, the total cost of this system will be €3,750,000. The total project cost would then be €15,750,000, representing a 5% increase over the original price.

This analysis demonstrates how sensitive the total cost of acquiring a building is to fluctuations in specific cost increases or decreases. Table 2 presents additional scenarios based on the same calculation logic. Notably, this table is general to any cost changes and could therefore be applicable to cost increases related to virgin materials, the cost impact of new legislation, or any other costs associated with an increase in project costs.

Table 2: Scenarios of how cost increases or decreases impact the total costs of an apartment building.

Project cost	Structural frame cost change	New project cost	Project cost change-%
€15,000,000	-25%	€14,250,000	-5%
€15,000,000	+25%	€15,750,000	+5%
€15,000,000	+50%	€16,500,000	+10%
€15,000,000	+75%	€17,250,000	+15%

Are these scenarios reliable and meaningful? First, the reviewed literature shows that construction costs vary significantly based on factors such as location, building type, materials used, project scale, labor costs, and regulatory requirements. Therefore, the estimates are mainly indicative.

Eklund et al. (2003) report on a Swedish case of new student accommodation constructed in 2001 in Linköping using reused elements. The project was 10%–15% more expensive than similar buildings constructed using conventional methods. Nevertheless, the contractors were confident that, through learning and larger-scale projects, the costs could be reduced to the level of conventional methods. This statement demonstrates the importance of economies of scale and the learning curve effect (Besanko et al., 2010) in driving down costs over time, thereby contributing to the increased adoption of PCE reuse. Furthermore, in this case, the Swedish government covered the costs with grants for developing new environmentally responsible construction methods (Eklund et al., 2003).

Some other projects reported in the literature indicate a variance in construction costs between approximately -80% – +60% when compared to alternative methods (Küpfer et al., 2023). However, as Küpfer et al. (2023, p. 23) state, these comparisons should be made with caution, as "computing methods, system boundaries, and hypotheses are heterogeneous." In the Swedish Återhus pilot project (2023), the costs have been comparable to or slightly higher than conventional methods, with the expectation of becoming directly economically beneficial once reuse is implemented with more standardized methods. For more information on cost levels, see also Salama (2017) and Huuhka et al. (2015).

The variance in reported costs for projects with PCE reuse demonstrates that costs can vary significantly depending on the type of project and its organization. On the other hand, annual construction cost fluctuations at the range of -10% – +10% are quite normal in the construction industry. Therefore, price changes reported in the literature and Table 2 may be considered moderate.

Given that most of these examples are dated, technological developments over the past 10–15 years have likely contributed to reducing the cost difference, bringing project costs closer to price parity with the use of virgin materials. Emerging technological developments could offer the potential for reduced costs in PCE reuse practices over time. For example, the efficiency and

https://doi.org/10.52202/080684-0018

quality of deconstruction and design operations could be enhanced with artificial intelligence, sawing and drilling operations could be automated using robotics, and material tracking could be managed through digital technologies and workflows (Brozovsky et al., 2024; Dervishaj et al., 2023a; Dervishaj et al., 2023b; Dervishaj & Gudmundsson, 2024). On the other hand, there is uncertainty regarding the maturity of these technologies and their cost impacts.

However, the accumulation of the learning curve effect and economies of scale (Besanko et al., 2010) is unlikely, as current projects are isolated pilot initiatives. Economies of scale — cost advantages that lead to a decrease in the average cost of production — would require repetition across multiple similar projects. The path toward economies of scale could begin with governmental support and investment, structured efforts within the innovation system, and legislative changes that prioritize PCE reuse over the use of virgin materials.

Furthermore, decision-making exhibit inertia (Engström and Hedgren, 2012), meaning that decision-makers are likely to choose options familiar to them. This is problematic from the PCE perspective and requires a change in attitudes, as well as efforts in research, education, and deliberate attempts to promote these new ideas and solutions to decision-makers. Pulkka and Junnila (2015) discuss a "gravitational slingshot analogy," suggesting that innovation systems can leverage change-driven momentum to shift trajectories toward desired system states. Furthermore, a shift in trajectories toward the larger adoption of PCE reuse practices would require improved legitimacy for these practices within the sector (Thomas and Ritala, 2022).

The learning curve effect, in turn, would require either repetition by the same actors to accumulate expertise or effective knowledge transfer between actors. On the other hand, many current projects are regional, and it is likely that future value chains will remain local due to high transportation costs and varying local norms and regulations (Svedmyr, 2024; Ghisellini et al., 2018). From the perspective of building buyers, this was particularly challenging in the observed pilot projects, as regional buyers rarely engage in the construction of new buildings. The knowledge fragmentation, a common challenge in the construction sector (Dubois & Gadde, 2002), generally hinders learning, innovation, and the scalability of new ideas.

Nevertheless, current projects, research and development, and education in circularity practices foster the learning curve effect and knowledge transfer within the sector. The development of theoretical frameworks and practical guidelines for PCE reuse is particularly important because building owners and other actors need novel frameworks for business development and to guide their sustainability initiatives (Nyoni et al., 2023). With further analysis, these frameworks could also be tailored for investors.

3.3 WHICH ECONOMIC FACTORS INFLUENCE THE PROFITABILITY OF INDIVIDUAL ACTORS WITHIN SUPPLY CHAINS AND THE SUPPLY CHAIN AS A WHOLE?

The third relevant question related to PCE reuse concerns profitability. Economic theory suggests that, over the long term, firms must remain profitable to avoid bankruptcy. Without consistent profitability, firms cannot cover operating costs, repay debts, or invest in necessary resources, ultimately leading to financial distress and potential insolvency. Profitability is a key driver of sustainability and circularity, as firms require profits to invest in new, sustainable technologies and methods. Additionally, these new technologies and methods must be more profitable than alternative options for firms to have the incentives to make the costly investments required for their adoption.

It follows that each actor in the supply chain, as well as the supply chain as a whole, must remain profitable over the long term. This has two implications for the unit of analysis. First, each firm must have incentives to invest in alternative methods, meaning the analysis should focus on firm-level incentives. Therefore, a firm reluctant to invest in new capabilities may slow down the development for others. Second, the way value is created and captured within the supply chain or broader business ecosystem sets the analytical focus at the system level (e.g., Harala et al., 2023; Riuttala et al., 2024; Sairanen et al., 2024; Vigren, 2024). Here, while value is created through the interdependent supply relations across the supply chain.

In the PCE reuse supply chain, the individual actors include building donors and sellers; architectural and engineering firms that make inventories of existing PCEs in buildings and conduct suitability tests and inspections; deconstruction firms; storage operators; transport firms; design firms; reconditioning facilities; and contractors and clients of the new building. Additionally, actors in the standard demolition supply chain include demolition firms, waste management and recycling facilities, while actors in the virgin material value chain include concrete suppliers, concrete manufacturers, and PCE factories. Furthermore, all supply chains rely on consultants, such as environmental consultants, have relationships with government agencies, and are indirectly connected to other supply chains, such as those involving other materials supplied for construction sites.

These operations may be organized by individual firms or vertically integrated firms that operate across multiple phases within the supply chain (Besanko et al., 2010). Nevertheless, effective operations present a major coordination challenge between people and workflows (Eriksson et al., 2019).

For most, time efficiency is a major profitability driver, as labor costs represent a large share of their operations. In this sense, learning new construction methodologies represents a challenge because it requires an investment of time, which may decrease the overall efficiency and profitability over an uncertain period. As a result,

https://doi.org/10.52202/080684-0018

innovation is not generally incentivized in the sector. Additionally, the construction sector is generally a low-margin industry, making it difficult to allocate resources for learning and innovation.

However, with PCE, there is limited knowledge about the cost structure within the value chain. First, the literature generally indicates that deconstruction costs are significantly higher with PCE reuse compared to standard demolition. An estimate from a representative of a precast concrete building systems provider suggests that dismantling a building for PCE reuse is 1.5 to 2 times more expensive, although there are significant differences between building types, such as office and residential buildings. On the other hand, demolition can be performed in various ways (Ghisellini et al., 2018), and salvaging other building materials on-site is becoming more common. This suggests potential synergies between deconstruction efforts aimed at salvaging PCEs.

Second, substantial cost savings arise from avoiding landfill fees and other expenses associated with standard demolition.

Third, additional savings are achieved through reduced material costs compared to using virgin materials. Some estimates suggest that salvaged panels can cost as little as one-third of the price of new ones (Huuhka et al., 2019; see also Küpfer et al., 2023).

Fourth, the storage costs of PCE reuse are significantly higher compared to standard demolitions, as reuse often requires both on-site and intermediate storage. These costs are closely related to the distance between the deconstructed building and the new building. Close proximity may reduce the required transportation and intermediate storage. Additionally, as Addis (2012) points out, inventory turnover is another important metric. Stored inventory accrues costs over time, as money, time, and other resources, such as space, are tied up in the PCEs. Therefore, the contractor owning the PCEs would only salvage items with a high likelihood of being quickly demanded for new construction, thus keeping storage costs to a minimum (Addis, 2012). Furthermore, the type of storage will impact the costs, such as the amount of protection needed from the weather.

Fifth, transportation costs are a major cost driver, and distance may also impact the environmental benefits of PCE reuse. With longer distances, transportation costs and environmental impact increase. Therefore, authors such as Ghisellini et al. (2018) and Svedmyr (2024), highlight that the circular economy in the construction and demolition sector is primarily a territorial activity.

Sixth, and finally, the increased availability of PCEs in the construction sector could lead to the creation of what might be considered an entirely new market, which is currently in its infancy. The development of new markets can have significant economic implications. First, markets serve as forms of coordination that promote efficiency and facilitate information exchange. Second, markets generate signals for investors, with growing markets being particularly attractive to them. Third, increased economic activity could stimulate further innovation, economies of scale, and learning curve effects, all of which may have substantial impacts for all actors in the supply chains in the future (Besanko et al., 2010).

4 TOWARDS AN AGENDA FOR RESEARCH AND INVESTMENT DECISION-MAKING

Current literature and analyzed sources highlight that PCE reuse has a long history and is supported by extensive contemporary studies (Table 1). However, surprisingly little attention has been given to economic analysis. This is notable, as the economic feasibility of any innovation is a major factor in its adoption.

Further economic analysis would make important contributions to research and would also be valuable for investors. In this context, investors broadly refer to those managers who decide to invest in human resources, such as new skills and capabilities, or in capital investments, including buildings, logistics capacity, storage capacity, production capacity, and machinery. These capital investments require long-term planning and financing. Currently, the PCE reuse market is in its infancy and faces significant uncertainty due to legal and economic factors. To alleviate these uncertainties, further analysis of economic feasibility and legal impacts is needed.

Based on economic theory (Besanko et al., 2010) and cost management perspectives (Kulmala et al., 2002; Paranko, 2012), this article has made an attempt to address these needs by outlining the economic principles for assessing the boundary conditions of the economic feasibility of PCE reuse. The use of economic theoretical concepts to identify key issues and interdependencies constitutes the article's contribution (cf. Tarafdar and Davison, 2018) and represents the first step in economic feasibility analyses. Our model (Figure 1) establishes the theoretical boundary for further analysis and comparison of different alternatives of demolition, reuse, and construction from virgin materials. Specifically, we contribute by analysing the key decision-making moments. This contribution may also be relevant to the wider circular economy and sustainability literature.

These economic principles hold regardless of context, but significant research opportunities exist in exploring specific cost items, such as the cost of PCE components across supply chains, or conducting case studies focused on particular phases within these supply chains. Further studies could also target the economic feasibility of individual buildings and projects. Additionally, research could examine the impact of regulatory changes on PCE reuse costs, investigate how supply chain dynamics influence economic feasibility, and explore the role of technological innovations in reducing costs across the value chain. Exploring the cost implications of different PCE reuse methods also offers valuable avenues for future research.

However, studies focusing on the economic feasibility of reusing structural steel (Yeung et al., 2017) show that the analysis is highly sensitive to context-specific factors,

https://doi.org/10.52202/080684-0018

such as labor costs, (de)construction methods, and the value of steel components. This challenge was also recognized in Swedish pilot project (Återhus Project, 2023), as it was difficult to have a comprehensive view of the economics of projects related to reuse, as well as how the value of reuse is communicated throughout the value chain. Stakeholders assess reuse in different ways, and the benefits and costs arising from reuse are allocated to different actors. Additionally, costs vary significantly between regions and countries (Svedmyr, 2024).

These variances serve as a caution against generalizing findings from case studies. Another caution pertains to generalizing findings from pilot projects, which often involve high exploration costs and high unit costs. Nevertheless, specific studies can be highly informative, especially if they demonstrate profitability and investment opportunities despite these uncertainties and higher costs. Furthermore, we encourage, along with others (Svedmyr et al., 2024; Küpfer et al., 2013; Kulmala et al., 2002), that data from public and private sources be made available to researchers and analysts for further research and more comprehensive analysis of cost structures. This also serves as a recommendation for industry analysts to monitor these developments closely. Systematic data on actors, costs, prices, and markets are prerequisites for informed investment decision-making.

This article also sets aside other important areas of economic research for future investigation. Küpfer et al. (2023) aptly point out that reuse has the potential to create new jobs and business models, and promote local sourcing of materials, thereby contributing to local job markets and economic activity — important topics for future research.

This article also does not focus on other societal perspectives, such as externalities of construction, which are an important area of future research (see also Återhus Project, 2023). Other relevant questions include: What new roles or actors may arise? How can the mediation or matching between deconstructed buildings and new constructions be facilitated?

Additionally, we have not addressed other possible mechanisms related to the value chains, including more specific categorization of tasks and processes (Crowston, 1997) involved in standard demolition, reuse, and the use of virgin materials in construction, as well as their combinations. For example, in reality, not all PCEs fulfill the criteria for reuse, and a certain percentage of materials used in new constructions would still need to be produced from virgin materials. For example, a precast concrete building systems provider stated that the largest economic potential lies in the reuse of floors rather than walls.

Finally, further research could investigate incentive structures related to tax incentives, grants, subsidies, compliance or avoidance of sanctions related to legislation, eligibility for government-sponsored sustainability initiatives, green financing opportunities, lower insurance premiums tied to sustainable operations, and enhanced market competitiveness through positive branding and alignment with corporate social responsibility goals.

5 CONCLUSION

In sum, the analysis concludes that:

Economic factors such as reduced waste disposal costs, potential revenue from salvaged components, contributions to sustainability goals, and possible associated economic incentives (e.g., tax incentives, grants, or green financing) provide building owners with compelling incentives to donate or sell PCEs for reuse. However, these incentives are highly context-dependent and require further data for comprehensive analysis.

Building buyers' decisions to choose reused PCEs over virgin materials are influenced by cost considerations. Costs have varied significantly in previous reuse projects, showing both cost savings and additional expenses compared to other methods. Therefore, economic feasibility is highly contextual. Future investments can already be directed toward the most promising opportunities. However, further data on costs and prices are needed. Expectations of economies of scale, learning curve effects, and technological advancements present opportunities to improve economic feasibility.

For the supply chain, the main cost categories in PCE reuse include higher deconstruction, storage, and transportation costs, while cost reduction drivers come from savings on landfill fees and material costs. Profitability depends on these costs, as well as the potential for new markets, economies of scale, and innovation, which can enhance economic feasibility in the long term.

Further economic feasibility research on the cost structures, regulatory impacts, technological innovations, and supply chain dynamics is necessary to inform better investment decisions in this emerging market.

ACKNOWLEDGEMENTS

This project has received funding from the European Union's Horizon 2020 research and innovation programme under grant agreement No 958200.

REFERENCES

Aarikka-Stenroos, L., Alkki, L., Harala, L., & Riuttala, M. (2021). Reuse innovation in construction industry: Value creation and the ecosystem. *In Conference proceedings of ISPIM Connects Valencia – Reconnect, Rediscover, Reimagine*. LUT Scientific and Expertise Publications.

Addis, B. (2012). *Building with reclaimed components and materials: a design handbook for reuse and recycling*. Routledge.

Agndal, H. and Nilsson, U. (2009). Interorganizational cost management in the exchange process. *Management Accounting Research*, 20(2), 85-101.

Al-Najjar, A., & Malmqvist, T. (2025). Embodied carbon saving of reusing concrete elements in new buildings: A Swedish pilot study. *Resources, Conservation and Recycling*, 212, 107930.

Besanko, D., Dranove, D., Shanley, M., & Schaefer, S. (2010). *Economics of strategy*. Fifth edition. John Wiley & Sons.

Brozovsky, J., Labonnote, N., & Vigren, O. (2024). Digital technologies in architecture, engineering, and construction. *Automation in Construction*, 158, 105212.

Circular Economy Stakeholder Platform. (n.d.). Cementing Europe's future: Buildings for a net-zero age. *Circular Economy Stakeholder Platform, a joint initiative by the European Commission and the European Economic and Social Committee.* https://circulareconomy.europa.eu/platform/en/news-and-events/all-events/cementing-europes-future-buildings-net-zero-age

Crowston, K. A. (1997). Coordination theory approach to organizational process design. *Organization Science*, 8(2), 157–175.

Dervishaj, A., Fonsati, A., Hernández Vargas, J., & Gudmundsson, K. (2023b). Modelling Precast Concrete for a Circular Economy in the Built Environment: Level of Information Need guidelines for digital design and collaboration. *41st Conference on Education and Research in Computer Aided Architectural Design in Europe*. Graz University of Technology, Faculty of Architecture.

Dervishaj, A., Hernández Vargas, J., & Gudmundsson, K. (2023a). Enabling reuse of prefabricated concrete components through multiple tracking technologies and digital twins. *European Conference on Computing in Construction and the 40th International CIB W78 Conference*. European Council for Computing in Construction.

Dervishaj, A., & Gudmundsson, K. (2024). From LCA to circular design: A comparative study of digital tools for the built environment. *Resources, Conservation and Recycling*, 200, 107291.

Dubois, A., & Gadde, L. E. (2002). The construction industry as a loosely coupled system: implications for productivity and innovation. *Construction Management and Economics*, 20(7), 621-631.

Eklund, M., Dahlgren, S., Dagersten, A., & Sundbaum, G. (2003). The conditions and constraints for using reused materials in building project'. In: R. Chini (Ed.), *Deconstruction and Materials Reuse*, CIB Publication, 287, 248–259.

Engström, S., & Hedgren, E. (2012). Sustaining inertia? Construction clients' decision-making and information-processing approach to industrialized building innovations. *Construction Innovation*, 12(4), 393-413.

Eriksson, K., Wikström, K., Hellström, M., & Levitt, R. E. (2019). Projects in the business ecosystem: the case of short sea shipping and logistics. *Project Management Journal*, 50(2), 195-209.

European Commission. (n.d.). What is the EU ETS? *European Commission.* https://climate.ec.europa.eu/eu-action/eu-emissions-trading-system-eu-ets/what-eu-ets_en

European Union. (n.d.). Fit for 55. *Council of the European Union.* https://www.consilium.europa.eu/en/policies/fit-for-55/

Fischer, M., Ashcraft, H. W., Reed, D., & Khanzode, A. (2017). *Integrating project delivery*. John Wiley & Sons.

Gartner, E., & Sui, T. (2018). Alternative cement clinkers. *Cement and Concrete Research*, 114, 27-39.

Geroski, P. A. (1997). What Can Economics Offer Strategy? A Review of: D. Besanko, D. Dranove and M. Shanley Economics of Strategy. *International Journal of the Economics of Business*, 4(2), 215-227.

Ghisellini, P., Ripa, M., & Ulgiati, S. (2018). Exploring environmental and economic costs and benefits of a circular economy approach to the construction and demolition sector. A literature review. *Journal of Cleaner Production*, 178, 618-643.

Harala, L., Alkki, L., Aarikka-Stenroos, L., Al-Najjar, A., & Malmqvist, T. (2023). Industrial ecosystem renewal towards circularity to achieve the benefits of reuse-Learning from circular construction. *Journal of Cleaner Production*, 389, 135885.

Huuhka, S., Kaasalainen, T., Hakanen, J. H., & Lahdensivu, J. (2015). Reusing concrete panels from buildings for building: Potential in Finnish 1970s mass housing. *Resources, Conservation and Recycling*, 101, 105-121.

Huuhka, S., Aarikka-Stenroos, L., Lahdensivu, J., Jonker-Hoffrén, P., Arnold, V., Stenberg, E., Blok, R., Gudmundsson. K., Teuffel, P., & Mettke, A. (2024). Recreating the Construction Sector for Circularity: Catalysing the reuse of prefabricated concrete elements. In Lehtimäki, H., Aarikka-Stenroos, L., Jokinen, A., & Jokinen, P. (Eds.), *The Routledge Handbook of Catalysts for a Sustainable Circular Economy* (pp. 42-66). Routledge.

Huuhka, S., Naber, N., Asam, C., & Caldenby, C. (2019). Architectural potential of deconstruction and reuse in declining mass housing estates. *Nordic Journal of Architectural Research*, 31(1), 139-179.

Kulmala, H. I., Paranko, J., & Uusi-Rauva, E. (2002). The role of cost management in network relationships. *International Journal of Production Economics*, 79(1), 33-43.

Küpfer, C., Bastien-Masse, M., & Fivet, C. (2023). Reuse of concrete components in new construction projects: Critical review of 77 circular precedents. *Journal of Cleaner Production*, 383, 135235.

Nyoni, V., Piller, W. B., & Vigren, O. (2023). Sustainability action in the real estate sector—An organizational and institutional perspective. *Cleaner Production Letters*, 5, 100049.

Paranko, J. (2012). Metallituoteteollisuuden kannattavuuskehitys: Pitkän aikavälin näkökulma.

 https://doi.org/10.52202/080684-0018

PhD Thesis. *Tampere University*. https://trepo.tuni.fi/bitstream/handle/10024/115318/paranko.pdf?sequence=2

Pulkka, L., & Junnila, S. (2015). Gravitational slingshot analogy of discontinuous sustainability innovation in the construction industry. *Construction Innovation*, 15(4), 409-427.

ReCreate Project. (n.d.). Recreate project: Circular economy and low-carbon technologies for construction. *Recreate Project*. https://recreate-project.eu/

Regeringskansliet. (2021). Lag (2021:787) om klimatdeklaration för byggnader. *Riksdagen, Landsbygds- och infrastrukturdepartementet*. https://www.riksdagen.se/sv/dokument-och-lagar/dokument/svensk-forfattningssamling/lag-2021787-om-klimatdeklaration-for-byggnader_sfs-2021-787/

Riuttala, M., Harala, L., Aarikka-Stenroos, L., & Huuhka, S. (2024). How building component reuse creates economic value–Identifying value capture determinants from a case study. *Journal of Cleaner Production*, 443, 141112.

Räsänen, A., & Lahdensivu, J. (2023). Quality assurance process for reuse of building components. *Life-Cycle of Structures and Infrastructure Systems*. CRC Press.

Sairanen, M., Aarikka-Stenroos, L., & Kaipainen, J. (2024). Customer-perceived value in the circular economy: A multidimensional framework. *Industrial Marketing Management*, 117, 321-343.

Salama, W. (2017). Design of concrete buildings for disassembly: An explorative review. *International Journal of Sustainable Built Environment*, 6(2), 617-635.

Silfwerbrand, J. (2020). Concrete and sustainability–some thoughts from a Swedish Horizon. *Nordic Concrete Research*, 63(2), 79-87.

Social- og Boligministeriet. (2024). Tillægsaftale mellem regeringen (Socialdemokratiet, Venstre og Moderaterne) og Socialistisk Folkeparti, Det Konservative Folkeparti, Enhedslisten, Radikale Venstre og Alternativet om national strategi for bæredygtigt byggeri. *Social- og Boligministeriet*. https://prodstoragehoeringspo.blob.core.windows.net/e479e1f4-4b8d-45b5-b69d-bc9e47689a69/Till%C3%A6gsaftale_om_stramning_af_CO2e-krav_til_bygninger.pdf

Suomala, P., Lahikainen, T., Lyly-Yrjänäinen, J., & Paranko, J. (2010). Open book accounting in practice–exploring the faces of openness. *Qualitative Research in Accounting & Management*, 7(1), 71-96.

Suchorzewski, J., Santandrea, F., & Malaga, K. (2023). Reusing of concrete building elements–assessment and quality assurance for service-life. *Materials Today: Proceedings*.

Svedmyr, K., Weckman, A., Suikkanen, T., Wium Pálmarsdóttir, J., Grønhaug, B., Damsgaard, J., ... & Jönbrink, A. K. (2024). Reuse, recycling and recovery of construction and demolition waste in the Nordic countries: An inventory of facilities and techniques in the Nordics including an outlook on other European countries. *Nordic Council of Ministers*.

Tarafdar, M., & Davison, R. M. 2018. Research in information systems: Intra-disciplinary and inter-disciplinary approaches. *Journal of the Association for Information Systems*, 19(6): 523-551

Thomas, L. D., & Ritala, P. (2022). Ecosystem legitimacy emergence: A collective action view. *Journal of Management*, 48(3), 515-541.

Vigren, O. (2022). Digitalization, sustainability, and ecosystems: An application of social network analysis to the real estate sector. PhD Thesis. *KTH Royal Institute of Technology*.

Vigren, O. (2024). Ecosystems in construction management and urban development: a comprehensive review of conceptualizations and contributions. *Construction Management and Economics*, 42(2), 162-181.

Vigren, O., & Eriksson, K. (2025). A multilayer network model for studying business ecosystems: insights from enterprise architectures in the real estate sector. *Journal of European Real Estate Research*, 18(1), 170-191.

Vigren, O., Kadefors, A., & Eriksson, K. (2022). Digitalization, innovation capabilities and absorptive capacity in the Swedish real estate ecosystem. *Facilities*, 40(15/16), 89-106.

Yeung, J., Walbridge, S., Haas, C., & Saari, R. (2017). Understanding the total life cycle cost implications of reusing structural steel. *Environment Systems and Decisions*, 37, 101-120.

Återhus Project. (2023.) UDI: Återhus - att bygga hus av hus - Slutrapportering Steg 2. *Återhus Project*. https://aterhus.nu/wp-content/uploads/2023/08/Slutrapport_Aterhus.pdf

LUND UNIVERSITY | LTH FACULTY OF ENGINEERING

DESIGN FOR THE FUTURE – VERSATILE, RELOCATED AND VERTICALLY EXTENDED TIMBER BUILDINGS FOR A CIRCULAR ECONOMY

Karin Sandberg[1*], Frida Görman[2], Janina Östling[3], Alar Just[4], and Ylva Sandin[4]

[1] RISE Research Institutes of Sweden, WoodTech

[2]IVL Swedish Environmental Research Institute

[3] IsoTimber Holding AB

[4] RISE Research Institutes of Sweden

*Corresponding Author: karin.sandberg@ri.se

ABSTRACT

Background and aim. Developing timber buildings suitable for deconstruction, reuse, and adaptability in practice is challenging and complex. The project "Design for the Future - Reuse of Timber Buildings in a Circular Economy" developed two concept buildings to be reused with preserved functionality. Focus was on environmental benefits and was obtained through collaboration within the circular value chain and according to real estate developers' requirements. One building featured industrially manufactured volumes designed to be relocated and rebuilt. The other was an adaptable building with planar elements, designed to be flexible, relocated and vertically extended with two added floors.

Methods and Data. The concept method, a co-creation process, was used that involved possible scenarios, construction, deconstruction, reconstruction, waste management and estimation of reusability. The method SimFORCE, Simulation for Future Oriented Reuse and Circular Economy, was developed. Evaluation of reusability and preserved functionality was conducted in cooperation with expert groups. The climate reduction potential of reuse was analysed using Life Cycle Assessments.

Findings. SimFORCE helps identify whether structures are designed for deconstruction or need improvement. Further, the results were useful in preparing and writing deconstruction and reconstruction guides. Climate calculations show a significant reduction in environmental impact when buildings are reused.

Theoretical/Practical/Societal Implications. With SimFORCE, two timber buildings were demonstrated as possibly being reusable with preserved functionality (structural, acoustics, fire resistance, etc.) with a considerably reduced climate impact. Assessments were based on profound knowledge and experiences of the building systems, deconstruction and testing. The actual buildings have not been deconstructed and rebuilt.

KEYWORDS: Adaptability, Co-creation, Design for deconstruction, Reconstruction, Reuse

1 INTRODUCTION

Global consumption of materials is expected to double in the next forty years (CEAP, 2020). The Circularity Gap Report (CGR, 2024) shows that the share of secondary materials is barely 7.2% in 2023, steadily declining since 2018. It also mentions that construction and demolition processes drive nearly one-third of all material consumption. Therefore, the total amount of materials consumed by the global economy is expected to increase, out of which most extracted materials entering the economy are primary. It can be concluded that there will be a material shortage if we do not leave the linear economy and make more use of the earth's resources. We also need to reduce emissions of greenhouse gases due to the climate impact. The built environment, including

https://doi.org/10.52202/080684-0019

housing and commercial buildings, is essential for our quality of life. About 40% of global greenhouse gas emissions can be attributed to buildings' construction, use and demolition (CGR, 2024). The European Commission adopted the new circular economy action plan (CEAP, 2020), one of the main building blocks of the European Green Deal, Europe's new agenda for sustainable growth. The EU's transition to a circular economy will reduce pressure on natural resources and create sustainable growth and jobs. It is also a prerequisite to achieving the EU's 2050 climate neutrality target. Building with wood has, therefore, become more critical since it is a renewable building material. However, it must be used efficiently and in accordance with the waste hierarchy and should be used as long as possible as a building material. A considerable amount of wood from the building stock can be available for cascading and second use (Nasiri et al., 2021) and recovered wood from the building stock could potentially be substituted into products (Höglmeier, 2013). Achieving this requires meticulous deconstruction of buildings and careful handling of materials. Research and development in recent years have increasingly transferred from a linear to a circular economy. The European project InFutUReWood investigated, for example, how we should build today to be able to circulate tomorrow and compiled findings on design as well as material (Sandberg et al., 2022). Several publications supporting the development of design for adaptability have been published (Ottenhaus et al., 2023) and constructions in circular economy (Çimen, 2021). Still, it is highly complex to manage the development of a fully circular building. It is used over a long period of time and consists of thousands of components. Sandin et al. (2023) support designers and industries applying Design for Deconstruction and Reuse and Adaptability (DfDR/A) to interpret ISO 20887:2020 by providing practical examples from case studies. Jockwer et al. (2020) mention the lack of existing methods to evaluate the performance of the dismantled elements before reuse as one of the reasons that the circularity concepts are not yet effectively established in timber buildings. This can also be due to considering buildings long-lasting and not anticipating disassembly and reuse of their elements.

1.1 AIM OF THE STUDY

The aim was to contribute to a deeper knowledge of how to build today to simplify future reuse and preserve the earth's resources, by developing concept buildings demonstrating reusable timber structures with preserved functionality. The intention was to create timber buildings adapted for increased circularity through Design for Deconstruction, Reconstruction and Reuse (DfDR&R). Environmental benefits and collaboration in the circular value chain were in focus. This was to be done by theoretical simulations to obtain more reusable designs in a process that relied on qualified estimates and calculations based on today´s knowledge and experience.

2 METHODS AND DATA

The work was part of the project "Design for the Future - Reuse of Timber Buildings in a Circular Economy" and two concepts with the following scenarios were investigated:
- The Modular Building: a timber building with industrially manufactured volumes designed to be relocated and reconstructed (elastic).
- The Adaptable Building: a timber building with planar elements, designed to be flexible and versatile, relocated and vertically extended with two added floors (elastic).

The concept method is described in section 2.1-2.6 and is illustrated in Figure 2. A workgroup conducted the concept studies in a co-creation process with the following assumptions and limitations. Anticipated scenarios and developed concept buildings are based on the knowledge of the project participants, and the processes are based on current industrial off-site timber-building techniques in Sweden. This involved reviewing technical solutions, theoretical and practical studies of building processes, testing and calculations, transport, storage and business models through work meetings, drawings and document studies, and building regulations and standards reviews. Requirements of real estate developers were included.

Definitions were used according to EN-17680:2023.
- Adaptability, is the ability of the object of assessment or part to be changed or modified to make it suitable for a particular use. Adaptability can be subdivided into functions of flexibility, versatility and elasticity of the building, part of or group of buildings.
- Flexibility is related to changing space distribution within the existing building unit.
- Versatility is related to changing the use of the building.
- Elasticity is related to changing the volume of the building space either outside the existing building unit or addition of a new building(s) within the site.
- Reuse is an operation by which products or components that are not waste are used again for the same purpose for which they were conceived or used for other equivalent purposes without reprocessing but including preparation for reuse.

2.1 CO-CREATION IN THE CIRCULAR VALUE CHAIN – THE TEAM

An essential part of the project was to engage the circular value chain for residential buildings, from procurement and planning and manufacturing to building and waste management, in a co-creation process to understand and cover the whole process of a building's life. More than 30 persons have participated to varying extents in the "Value -Chain-Team". The two concept buildings are based on the needs of the project participants in the role of clients, property developers, owners and managers, and the municipality as an authority. Implementation and assessment of the structure's functions are based on existing knowledge of multistore timber buildings, off-site manufacturing methods, building at the construction site, component suppliers and transportation. Co-creation and a common goal were created through regular documented meetings with presentations and discussions and several workshops using visual work platforms (i.e. Mural) with a digital whiteboard.

2.2 SCENARIOS AND REQUIREMENTS

The results depend on the anticipated scenario. The assumed life cycle impacts the reusability of the building or the building component suited for its purpose. The project discussed what to reuse, the entire building or a structural part of it, and for how many times. The building's functional performance, required by the client, a user, or by regulations, also affects the outcome. Therefore, the anticipated scenario must be described for the building and the client's requirements (procurement) must be documented. Moreover, an execution plan for the simulation process and competence requirements for the evaluation process and the Expert Team (see section 2.3) are needed. Additionally, the boundaries and system limits that apply in the LCA must be specified. Scenarios in this project were determined through several workshops. The scenarios for the buildings were summarized in PowerPoints, presented and discussed at meetings, and thereafter reviewed in Word and Excel.
Questions that the Value-Chain-Team arose in the process were for example:

- How should we design for reusability, to maintain value and functional qualities for optimal reuse, deconstruction, and reconstruction to accommodate reapplication for the same purpose or adaptability?

2.3 SIMULATED DECONSTRUCTION, RECONSTRUCTION AND ADAPTABILITY

The project developed a method for a theoretical simulation of the possible reuse of a building, i.e. SimFORCE - Simulation for Future Oriented Reuse and Circular Economy. The method is based on today's knowledge and consists of several steps as described in section 2.3.1. The simulation method assumes an initial building design (Design 1) to be assessed and developed into an improved building design (Design 2) optimizing the initial building (Phase 1) for deconstruction and reconstruction (Phase 2). To help structure the complex and iterative work answering the questions and scenario while developing Design 2, the SimFORCE method was used. With the purpose of finding an improved structure adapted for deconstruction, the method is based on the already existing 'case study method' (Sandin et al., 2022). Within this project, the 'case study method' was complemented with a functional analysis (i.e. assessment of preserved functionality) to predict the outcome of deconstruction, relocation and reconstruction at a new site, but also an assessment of possible adaptability.

The method is based on the collective assessment of reusability and functionality by an Expert Team with diverse competencies and extensive experiences about the building system to be evaluated. In this project a profound knowledge in timber buildings were present, with experience from construction, deconstruction and reconstruction. The members were Quality and Product Engineers, R&D Managers, Designers, Structural Engineers, Constructors, Production Managers (including building planners), Sustainability Managers and Research Engineers in Wood Technology. Simulations were done for the two concept buildings, with different Expert Teams of 4-7 participants per session. Verifying new, improved solutions may require practical tests and lab experiments if functional performances are unknown and difficult to estimate. The estimates of material consumption, energy consumption, etc, have been used in the LCA calculations.

2.3.1 Simulation by the SimFORCE method

To make the work logical, Excel sheets could be used that are prepared by a process leader experienced in building techniques, leading the work and asking supplementary questions to the Expert Team. The Excel should consider topics such as those specified in Figure 1.

2.4 ENVIRONMENTAL EVALUATION

To evaluate the potential from circular construction and reuse a life cycle assessment was carried out for each of the concept studies in this project. The assessment was conducted to gain knowledge of potential benefits as well as to identify climate driving factors.
The climate calculation using LCA-methodology was based on the chosen scenarios and collection of data formed in mentioned process conducted by suppliers, manufacturers and architects. The life cycle stages assessed are cradle to gate (A1-A5) as well as energy use for deconstruction (C1). The assessment includes the entire building from the foundation to its insulation. The LCA were based on the following standards: EN 15978:2011 for buildings and EN 15804:2019 for building products. The calculation was performed using the Building Sector's Environmental Calculation Tool (BM 3.0). The tool contains a database with generic LCA data representative of the Swedish construction market, as well as generic data for waste and transport.
The result is reported in global warming potential (GWP), measured in kilograms of carbon dioxide equivalents (kg CO_2e), and includes the greenhouse gases carbon dioxide (CO_2), methane (CH_4), and nitrous oxide (N_2O). Scenarios for reuse and reconstruction assumed to occur in the future was calculated based on current knowledge. This means, for example, that future scenarios for the climate impact of building materials have not been applied.

https://doi.org/10.52202/080684-0019

Step 1. Scenario and background.	
Define the background and the scenarios to anticipate the clients' brief/requirements and demand levels, whether to investigate on a system level or element level.	
Step 2. Suggestion of initial design suitable for customer requirements (Design 1) for the first life cycle (Phase 1).	
Describe the building system, how it is assembled (drawings) with text describing crucial components such as junctions, wall elements and other technical details that are important for Design 1 for the first life cycle (Phase 1).	
Step 3. Simulation of deconstruction.	
Identify (based on the structure in Step 2) and select possible areas to improve for an efficient deconstruction.	
Column:	**Activity**
1	**The assessment starts from top of the building**, going downwards the structure (material, component, structural) one Excel row for each part to be analysed.
2	**Comments.**
3	**Tools and vehicles needed.**
4	**Handling.** What happens to the part? Damage that occurs to components and materials from handling during deconstruction and reconstruction.
5	**Need for reconditioning, repairs and inspections.**
6	**Foreseeable problems with transportation or intermediate storage.**
7	**Predict material loss and waste.**
8	**Personal risk assessment.**
9	**Environmental risk assessment.**
Add columns for "Measuring performances" depending on scenario/client, for example from ISO 20887:2020. Describe the process, building, structure, how to deconstruct for reuse. Identify areas to improve. Necessary prework to perform before deconstruction etc.	
Step 4. Improved design solution (Design 2).	
The same layout and context as in Step 3, but with the modified and improved building/structure and solutions implemented (Design 2).	

Step 5. Assessing preserved functionality of Design 2 for the second life cycle (Phase 2).	
The initial building (Phase 1) has now theoretically been deconstructed, relocated and reconstructed. Based on previous work (Steps 2-4), now estimate, calculate, test the preserved functionality required for reuse. Describe how the functionalities are achieved and verified, or how to restore them, for Phase 2.	
Column:	**Activity**
1	**Functionalities.** List those identified as crucial, such as Airtightness, Stability, Acoustics, Fire resistance for structures etc
2	**Requirements.** List those related to the specific functionality.
3	**Class.** Specify if relevant.
4	**(A) Experiences.** Indicate facts/knowledge based on previous experience, competence, test
5	**(B) Evaluations/Calculations.** Estimate and/or calculate preserved functionality required for reuse based on standards etc.
6	**(C) Required lab tests.** Identify whether lab tests are required to ensure the functionality in Phase 2.
7	**Measures to consider.** Specify foreseen measures to be taken at the deconstruction and/or reconstruction to preserve or restore the functionality. Repair or replacement of components might be necessary.
8	**Assessment of functionality**. The Expert Team concludes whether the functionality can be preserved or restored (explain how in such case).
Step 6. Summary.	
Summarize the improved building design (Design 2).	

Figure 1: The structure of the SimFORCE method.

2.5 GUIDE FOR ADAPTABILITY, DECONSTRUCTION AND RECONSTUCTION

Crucial information obtained by the Expert Team in the simulation process (SimFORCE) was transferred to a guide for adaptability, deconstruction and reconstruction.

2.6 BUILDING DESIGNED FOR REUSE

The process of the concept method continued iteratively until conformity was reached within the Value-Chain-Team and the goals were complied. The documentation should be consistent in accordance with stated requirements. Any deviations from the requirements should be described in the documentation with clarifying explanations of why.

3 RESULTS AND FINDINGS

The main result was a concept method (3.1) to be used when designing new buildings. It was applied to two concept buildings (3.2, 3.3) and demonstrated useful for deconstruction and reuse scenarios. Theoretical studies were carried out for two Swedish industrially manufactured building systems based on the project members' requirements. It included determining possible scenarios, processes, and logistics, evaluating improved solutions, possible service life, waste management, simplified structures/components and material efficiency.

3.1 CONCEPT METHOD

The development of the concept took place interactively in loops during the project period and many companies have been involved in the process, see Figure 2. The focus was on reusability and to cause as little damage as possible during the deconstruction, relocation and reconstruction of the load bearing structure. Also, to keep functionality from the first life cycle (Phase 1) to the second life cycle (Phase 2). The analysis steps and loops enhanced an increased understanding of whether the technical functionalities were preserved or how they could be restored.

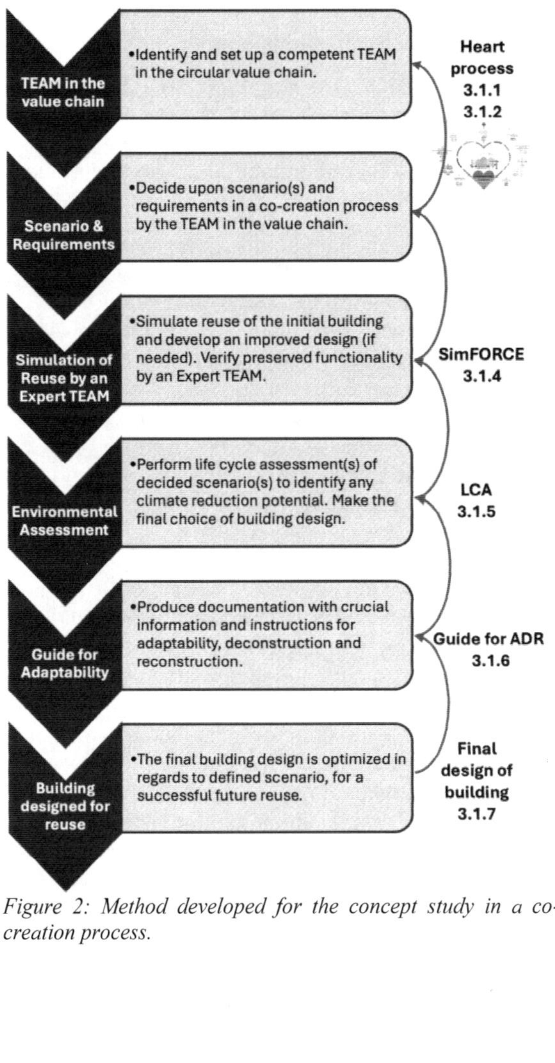

Figure 2: Method developed for the concept study in a co-creation process.

Industrial timber building process in Sweden

There are several ways to build in Sweden. The most common are frame structures of timber studs and/or lightweight beams (e.g. I-joists). Solid timber structures can be cross-laminated timber (CLT) or a post-and-beam structure made of glulam or laminated veneer lumber (LVL). IsoTimber is a semi-massive timber structure with a combined load bearing and insulating function. The building systems can be delivered to the building site as planar elements (panels), usually as walls and floor panels, or as 3D volumes (modules), which form entire rooms or apartments. Transport is usually conveyed by trucks. The industrially produced panels are assembled at the building site by contractors and completed to a building at a system level with plans for installations etc. The modules are built in the factory under the manufacturer's name and delivered to the building site, where the manufacturer's builders complete them (see Figure 3).

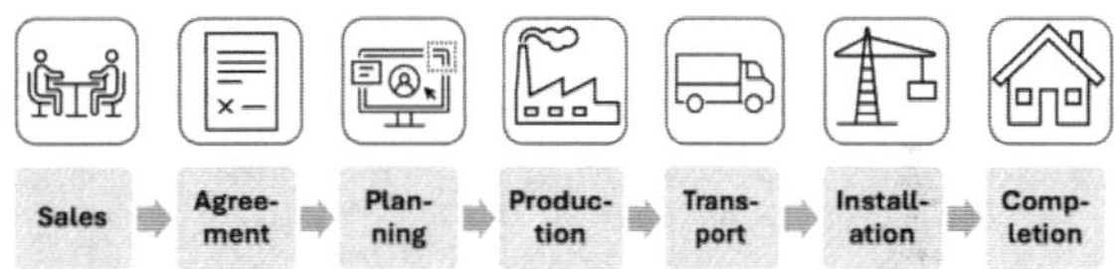

Figure 3: Industrially manufactured buildings in a schematic process of today.

3.1.1 Business in the Future Circular Value Cycle

The process for the first life cycle is known, but how will the deconstruction, relocation, and reconstruction of a building be managed in the future? The Value-Chain-Team identified that the process can be illustrated for a truly circular and sustainable building as in Figure 4. The complex process of finding a resource-efficient use of a building in its built environment, including its material use at any time, is explained in stages 1-10. Several new actors will be involved; for example, digital trading platforms and digital data management systems are under development. Therefore, the emergence of new actors will continue to enable the circular process.

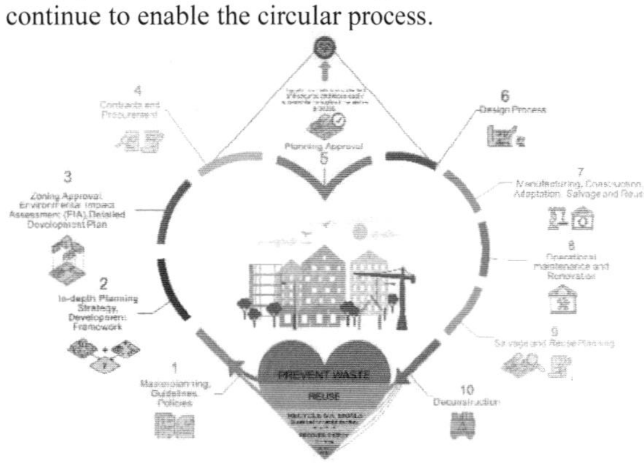

Figure 4: Circular value cycle process (the heart process) in 10 stages before the decision in the small heart: Prevent waste, Reuse, Recycle materials, Recover energy, and lastly Dispose.

https://doi.org/10.52202/080684-0019

3.1.2 Identify the Value-Chain-Team and prepare for the concept study

The concept study was based on co-creation by a team in the value chain. It was important to identify members in the value chain who understood the challenge and the tasks. Relevant topics to discuss were, for example as listed below. The discussions led to agreeing on scenarios, client and general requirements, see examples in Figure 5.

- Contribution to circularity: Reusability was a priority, and the elements should be able to be reused after deconstruction. However, in some cases, it was preferable to repair or refurbish parts of the component after a relocation but before reconstructing the building so that it would last for many more years. An example of this was the sealing tape at element joints.
- The level of assessment, material, component or the entire building: The reuse was primarily on element and volume levels. Therefore, the focus was on element connections and identifying and finding important structural intersections to address in the guide for deconstruction.
- Management of the building foundation: Should it be relocated or not? Both options were explored.
- Relevant regulations and standards, both current and upcoming, and how to handle them in the simulations: The concept buildings are expected to perform with the same functionality after reconstruction as they do today, and the requirements are expected to be the same.
- Environmental and building requirements for new construction versus requirements for renovation/relocating the building and adaptation to new users.
- Verification of functional requirements: Since the Expert Team had a lot of experience in constructing timber buildings, but also deconstruction of buildings, transportation, etc, the requirements are based on existing knowledge, tests and calculation methods.
- Verification of the technical lifespan of materials: Systems are based on current communication and guarantees. Hence, if possible, environmentally friendly materials with a long technical lifespan, preferably more than 50 years, were chosen.
- The extent of documentation required, but also what kind of information should be saved for the future and who is responsible for archiving the information: Documentation as constructions and building documents, drawings, operational and maintenance instructions, material specifications, and structural documents.
- Content in a guide for adaptability, deconstruction and reconstruction of the building.

FUNCTIONALITIES TO CONSIDER

CATEGORY OF FUNCTIONALITIES	
	Structural capacities, such as vertical and foundations load bearing capacity etc.
	Healthy indoor environment, such as acoustics, emissions, temperature etc.
	Climate impact and other **environmental** impact, such as energy consumption etc.
	Certifications, such as today´s Miljöbyggnad silver, Svanen etc.
	Regulations by law, national, municipal and other local **recommendations**, such as and EKS and BBR in Sweden, and municipal strategies etc.
	Market requirements, such as an aestethic architecture etc.
	And many more functionalities and requirements that the TEAM in the value chain finds are important for the specific building to be analysed.

Figure 5: Other categories of requirements.

3.1.3 SCENARIO AND REQUIREMENT

The Value-Chain-Team defined and agreed upon clear scenarios and which requirements should apply. Scenario and requirements depend on each specific concept study. See section 3.2.1 for The Modular Building and section 3.3.1 for The Adaptable Building.

3.1.4 SIMULATION OF REUSE AND FUNCTIONALITY ASSESSMENT

Technical functionality requirements are very important to verify when reusing a building, and therefore, Expert Teams have worked with this task. The method is based on the collective assessment of reusability and functionality by this Expert Team with diverse competencies and extensive experience with the building system to be evaluated; see section 2.3 for the Expert Team in this project.

The development and evaluation process in the two concept studies followed the SimFORCE method described in section 2.3. Each study ending with an improved building design (Design 2) assessed its' preserved functionality to predict the outcome of deconstruction and reconstruction at a new site, but also an assessment of possible adaptability. See Figure 6 for a schematic overview of the functionality and fire safety assessed in the project.

Which functionalities should be assessed depends on the defined scenario and requirements; see sections 3.1.2 and 3.1.3. Each functionality should be assessed and documented according to the steps in the SimFORCE method. Three methods were considered: A: Experiences,

B: Evaluation/Calculation and C: Lab test required. The Expert Teams decide which method to use or a combination of methods. However, the verification must be documented and clearly indicated in the final statement of the functionality assessment, along with any presumptions being made.
The project focused on reusability. Therefore, were minor damages and preserving the functionality from the first life cycle (Phase 1) to the second life cycle (Phase 2) assessed as the most favourable outcome. The standard ISO 20887:2020 provides examples of assessment criteria, see Annex C – Measuring performance, where C.5 "Ease of access to components and services" with a relative rating scale and C.9 "Supporting reuse (circular economy) business models" are valid for the evaluation.

EXAMPLE OF FUNCTIONALITY ASSESSED:
FIRE SAFETY

REQUIREMENTS	Sufficient fire resistance (REI30, 60, 90) according to EKS.
STANDARD/ CLASS	EN 13501-2
	EN 1995-1-2
A:EXPERIENCES	Personal: Participating in fire tests and analysing building parts.
	Facts: Fire stops can loose their functionality.
B: EVALUATION/ CALCULATION	Fire resistance calculations according to listed standards.
C: LAB TEST REQUIRED	EN 1363
	EN 1364
	EN 1365
MEASURES TO CONSIDER AT THE TIME OF REUSE	Evaluate protective boards (gypsum, plaster, wooden boards) – is there any presence of cracks, is the board thickness not valid or is there a presence of fasteners? If so, exhange the board.
	Check all fire stops, are they tightly fit to the joints? If not, replace the fire stop.
	Load bearing timber studs, are they of sufficient dimensions according to fire specifications? If not, strengthen the specific load bearing structural part.
ASSESSMENT OF FUNCTION 2:ND USE	The Expert TEAM assess that the building design will fulfill specified functionality at a future reuse.
	The assessment is based on A and B above, see (references) and presumes that indicated measures above are performed.

Figure 6: Examples of one functionality assessment, fire safety.

3.1.5 ENVIRONMENTAL ASSESSMENT

The environmental assessment findings are described under respective concept building; see sections 3.2.3 and 3.3.3.

3.1.6 GUIDE FOR ADAPTABILITY, DECONSTRUCTION AND RECONSTRUCTION

Valuable information is extracted using the SimFORCE method, as described in Figure 1. The information could be used to develop guides to facilitate the deconstruction and reuse of modules and panels, see Figure 7. Describing step by step, from securing walls and floors to fire safety and structural integrity throughout the deconstruction process to detailed descriptions of how to disassemble and remove installations, joints, elevators, balconies, stairs, access balconies, access balconies, roofs, facades, and finally lifting off the roof and then the volumes or planar elements for transport and storage. The deconstruction guide could also be complemented with a guide for reconstruction based on the functionality assessment. See the example in Figure 6.

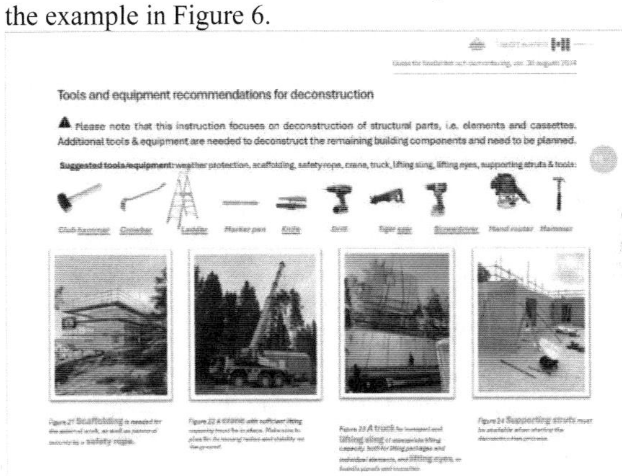

Figure 7: A guide for adaptability, deconstruction and reconstruction was developed for the Adaptability building.

Understanding the deconstruction process requires insight into the various steps outlined in Figure 3. During the initial building phase, there is a multitude of information about the building, including planning documents, drawings, and assembly instructions at the building site. However, this knowledge is often held by different people or departments. To learn more, meetings and discussions were conducted with individuals experienced in planning, industrial manufacturing, transport, and assembly at the building site, as well as the deconstruction of modules and planar elements. The assembly of the different parts and elements is planned during the initial design phase. Still, the deconstruction can be affected by the mounting and assembly at the building site and subsequent renovations and should be documented.

3.1.7 FINAL DESIGN OF BUILDING

The process, as described according to Figure 2, was completed. With quite an intense iterative looping, the project agreed upon two concept buildings that fulfilled the scenarios and requirements decided; see section 3.2 Concept - The Modular Building and section 3.3 Concept - The Adaptable Building.

https://doi.org/10.52202/080684-0019

3.2 CONCEPT - THE MODULAR BUILDING

The Modular Building is based on the real estate developer Folkhem's planned five-storey high buildings at Klockelund in Farsta, Stockholm, Sweden, see Figure 8. The buildings should be certified according to the Nordic Swan Ecolabel Buildings. The requirements promote resource efficiency, reduced climate impact, circular economy and conservation of biodiversity.

Figure 8: Folkhem's proposal of a modular building in Klockelund in Farsta. Illustration by In Praise of Shadows.

3.2.1 Scenario, requirements and boundaries

The building should be manufactured with a frame structure of timber studs as 3D volumes, fully equipped with a kitchen and bathrooms delivered from the factory, assembled and completed at the building site with installations, roof and elevator shaft. The Value-Chain-Team concluded that The Modular Building was developed with the reuse scenario. That results in two phases:

- Phase 1 - Initial building, five-storey.
- Phase 2 - Deconstruction and relocation of the building to a new site (reuse). The foundation of the building is not relocated.

3.2.2 Simulation of deconstruction and reconstruction by SimFORCE

The building system is based on existing building systems from manufacturers of multistorey modular timber buildings in Sweden, Lindbäcks Bygg, Derome and OBOS. They agreed on one joint building design. Studies on design for reuse, separation, sorting, and handling of reclaimed timber were conducted in collaboration with personnel knowledgeable about the issues from various companies in the project. This was made in many sessions dedicated to the overall concept study process, as described in Figure 2. Two sessions were performed with Expert Teams in the SimFORCE process, see Figure 1, to develop improvements and evaluate functionalities. The manufacturers have deep knowledge and experience of their building systems and have the competence to assess improvements and functionality.

Findings of improvements of the building structure

Identified improvements were, for example, prefabricated roof cassettes. The reuse process and transport are more efficient if the roof structure is constructed in sections. Improved connections between volumes were developed for easier deconstruction and reconstruction. It is important to balance the 3D volumes precisely when lifting them at reuse, and a device was identified to get hold of and place the lifting slings easily.

3.2.3 Climate calculation and collection of data

The chosen scenario for the life cycle assessment considers the two phases of the modular building, which are considered two separate life cycles. The activities included in the two phases are:

- Phase 1: Initial construction. The assessment includes using primary materials and energy for transport to the construction site and the construction.
- Phase 2: Change of location (100 km) includes reuse of materials from Phase 1, energy use for deconstruction, transport and reconstruction and new materials used for parts that need to be replaced.

Material used for the foundation, frame structure, façade, roof and frame completion were determined and quantified through the project planning document. For interior surfaces and room completion, as well as installations (Technical installations (not solar cells) of timber building apartments), standard values were used (Malmqvist et al., 2021). Data on waste quantities and additional materials, as well as estimates of waste and energy consumption, have been provided by all three suppliers based on their building systems. However, when the building is reused in the second phase, the foundation, installations and technical equipment are assumed not to be reused. The interior surfaces and room completion are also assumed to have to be remade. These choices were made since the foundation cannot be removed and reused and not to overestimate the potential of what can be reused. The results from the climate calculation for Phase 1 and Phase 2 are shown in Figure 9. A considerable climate benefit, more than 50%, can be obtained in the product stage (A1-A3) by reusing materials from the initial building at a change of location.

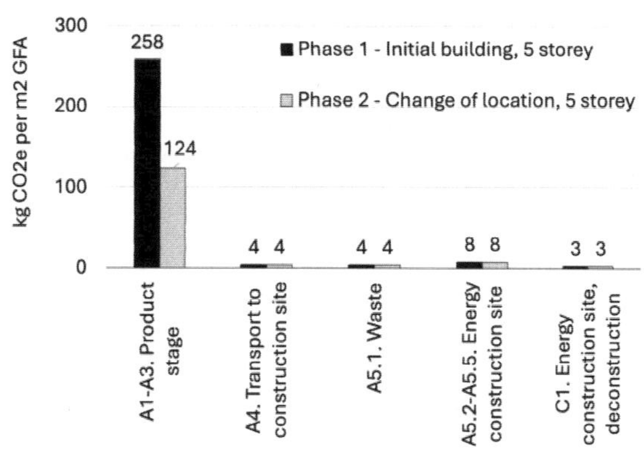

Figure 9: Climate calculations (kg CO2e per m2 Gross Floor Area (GFA)) for the initial building (Phase 1) and the relocated building (Phase 2), of the Modular Building.

3.3 CONCEPT - THE ADAPTABLE BUILDING

The Adaptable Building was developed based on the indicated needs of Skellefteå Municipality and Kiruna Bostäder AB (owned by Kiruna Municipality). They own buildings and rental apartments. Skellefteå Municipality is also an authority for building and demolition permits. They are in an expanding region in the north of Sweden and urgently need housing for entrepreneurs. However, it was identified that the property owner's needs will most likely change. In this case, from entrepreneur housing to student housing, or tourist accommodation. IsoTimber (supplier of external wall elements), Masonite Beams (supplier of floor/roof elements), and ETTELVA Architects jointly developed the new concept building. Figure 10 shows an illustration of The Adaptable Building.

Figure 10: The Adaptable Building was studied for different phases during its lifetime, from a new initial building to changes in layout (flexible and versatile), relocation and adding two floors (elasticity). Illustration by ETTELVA Architects.

3.3.1 Scenario, requirements and boundaries

Design for Adaptability (DfA) means both a flexible change, as in changing the space (layout) within the existing building, and versatility, i.e. changing the function or the use of the building, from larger apartments to tourist accommodation, for example. Further, the building can be adapted by adding new floors vertically (elasticity). The Adaptable Building should enrich its surroundings.

Therefore, it was assumed that the building's layout and function change could be implemented in 20 years. A deconstruction and relocation of the building, for example, due to changes in infrastructure and new roads, is more likely to occur in a longer perspective, e.g. in 50 years. Hence, the building should also be designed for deconstruction, reconstruction and reuse (DfDR&R), with a foundation that can be relocated and reused. Based on these assumptions, the Value-Chain-Team concluded that the concept for The Adaptable Building should be developed regarding a scenario with four different life cycles (phases):

- Phase 1: The initial building, two-storey.
- Phase 2: Change of function and layout (flexibility and versatility).
- Phase 3: Deconstruction and relocation (100 km) of the building to a new site (reuse).
- Phase 4: Extension, from two to four-storey building (elasticity).

3.3.2 Simulation of deconstruction and reconstruction by SimFORCE

As mentioned earlier, general studies and co-operative learning occurred in the project, as described in Figure 2. Dedicated sessions with Expert Teams in the SimFORCE process, see Figure 1, took place 6 times for this concept building. Improvements were developed, and the Expert Teams assessed preserved functionalities.

Architectural design with flexibility in mind

The first life cycle (Phase 1) is planned as a two-storey multi-family house, with 2-room and 4-room apartments, that can be used as shared contractor accommodations. Design features included to maximize flexibility:

- Kitchens and bathrooms are concentrated around a combined shaft in the building's core to free up floor and facade space for varying rooms requiring daylight.
- The facade is well planned with generous, general, and repetitive window placements to divide rooms or move walls in a maximum number of different positions as needs change.
- Load bearing interior walls are not required as the flexible building systems from Masonite Beams and IsoTimber handle the spans through load bearing in the floor elements and outer walls.

At a later stage, if a change in layout is needed due to a change in functional requirements (Phase 2), the building can be adapted. This is possible by opening and partially removing the apartment-separating wall between the units while maintaining shafts and wet rooms, stabilising the building, and facades/windows in the same positions. According to the decided scenarios, the building should be able to be relocated (Phase 3) and extended with two floors (Phase 4). The building's climate shell and load bearing structure were specified, while the interior and technical installations were not specified or quantified and were included in the climate calculations by standard values.

https://doi.org/10.52202/080684-0019

Findings of improvements of the building structure
The main improvement developed was a new connection detail at the junction between the external wall made of IsoTimber and the floor element built with I-joists of Masonite Beams. A consulting company assessed the acoustic performance of the new solution, and the fire safety performance was calculated and assessed as sufficient by members of the Expert Team.
It was identified that roof cassettes would minimize the need to remove the under-roof sheathing and parts of the roof and, therefore, would be preferable. The elements are easier to lift down, transport, and reuse. The lifting slings are left in place to indicate where to lift, but it is recommended to replace them if they are old.
The joints between elements are screwed, and that is a well-tried procedure and tested function. A problem might be finding the screw's right positions and uncovering them. There are different solutions depending on the joints. They can be covered with wear-layer or tape removed at deconstruction and replaced at reconstruction. It was suggested that panel joints can be clearly marked with a colour that is easy to find. The choice of screws can also be of importance, dimensions of screw-head and or replaced by wood screws. However, the performance needs to be tested and calculated before use.

3.3.3 Climate calculation and collection of data
The chosen scenario for the life cycle assessment considers the four different phases of The Adaptable Building which are considered for separate life cycles. The activities included in the four phases are:

- Phase 1: Initial construction. The assessment includes the use of primary materials and energy for transport to the construction site and the construction.
- Phase 3-4: Includes reuse of materials from Phase 1, energy use for deconstruction, transport, and reconstruction, as well as new materials used for parts that need to be replaced.

Material use for new materials for each phase was determined from the architectural drawings made by ETTELVA Architects, where amounts for the foundation, frame structure, façade, roof and frame completion were quantified. For interior surfaces and room completion, as well as installations (Technical installations (not solar cells) of timber building apartments), standard values were used (Malmqvist et al., 2021).
Data on waste quantities and additional materials and estimates of waste and energy consumption are the same as for The Modular Building.
When the building is relocated in Phase 3, installations and technical equipment are assumed not to be reused. The interior surfaces and room completion are also assumed to have to be remade. These choices were made not to overestimate the potential of what can be reused. However, the foundation can be reused compared to The Modular Building.
The results from the climate calculation for Phase 1 to Phase 4 are shown in Figure 11.

A considerable climate benefit, more than 50%, can be obtained in the product stage (A1-A3) for all scenarios (change of layout, change of location and extension).

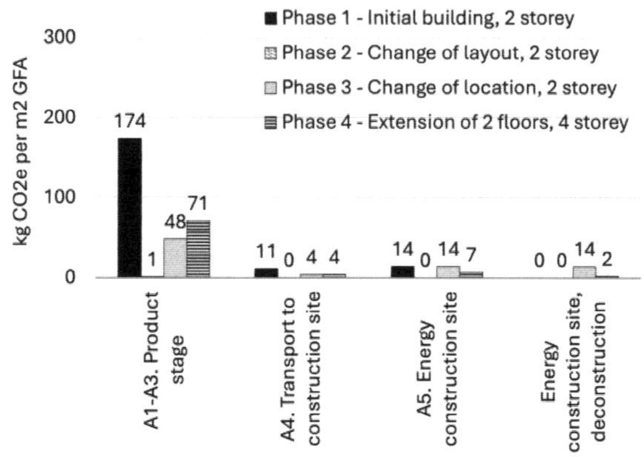

Figure 11: Climate impact (kg CO2e per m2 Gross Floor Area (GFA)) per respectively life cycle, i.e. Phase 1-4, for the Adaptable Building with planar elements, designed to be flexible, relocated and vertically extended with added floors.

4 DISCUSSIONS
Regarding results obtained by using SimFORCE, see 3.1.4 (Simulation of reuse and functionality assessment), the Expert Team must make sure to transfer any measures identified to be advantageous during reuse. The assumedly preserved functionality might be ruined if identified measures are not practically performed at reuse. For example, if the planar elements or volumes are not deconstructed carefully or connections are not exchanged as indicated, they might risk diminishing load capacity. The project identified the guide for adaptability, deconstruction and reconstruction as the best place to keep this kind of information today. Digital product passports are under development and might be a place to save information in the future.
To understand the complexity, the study identified that the SimFORCE method for a specific building with its defined scenario, covers mainly the right side of the circular value cycle process (heart process), see Figure 4. However, the scenario for the building also depends on the left side of the heart, for example procurement and building permits, but also the lower part (the small inner heart) as waste management and demolition plans. The circularity depends on the possibilities to implement the circular strategic 10 R's (Potting, 2017) and the cascading hierarchy.
Construction and assembly details and their practical implementation on the building site must be verified during renovation and deconstruction. The wear and tear of different parts after 50 years of use depends on the components' quality and position. Therefore, the deconstruction plan must vary depending on the building's structure and describe in chronological order what should be done, potential risks and tools needed. For deconstruction and future reuse, the components of the

building must be thoroughly documented before construction. Describing elements, components, materials, weight, dimension, connection position, type of connections etc. The documentation should be digital as well as physically marked on the building elements. This is standard procedure in the prefabrication process used by current suppliers of panels and roof cassettes today.

The findings of this work indicate very positive outcomes of reusing timber buildings designed for deconstruction and reconstruction. Looking at the climate evaluations for the two concept buildings, it was clear that the largest climate benefits can be made by reusing the materials of the initial building in a second building. The climate impact in the product stage, according to A1-A3 in standard EN-15978:2011, could be reduced by 50% or more in CO2e/m2 GFA.

New building regulations are expected in Sweden in the coming years. Those will set limits for the CO2 emissions to be declared in climate declarations for the structural and building envelope. The results from the concept buildings are below today's anticipation of the coming limit values. For the Modular Building the assumed scenario was a change of location. As mentioned, the largest climate benefit is obtained by reusing material, see Figure 9. The impact from construction, reconstruction, transport, and deconstruction is less significant than the material impact. Regarding the Adaptable Building, four scenarios were considered. See Figure 11. It was clear that the largest climate benefits can be made by reusing planar elements and other large building parts to save material. Phase 2 (change of layout) and Phase 4 (extension) apply most reuse, even though Phase 2 shows the lowest values since very little material is added combined with low energy use. Phase 4, on the other hand, has added new material for two more floors which is why the impact is much higher than in Phase 2. Both Phase 1 and Phase 4 add material equal to two floors. However, Phase 4 shows less than half the impact from material use. This since both the foundation and the roof can be reused in Phase 4, while this is considered new material in Phase 1. Phase 3 shows less impact than Phase 4 but with a higher impact from energy use. It is also shown that the impact from construction, reconstruction, transport, and deconstruction is less significant when compared to the material impact. The lowest contribution is from the transportation of new and reused materials.

5 THEORETICAL/PRACTICAL/ SOCIETAL IMPLICATIONS

The concept method can be a strategic help to guide the process of defining which scenario(s) to aim for in practical cases when developing new circular buildings. This would encourage the industry to produce more buildings adapted for more efficient future reuse while making well-advised choices to keep the climate impact low. In the same way the method could guide buyers, such as real estate developers or municipalities, to assess which scenario(s) would best suit their situation and future.

The development of the concept buildings was based on the requirements of clients in the project and done by a team representing the value chain. Evaluation of preserved functionality for a second life cycle was based on the Expert Team's experiences of the building systems, building deconstructions, testing and modelling of structural engineering, fire safety and acoustics. The concept buildings have not been deconstructed, rebuilt, or tested in a laboratory. Even so, the method of simulating reuse, SimFORCE, was a valuable tool when developing buildings and a positive implication for society.

SimFORCE provides valuable understandings of the context and results useful when writing guides for DfDR&R&A, see Figure 7, which is practically useful for future building reuse.

A building consists of many different products and components with varying lifespans. The performance of reused building elements depends on components, the usage phase length, the construction type and load cases. Maintenance and renovation can also affect the performance. Potential damages depend on deconstruction or demolition, but also storage and transport. This affects the outcome, such as quality and maintained value. SimFORCE and the heart process (Figure 4) help to structure evaluation of the functional requirements for deconstructed, relocated and reused structures that need to be defined according to regulations and standards.

Choosing a suitable building system according to procurement and the business model is important. Consulting with industrial structural suppliers and contractors at an early stage can save time, materials, and costs, as they know their building systems' capabilities.

6 CONCLUSIONS

For nature and the public good, it is crucial to respect a building and its materials, retaining functionality and reuse. This must be the focus of the design process of new buildings today. The concept method is a practical tool in steps to obtain more circular timber buildings:

- Collaboration by co-creation in the building's value chain and sharing of the team's knowledge were essential for successful development.
- The circular value cycle process (the heart process) provides an understanding of the context and results affecting circular buildings.
- Defining scenarios and requirements of the building to be designed and assessed regarding its potential climate impact reduction.
- The method SimFORCE (Simulation of Future-Oriented Reuse for a Circular Economy). It was valuable for improving designs and assessing preserved functionality in circular buildings, although the actual buildings have not been deconstructed, reconstructed or tested in a laboratory. A competent Expert Team is required for the assessment.

https://doi.org/10.52202/080684-0019

- It gives information to guide documents for adaptability, deconstruction and reconstruction.

The concept method was applied to two concept buildings, designed to be adaptable and reusable with preserved functionality with environmental benefits:

- The Modular Building: a timber building with industrially manufactured volumes designed to be relocated and reconstructed (elastic).
- The Adaptable Building: a timber building with planar elements, designed to be flexible and versatile, relocated and extended with two added floors (elastic).

Environmental evaluations compared the first life cycle (initial building) to the second life cycle of the respective building. In all scenarios, the reuse of timber buildings shows a substantial potential to reduce the climate impact, in the order of 50%, in the product stage (A1-A3).

The results demonstrate that various scenarios can be considered to adapt to future needs. The concept method can be used to define strategies for clients and authorities. The results of the concept of buildings demonstrate that buildings can be designed for adaptability while keeping the climate impact low of the initial building. Guides for adaptability, deconstruction and reconstruction were formed to assist this future building transformation. This is valuable for real estate developers.

Future work is to continue the development of the SimFORCE method and interpret the complex process of designing buildings following the circular value cycle.

ACKNOWLEDGEMENTS

Thanks to the founder Vinnova Sweden's Innovation Agency. Thanks also to the industry partners for contributing with great commitment and knowledge in alphabetic order: Derome, ETTELVA Architects, Folkhem Trä, IsoTimber Holding, Kirunabostäder, Lindbäcks Bygg, Masonite Beams, OBOS, Ragn-Sells Recycling, Skellefteå municipality, Sortera, Swedish Wood, SSAM, Veidekke Entreprenad.

REFERENCES

CEAP 2020. Circular Economy Action plan for a cleaner and more competitive Europe #EUGreenDeal, PDF ISBN 978-92-76-19070-7, doi:10.2779/05068.

CGR, 2024. Circularity Gap Report, CGR+Global+2024+-+Report.pdf, Retrieved from website in January 2025, CGR 2024.

Çimen, Ö. (2021). Construction and built environment in circular economy: A comprehensive literature review. *Journal of cleaner production,* 305, 127180. https://doi.org/10.1016/j.jclepro.2021.127180

EN-17680:2023 (2023). Sustainability of construction works- evaluation of potential for sustainable refurbishment of buildings, Edition 1.

EN-15978:2011 (2011). Sustainability of construction works - Assessment of environmental performance of buildings - Calculation method, Edition 1. SIS/TK 209

EN-15804:2019 (2019). SS-EN15804:2012+A2:2019. Sustainability of construction works - Environmental product declarations - Core rules for the product category of construction products, Edition 1.

Höglmeier K., Weber-Blaschke G., Richter K. (2013). Potential for cascading of recovered wood from building deconstruction- A case study south-east Germany, *Resources, Conversation and Recycling*, 78 (2013) p 81-91).

ISO 20887:2020 (2020). Sustainability in buildings and civil engineering works -- Design for disassembly and adaptability -- Principles, requirements and guidance, Article Number STD-80019435, Edition 1.

Jockwear, R., Goto, Y., Scharn, E., & Crona, K. (2020). Design for adaption–making timber buildings ready for circular use and extended service life. In *IOP conference series: earth and environmental science* (Vol. 588, No. 5, p. 052025). IOP Publishing. http://dx.doi.org/10.1088/1755-1315/588/5/052025

Malmqvist, T., Borgström, S., Brismark, J.Erlandsson, M. (2021). *Referensvärden för klimatpåverkan vid uppförande av byggnader*, ISBN: 978-91-7873-954-7 https://kth.diva-portal.org/smash/record.jsf?pid=diva2:1626114

Nasiri B., Piccardo C., Hughes .M (2021). Estimating the material stock in wood residential houses in Finland, *Waste Management* 135 (2021) p. 318-326.

Ottenhaus L-M., Yan Z., Branders R., Leardini, Fink G., Jockwer R. (2023). Design for adaptability, disassembly and reuse- A review of reversible timber connection systems, *Construction and Building Materials 400 (2023)* https:/doi.org/10.1016/j.conbuildmat.2023.132823.

Potting, J., Hekkert M.P, Worell E., Hanemaaijer, A. (2017). *Circular Economy: Measuring Innovations in the Product chain,* Policy Report. PBL Netherlands Environmental Assessment Agency, The Hague, 2017 PBL publication number: 2544.

Sandberg K., Sandin Y, Harte A, Shotton E, Hughes M., Ridley-Ellis D., Turk G, Íñiguez-González G., Risse M., Cristescu C, (2022). *Summary report InFutUReWood- Innovative Design for the Future – Use and Reuse of Wood (Building) Components*, RISE Report 2022:08, DOI: 10.23699/p41e-ae46.

Sandin Y., Cramer M., Sandberg K. 2023. *How timber buildings can be designed for deconstruction and reuse in accordance with ISO 20227*, Proceeding of 13th World Conference on Timber Engineering, WCTE 2023, Oslo Norway, pp. 3558-3567, https://doi.org/10.52202/069179-0463,

Sandin, Y., Shotton, E., Cramer, M., Sandberg, K., Walsh St J., Östling, J., Cristescu, C., González -Alegre, V., Íñiguez-González, G., Llana, F.D., Carlsson, A., U í Chúláin, C., Jackson, N., García Barbero, M., Zaba la Mejia, A. (2022). *Design of Timber Buildings for Deconstruction and Reuse: Three methods and five case studies,* RISE Report 2022:52, https://urn.kb.se/resolve?urn=urn:nbn:se:ri:diva-59357.

Proceedings of the Circular Building Sector Conference 2025, Lund, Sweden, 1-3 June 2025

TOWARDS DOUBLE DESIGN: CREATING DURABLE AND ADAPTABLE BUILDINGS

Michael W. A. Cassidy[1] and Pieter de Wilde[2*]

[1]University of Plymouth, UK

[2]Lund University, LTH, Division of Energy and Building Design, Lund, Sweden

*Corresponding Author: pieter.de_wilde@ebd.lth.se

ABSTRACT

Background and aim: Long-life and long-usefulness are to be achieved by recognising the processes of functional obsolescence and structural degeneration and embracing uncertainty as an essential component of the future. Applying the Double-Design concept envisages designing buildings that will not only last a long time but, by incorporating adaptability and flexibility, continue to be useful for as long as they last. The exploration addresses a research gap in that while there are several studies of flexibility and adaptability, there have been no efforts to expand their scope to the limit.

Methods and Data: The work explores the possibility of designing for multiple uses over time with a distinction made between "hard" compatibilities between different clusters of activities (uses) and "soft" compatibilities, which relate to each successive transformation of function within a Double-Design framework. The analysis of hard compatibilities is summarised, while the significance of architectural and engineering design in managing uncertainty is supported by a detailed longitudinal study of a university in UK.

Findings: The exploration confirms the feasibility of implementing Double-Design regarding resource conservation. It is consistent with a movement towards high-performance buildings that invite greater user engagement.

Theoretical / Practical / Societal implications. Architecture, the construction industry and Architectural education need to emphasise a building's lifetime rather than just its first day of use. The public interest regulations guiding design must cover ethical principles embracing resource use and the environment. The concept is physically feasible, but several aspects of the professional and social mindset must change.

KEYWORDS: architectural design, building life expectancy, building performance, sustainability

1 BACKGROUND

Architecture carries a banner for the values and needs of its custodians and finds itself out of touch when those values and needs change. With some ingenuity, the inherited estate may sometimes be turned to good use to serve the incoming requirements and the changing values they represent. Yet a strategic choice remains: is it better that new buildings should be designed and built for their initial purpose only, to be killed off, demolished, abandoned as soon as that purpose has run its course (Cairns & Jacobs, 2014), or should they be designed to last and to be used productively well beyond the first use so that architecture may serve more readily the changing needs of society over an extended period (Kincaid, 2002)? Considering the common-sense aim of avoiding waste, the built environment must be designed to last as long as possible and guarantee its functional usefulness for as long as it lasts physically. This approach is called Double-Design which is envisaged as a response to the changing demands made upon architecture. Rather than designing for a single use, Double-Design allows for multiple changes of use, and while these principles are developed for application to new-build projects, they apply equally to reuse projects.

Physical compatibilities among the spatial needs of different activities can be identified, and designs can be based upon the highest common factors arising from this analysis. The implementation of Double-Design would affect the way design and building are undertaken in the future. Each new building would be required to respond to changes within its initial use and to accommodate different future uses.

This paper summarizes some key insights gained from a full PhD thesis which in itself reflects on a lifetime spend in architectural design practice. The structure of the paper

 https://doi.org/10.52202/080684-0020

is that of an exploratory essay that visits some of the key findings (Cassidy, 2023) (Cassidy, 2025).

2 FUNCTIONAL OBSOLESCENCE AND STRUCTURAL DEGENERATION

All buildings are subject to the powerful operatic processes of functional obsolescence and structural degeneration that apply respectively to the uses to which they are put and to the materials of which they are made.

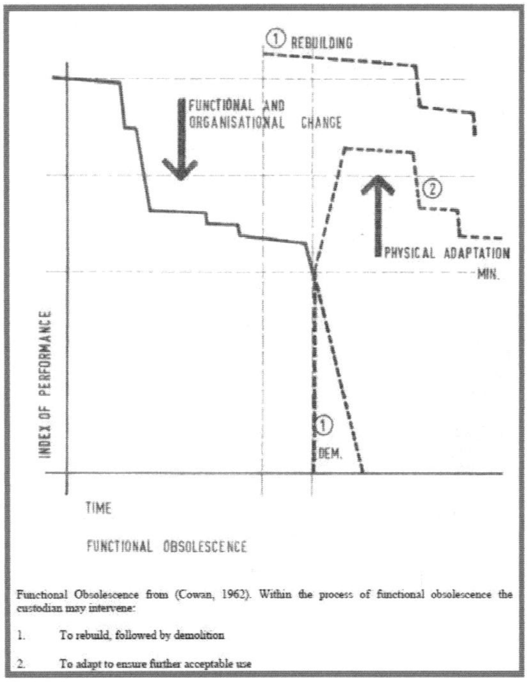

Figure 1: Functional Obsolescence (Cowan, Peter, 1962).

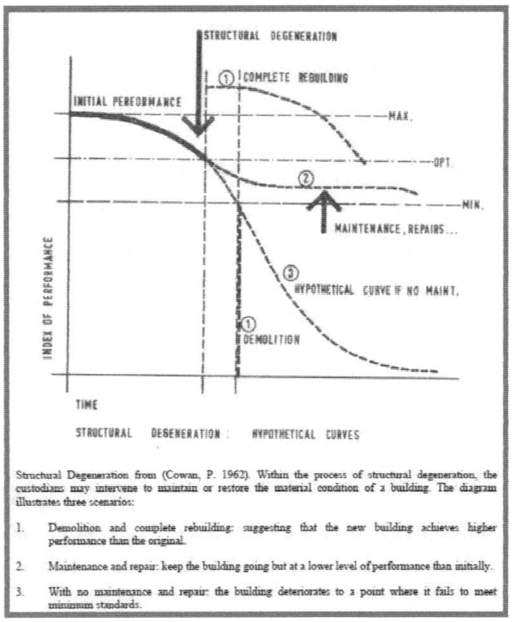

Figure 2: Structural degeneration (ibid.).

There is a renewed interest in adaptability and flexibility as design concepts that will contribute to a *longevity of usefulness* to complement a *longevity of physical lifespan*. The ease with which some older, more generously proportioned buildings are readily reused provides an obvious clue to how design needs to change. Some traditional forms of construction, stone for example, long considered too expensive, may also facilitate long life and return to serious consideration.

The alternative must also be considered. Can buildings or some of their components be designed with intrinsically sustainable materials that do not need to last so long? The counter-argument to Double-Design, is to deploy sustainable or recyclable materials and demolish/reuse them when structural degeneration or functional obsolescence kicks in. This approach might not achieve the smooth transition enabled by Double-Design from one use to another within a long-lasting space. It would rely upon comprehensive reconstruction to change use rather than upon interior adaptation, as with Double-Design. Indicative costs for buildings in use suggest that the longer-life options represent better value for money.

Schmidt and Austin provide a far-reaching analysis of adaptability that starts from a belief that: "a chasm remains between a perception of what architecture wants to be (in isolation as a finished and static sculpted work) and the reality of what architecture is (continually shifting in form and purpose to accommodate changing needs)" (Schmidt & Austin, 2016, p. xx). The Open Building movement led by Habraken has also laid the foundations for this exploration (Habraken, N.J., 2011). The work upon which this paper is based goes beyond current published research in examining the possibility of designing for very long-life buildings that would be able to accommodate many different uses with easy transitions between them. Each use would be able, by virtue of the inherent flexibility and adaptability, to deploy the best information and advice to support each new fit-out design. Although current research covers to some extent some of the issues considered here, architectural practice remains firmly within a short-term cost-based environment. This will need to change if Double-Design is to succeed.

The distinction proposed by Groak between adaptability (capable of different social uses) and flexibility (capable of other physical arrangements) is helpful. Both play a part in helping to enable buildings to last longer in productive use (Groak, 1992, p. 5).

Flexibility has been explored by the Open Building movement in the USA and internationally. Habraken's separation of the supports of a building from the infill is an essential contribution (Habraken, 2011). As Kendall argues, "Buildings are increasingly complex. Social change is accelerating. Given these circumstances, it is important to design and construct multi-unit buildings to avoid conflict, reduce dependencies among and between parties [...] and thus achieve maximum autonomy or freedom of decisions for each individual unit" (Kendall, 2004, p. 1).

Differing attitudes to the expected life and value of architecture characterise the sustainability debate, yet space itself is rarely mentioned. The buildings for which architecture is responsible comprise both space and materials. While it is taken for granted that some existing buildings can be reused productively, this cultural phenomenon has not influenced the design of the new stock. There have been few studies looking for the characteristics of buildings that render them suitable for productive reuse.

Forensic architecture is concerned primarily with the avoidance of decay and deterioration (Harris, 2001: Richardson, 2001: Ransom, 2002; Douglas, 2006: Watt, 2007) and through the creative analysis of positive interventions to achieve reuse (Kincaid, 2002: Wong, 2017). In addition, concern for the treatment of historic buildings provides a further, more specialized motivation (Grimmer, 2017).

Environmental concerns have influenced the development of high-performance buildings in which the quality of materials may be selected on the basis of long-term value.

3 PROBLEMS AND CHALLENGES

The work that architects undertake has a long-lasting impact, yet the focus of their attention on designing and the focus of their training in preparation for a professional career lies with satisfying requirements defined at the start of a project, with little consideration for long-term functionality and scant recognition of the inevitability of change. Hence, time and space are central to what architecture is about. Double-Design is especially relevant because architecture must be seen as occupying time as well as space. If Double-Design is to be fully implemented, the time dimension has to be central to the commissioning of buildings. As well as forming part of the evaluation/testing of new designs, the themes of change and growth and re-use must play a much more critical role in the briefing for new buildings and in evaluating the suitability for reusing existing buildings. The idea that buildings can be designed for multiple future uses recognises that the political, social and economic context of architecture changes over time. The distinction between place and space, together with the philosophical interdependence of space and time show that while the design process produces a fixed place, this place is the container for activities that are far from fixed and subject to varying degrees of uncertainty. The design process is focused on producing something spatial that is finite at its time of inception and construction but, thereafter, subject to the exigencies of use and transformation. The dichotomy is that of a building as object, fixed in time, and of a building as a container of human activities that occupy time as well as place.

Most of the commentaries regarding architecture and its place in the world have been written from an exo-architectural perspective, from outside looking in, and in many cases, from the outside looking in and back. It has proved difficult for journalists, architectural historians, and even architects to make the long-term use of buildings as compelling to the public as an iconic image. If architecture is to be improved, it is by understanding better the endo-architectural processes, what happens within the design process itself. The way in which architects navigate the information that guides design decisions is especially important.

Buildings are replaced over time. Despite the suggestion that the city is going to benefit from the additions of some kinds of spaces more than others, there does not seem to have been any attempt to ensure that this message gets through to those with the power to commission new space. The market-driven decentralized commissioning process relies upon the custodians and their architects to take into account the potential contribution that space can make to future activities. There are currently no incentives for custodians to look beyond their immediate and known requirements when starting a new project.

4 UNCERTAINTY

Considering the extraordinary diversity of changes of use observed throughout the building stock and throughout the world, it is tempting to be overwhelmed by the uncertainty that inevitably attends the start of a project. But while changes and the sequence of their occurrence cannot be forecast with accuracy, a range of possible changes in use could be suggested and, given that for each of those there is a set of requirements that can be defined, an environment could be designed to accommodate different activities throughout the physical life of the building.

Uncertainty is a condition confronting organizations and institutions, yet awareness and perception of the condition are experienced, communicated, and reacted to by individuals. Therefore, it is a surprise to find very little understanding of the interdependence of individual and institutional uncertainty. Anderson et al. address this question, suggesting that: "Uncertainty is fundamentally a mental state, a subjective, cognitive experience of human beings rather than a feature of the objective, material world. The specific focus of this experience, furthermore, is ignorance – i.e., the lack of knowledge. It is a higher-order metacognition representing a particular kind of explicit knowledge – an acknowledgment of *what* one does not know, but also *that* one does not know" (Anderson et al., 2019, p. 2). Importantly, far from the threat usually described, they show that uncertainty can have positive benefits (Anderson et al., 2019, p. 7). From a wider perspective, Kelly suggests that it is: "Impossible to be certain of anything except that everyone suffers as a consequence of being born. What is usually overlooked is that uncertainty, when consciously faced and perceived in the context of life's totality, is the creative aspect of being [...]The process can last for many years, even a lifetime, but with the knowledge that the uncertainty of living is

https://doi.org/10.52202/080684-0020

gradually being transformed to a higher octave of truth" (Kelly, 2018).

Designing for change brings organizational benefits if moves and disruption can be avoided. For some organizations working in exceptionally competitive environments, the speed at which a change can be affected may be critical to their survival. Looking at businesses, it may be imperative for them to introduce innovations quickly so as not to have to go through an elaborate change of use process. This accords with business models of decision-making in dynamic organizations (Lyneis & Sterman, 2009).

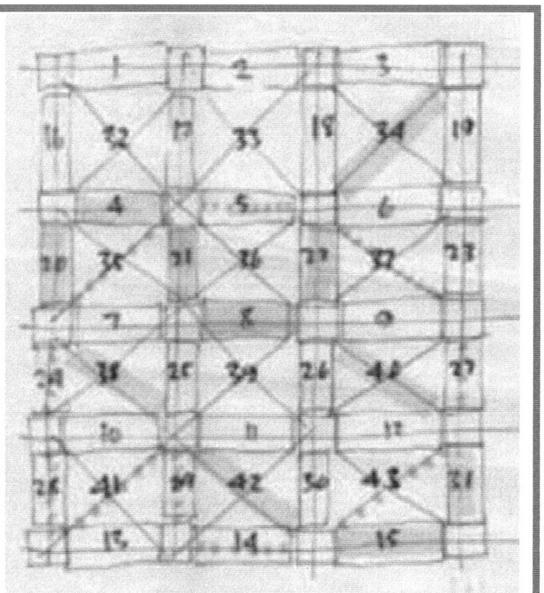

Theoretical diagram of grid design for university science complex showing possible interconnections between buildings. Narrow buildings can be multi-storey while 'courtyards' can accommodate 'lumpy' buildings like workshops and lecture theatres.

Figure 3: Diagrammatic design grid (YRM Architects).

A longitudinal study of a large and successful university building provided dramatic evidence of the way in which uncertainty affects the way it is necessary for buildings and their occupants to respond to change. The diagrammatic grid planned initially was intended to allow for both growth with connectivity and a high degree of internal flexibility. The unexpected events arising within the institution included:

Recruitment. The quality, effectiveness and ambition of employees are influential. The need to respond, sometimes very rapidly, to opportunities arising from the availability of special people and special money (investment, research funds, etc.) has a significant impact on campus development. There was a regular assessment of the academic marketplace regarding national and regional interests, which inevitably informed decisions about priorities. Opportunities for merging with other existing institutions arose when the momentum of the new institution was recognized.

The outcome of disputes. The refusal of newly appointed senior staff to respect the provisions for growth that were already part of the campus plan significantly impacted the connectivity of departments as the university expanded.

Changes in the administrative setup and decision-making machinery influenced changing priorities through patronage and funding. With campus growth, the mechanisms by which functional requirements are identified and communicated were divided into two parallel processes with separate teams responsible for space allocation and space procurement. This sophistication is matched by a changing balance between centrally timetabled space and locally controlled space. The allocation of space to solve short-term problems leads to complications when the temporary occupants demand changes to the fabric and service provision of their "temporary" home. The "host" is forced sometimes to struggle to get back their "lost" territory over decades. There were several examples in which changes of use took place in response to unexpected demands.

Technical decisions were made in light of the best available knowledge at the time. The central computer facility was initially located less than 200 yards (183 m) from places it served. As soon as technical advances outgrew this constraint, the space occupied was re-allocated to a succession of other uses. Space and environmental services needed to be updated as equipment was replaced.

Unexpected events arising from aspects of the external environment outside the control of the institution include:

Finance. To a large extent, the development of the campus reflects the timing of funding and the control exercised by the funding authority. Since the funding authority is itself subject to national financial allocation, the campus development was frequently at the mercy of what seemed to be arbitrary investment cuts and delays. The lack of funds at critical times led, in extreme instances, to staircases and toilets being converted to offices and laboratories. The change from being wholly publicly financed to being reliant upon diverse sources of finance affected every aspect of campus growth. Opportunities for private investment in campus buildings could not be overlooked.

Land and town planning. The need to assemble land from different donors and achieve development approved by local planning authorities influenced campus growth and traffic and pedestrian movement patterns.

Implementation. Many factors may influence the implementation of projects. These include design issues, contractor performance and financial stability, strikes, material availability and so on.

Regulations. The retrospective application of improved standards of health and safety affected both space and services provision.

Despite the turbulence, these institutional buildings have continued to work and it is not difficult to see that the unpredictability experienced would apply in some measure to many other projects, public and private, residential and commercial. It is essential to recognize the interdependent impact of these factors. The changes in university funding during the 1980s, referred to by Troiani and Carless (Troiani & Carless, 2021), encouraged an opportunistic approach to campus planning that was not consistent with continuing support for an established planned pattern, however rationally that was based upon a sensible appreciation of needs.

The example of Warwick University science buildings is not proposed as a prototype for Double-Design; rather it supports the basic idea that building morphology has an important role to play in establishing longevity.

5 KNOWLEDGE / RESOURCES / ENVIRONMENT

The design process must be fully understood if it is to be improved as a mechanism for society to manage uncertainty. This is the case whether decision-making for design assumes rationality (Simon, 1972), acknowledges complexity (Webber & Rittel, 1973) or relies upon regulatory prescription (RIBA, 2013). Given the legal and moral obligations of the architect, the importance of information and the recognition of its significance and its limitations cannot be over-emphasised.

Design professionals receive information from their clients and from their own searches. This traditional pattern of GIVEN and TAKEN information is disrupted when sources of information and the associated guidance are suspect. Professions that sign up for independent and honest service to the public will need, in such circumstances, to find a more robust and ethical basis for decision-making.

The politicisation of the narrative concerning news and priorities, its control, and its communication and promotion through mass media provide an unreliable information environment (C. P. Smith, 2021). The extent to which the objectivity of science is subverted by sponsorship further damages any hope for objectivity (Wall Street Journal, 2024) (Funding Sponsorship Bias, 2023). In the face of evidence supporting and opposing several current environmental themes, an initial approach has been to seek higher-order heuristics and, thus, find a more secure basis for decision-making in ethical principles. Yet the issues here are not related solely to the environment or to design but to the nature of the world we live in and the world of information we inhabit. Navigating these treacherous waters to seek truth needs to be approached with an open mind. The oversimplified dualism of correspondence-based philosophy is being challenged by those who see coherence in the mechanisms of perception and what is being perceived. The most persuasive and pragmatic solution relies upon a more open and holistic worldview in which different realms of knowledge, while overlapping and complementary, can nevertheless be applied to decision-making on an everyday basis. The intellectual context for this is suggested by McGilchrist, who argues that "(i) ancient spiritual truths, (ii) neuroscience, (iii) physics and (iv) the best kind of philosophy all lead us towards a world that makes sense as a whole: they bring things together, not drive them apart into their separate silos again. We need, he argues powerfully, to start seeing tables, mountains, nature, the cosmos and ourselves as facets of some ultimately connected, not sundered, state of affairs" (Read, 2022, p. 10).

Envisaging human society as an organism seeking its own sustainability creates an attractive metaphor for viewing the pursuit of truth. This is the idea behind wild systems theory, which, according to Jordan: "reconciles scientific and cultural narratives by first asserting that all of reality is inherently interrelated. Meaning, therefore, is this ubiquitous web of interrelations; choice is the means by which we navigate it, and selves are the patterns of interrelations we embody and manage over the course of our lives. Because such meaningful selves emerge step by step out of the trajectory of lived life, they are story-like; that is, they are narratives. And because these narratives always reflect a constellation of choices and chance, they are wild. In short, we are wild narratives" (Jordan, 2024). Here, we have a philosophical framework able to accommodate all forms of truth. It must be mapped to navigate the terrain of design. Our spectrum of cognitive abilities needs to match the many different domains of knowledge so that choices can be made with the confidence arising from a comprehensive understanding of the system whose interconnectedness strives for survival. In navigating successfully, we must be prepared to use the best information available in each situation and, recognising uncertainty and change, still be ready to make choices. We need both left- and right-hemisphere brain function but need to end with right-hemisphere holism "which has been dangerously eclipsed by left-hemispheric mono-maniac reductionism") (Read, 2022, p. 2).(Pinto et al., 2017) (Enns, 1997).

5.1 ENGAGEMENT/PARTICIPATION

While an allowance for user participation may help to humanise the experience of architectural space, it may also be necessary to dramatise and symbolize the differentiation of urban forms. The demands made of architecture go beyond the "purely functional" and must include other forms of satisfaction (de Botton, 2007). The application of "Double-Design" must not preclude the experienced pleasures that attend an architecture of variation (Spuybroek, 2009) as well as an architecture of eccentric intervention (Maudlin & Vellinga, 2014). As Vischer suggests, in seeking to develop a user-centred theory of the built environment, psychological comfort is included in the rating of how well the built environment performs as well as physical comfort and functional comfort (Vischer, 2008). These ideas support the value of manifest occupancy which could become an important element in the implementation of Double-Design.

Considering the central importance of interaction in any understanding of what architecture is about, it seems surprising that little attention has been paid to the active

https://doi.org/10.52202/080684-0020

encouragement of "user participation" concerning completed buildings. If we are to listen to the interests and wishes of building users, perhaps there need to be limits to the decisions left to the architects. Gone are the days when great architects designed everything in a building, from the door handles to the curtain rails. The architect Candilis put it well: "It is impossible for each man to construct his house for himself. But the architect must make it possible for each man to make his house his home. We must design the habitat only to the point at which man can take over" (Candilis, 1962, pp. 559–602). But how to establish exactly where that point is? A starting point is to assess the potentiality for participation for different building types.

There is evidence that offering users more control over their local environmental conditions brings a wide variety of benefits, not least in the current context of concerns about energy consumption. However, it would be ironic if the ready availability of control devices gave rise to the sacrifice of personal autonomy and the handing over of absolute control to the manufacturers of the devices. Studies are already identifying the public concerns and lack of trust in such technologies. As Wilson et al. say, in their analysis of the benefits and risks of smart home technologies: "Both prospective users and actual early adopters also express caution towards ceding autonomy and independence in the home for increased technological control. These broader sociotechnical risks are perceived more strongly than the privacy and data security concerns that have affected smart meter rollouts in the EU" (Wilson et al., 2017, p. 82).

Empowering the users of buildings to control their own comfort and environment is an intrinsically good thing with obvious benefits to the users themselves, to their employers and to the manufacturers of all the devices that support that empowerment. The involvement of users with the fabric of the building, with the local environment and with the furniture arrangements are all seen as helping to prolong the usefulness of the building so that it lasts functionally as long as it lasts physically.

6 PUBLIC INTEREST

Most countries in the world seek to protect their citizens from harm and to keep them safe, and they try to achieve this by means of regulation "in the public interest". Recent tragedies in the UK have demonstrated what happens when these regulators are weakened or compromised (Waite, 2022). For Double-Design to be fully implemented, against a backdrop of liberal economics and short-term thinking, it will need to be required by law so that all development takes place on a level playing field. The public interest will need to be redefined to accord with today's priorities.

As Arendt pointed out: "If the world is to contain a public space, it cannot be erected for one generation and planned for the living alone; it must transcend the life-span of mortal men" (Arendt, 1958, p. 55). Even before the onset of environmental concerns, the idea of societal altruism was commanding scholarly attention. Arguing that sociological theory had provided uncritical support for economic concepts like the rationality of self-interest, Monroe introduces a search for an alternative approach to the classical microeconomics of Adam Smith. She suggests that: "Only by understanding how people see themselves in relation to others can we begin to build a science of politics that allows for the complex interrelationship between the human needs to protect and nurture our self-interest *and* the needs for human sociability. Political science is a discipline looking for a new paradigm, a discipline ready for a new paradigm. Psychology and identity provide that paradigm through a theory of perspective on self in relation to others. (Monroe, 2001, p. 166). Other studies focus on the incompatibility of economic growth and sustainability and argue for a new approach to education that will emphasise this as a factual starting point. As Kopkina & Bedford say: "Just as the civil rights movement and rejection of racism and sexism have become mainstream in education in most institutional contexts across the world, so can an understanding of the need to halt environmental destruction be understood and widely shared and supported by both social movements (e.g., environmentalism, animal welfare/rights) and translated into the curriculum" (Kopnina & Bedford, 2024, p. 10). It is important that a clear academic understanding is emerging that accommodates the urgent redefinition of 'public interest'. The environmental argument for change is set out in an activist blog: "Sharing things and helping other people may damage the economy, but it's a great way to decrease our environmental footprint. Since the earth's resources are finite, competing to out-consume one another is a self-destructive course of action. This, however, is the natural outcome of capitalism, with its focus on money at the expense of all else ('Environmentalism & Altruism,' 2020).

7 HARD AND SOFT COMPATIBILITIES

The successful development of Double-Design requires an assessment of compatibilities covering practical criteria like floor-loading, floor-to-floor heights and plan depths. These physical compatibilities supporting functional changes of use are hard and the compatibilities supporting other transformations (which may be within an existing use or to secure a different use) are soft. The hard compatibilities cover the physical features of buildings, while the soft compatibilities allow for the essential manifestations of occupancy that contribute to the value of the experienced environment. The fundamental distinction between hard and soft compatibilities is that hard compatibilities are established at once and last for a long time. In contrast, soft compatibilities can be allowed, even encouraged, to merge, compete with, replace, and complement their predecessors without disrupting the long-term built infrastructure. While the application of the hard achieves the heavy lifting, the soft speaks of serving

https://doi.org/10.52202/080684-0020

the needs of specific uses, of meaning, feelings and the symbolic expressions of occupancy.

The hard compatibilities between different building uses have been assessed by analysing their basic physical requirements and assessing how much "spare" capacity would need to be incorporated to allow for other uses. Input data to this assessment has been taken from government guidance as well as published advice.

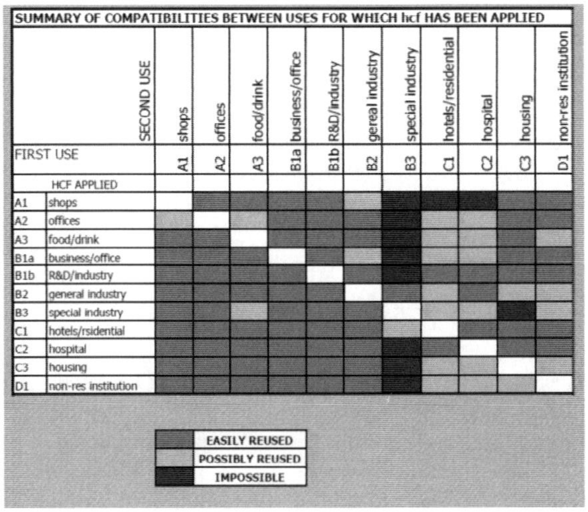

Figure 4: Summary of the outcome of physical compatibility analysis

7.1 RECONCILING HARD AND SOFT COMPATIBILITIES

Figure 4 indicates the potential value of a physical environment that incorporates the highest common factors covering floor-loading and the like. Having established a designed capacity to accommodate future changes of use, it is possible to factor in stated concerns about material conservation. With resource conservation as an unambiguous driver of the Double-Design idea, it is possible to envisage a framework that neatly encapsulates important aspects of the analysis:

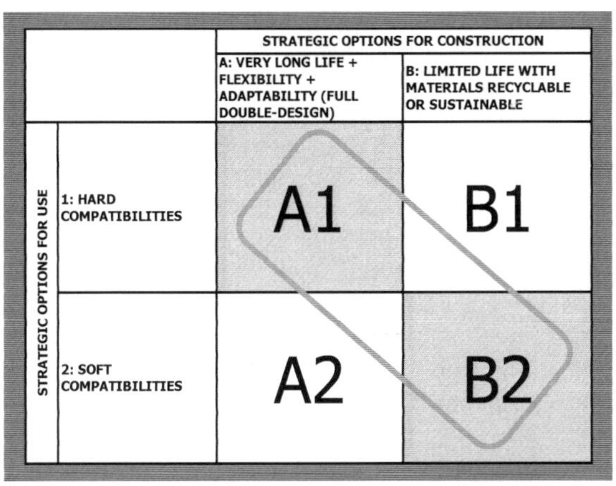

Figure 5: Strategic options for use and construction

7.1.1 A1 PLUS B2:

while A1 provides the long-lasting infrastructure securing compatibility of physical factors (floor loading, floor-to-floor heights and the like), B2 can provide the shorter-term interior design environment, the design for which can incorporate feedback and research intelligence specific to a particular use. The strategic combination of A1 and B2 is indicated as the optimum arrangement providing a robust 'infrastructure' within which soft requirements may change. This approach suggests that materials with an intrinsically long life must be deployed to achieve the desired longevity while shorter-life materials are used to match shorter-life functional and psychological requirements.

7.1.2 B1 alone

Limited life construction options using recyclable and sustainable materials may meet some design criteria but make no contribution to the overall lifetime-materials equation nor the ease of transformation from one use to the next.

7.1.3 A2 alone

It is possible to imagine examples, like some works of religious significance, in which soft compatibilities (the indicators of manifest occupancy) may be achieved through long-life construction. These will be exceptions to the general expectation that psychological needs will be met using shorter-lasting solutions.

8 DOUBLE-DESIGN SUMMARY

Figure 6 illustrates how the flow of given and taken information affects the management of functional obsolescence and structural degeneration over building life.

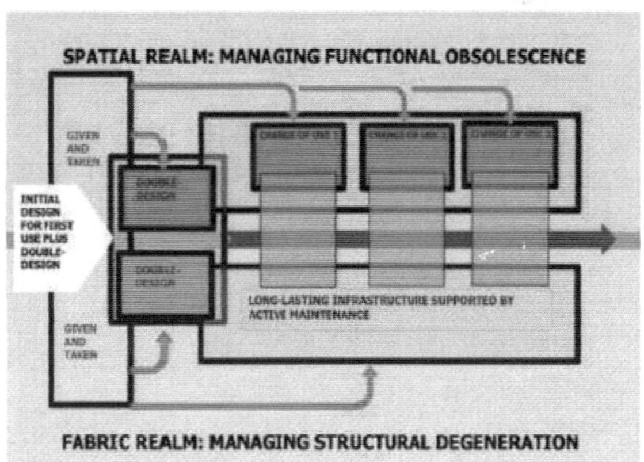

Figure 6: Managing Double-Design

Double-Design suggests that buildings should:

- last as long as physically possible using intrinsically long-lasting materials for the supporting infrastructure

https://doi.org/10.52202/080684-0020

- be designed with adaptability and flexibility to be useful for as long as they last physically
- allow for a succession of different uses
- use possibly short-life and/or recyclable materials for fit-outs as uses change
- allow for growth and change
- allow for uncertainty
- allow for the best information and advice to support each successive change of use
- allow for each successive use to express its occupancy

By designing from the start for future changes of use, fewer resources will be consumed over the life of the building, there will be less waste of material and that transformations of use in response to changing needs will be achieved efficiently and without wasting time. Architecture would be designed to accommodate unknown future uses and the custodians and users of buildings would be empowered and enabled to play their full part in ensuring the usefulness of buildings for as long as they last physically.

Double-Design is intended to achieve interventions that will be beneficial to future custodians and users, whoever they turn out to be (Harvey, 1996). The long-term value of an increasing percentage of built space incorporating flexibility and adaptability will contribute to the democratisation of space, and of cities.

REFERENCES

Anderson, E. C., Carleton, R. N., Diefenbach, M., & Han, P. K. J. (2019). The Relationship Between Uncertainty and Affect. *Frontiers in Psychology*, *10*. https://doi.org/0.3389/fpsyg.2019.02504

Arendt, H. (1958). *The human Condition*. University of Chicago.

Cairns, Stephen & Jacobs, Jane M. (2014). *Buildings Must Die*. MIT Press.

Candilis, G. (1962, December). *The Role of the Architect in Community Building,*. Architectural Design.

Cassidy, M. (2023). *Spatial Robustness: Double-Design and the Democratisation of Space* [University of Plymouth].https://pearl.plymouth.ac.uk/ada-theses/9https://pearl.plymouth.ac.uk/ada-theses/9; https://pearl.plymouth.ac.uk/cgi/viewcontent.cgi?article=1008&context=ada-theses

Cassidy, M. (2025). *Double-Design for Durable Architecture*. Routledge.

Cowan, Peter. (1962). Studies in the Growth, Change and Ageing of Buildings. *The Transactions of the Bartlett Society*.

de Botton, Alain. (2007). *The Architecture of Happiness*. Penguin Books.

Enns, J. T. (1997). Hemispheric Coordination of Spatial Attention. In *Advances in Psychology* (Vol. 123, pp. 197–231). Elsevier. https://doi.org/10.1016/S0166-4115(97)80074-X

Environmentalism & Altruism. (2020). *ENVIRONMENT AND ECOLOGY*. http://environment-ecology.com/environment-writings/508-environmentalism-a-altruism.htmlhttp://www.altruists.org/ideas/one_world/environmentalism/

Funding Sponsorship Bias. (2023, February 9). *The Embassy of Good Science*. https://embassy.science/wiki/Theme:B962d39b-ee34-4562-951c-5193700beff6

Grimmer, A. E. (2017). *THE SECRETARY OF THE INTERIOR'S STANDARDS FOR THE TREATMENT OF HISTORIC PROPERTIES WITH GUIDELINES FOR PRESERVING, REHABILITATING, RESTORING & RECONSTRUCTING HISTORIC BUILDINGS*. U.S. Department of the Interior National Park Service Technical Preservation Services Washington, D.C.

Groak, Steven. (1992). *The Idea of Building*. Spon.

Habraken, N.J. (2011). *Supports*. The Urban International Press.

Harvey, D. (1996). *Justice, Nature & the Geography of Difference*. Blackwell.

Jordan, J. S. (2024). *Wild Systems Theory Non-Reductive Naturalism and Narratives of Meaning, Choice, and Self.* World Scientific Publishing Company,. https://doi.org/10.1142/13635 |

Kelly, L. (2018). *The Call of the Spirit*. https://lancekelly.co.uk/a-step-closer-to-reality/

Kendall, D. S. (2004). An Open Building Strategy for Achieving Dwelling Unit Autonomy in Multi-unit Housing. *Housing and Society*, *31, 1*, 12.

Kincaid, David. (2002). *Adapting Buildings for Changing Uses*. Spon.

Kopnina, H., & Bedford, T. (2024). From Pseudo to Real Sustainability: Ecopedagogy in Business and Management Education. In *The Palgrave Handbook of Global Social Change* (pp. 1–15). Springer International Publishing. https://doi.org/10.1007/978-3-030-87624-1_369-1

Lyneis, J. L., & Sterman, J. D. (2009). *Green, Simple, and Profitable: The Paradox of Failed Best Practices in University Building Maintenance*. 26.

Massey, D. (2005). *For Space*. Sage.

Pinto, Y., De Haan, E. H. F., & Lamme, V. A. F. (2017). The Split-Brain Phenomenon Revisited: A Single Conscious Agent with Split Perception. *Trends in Cognitive Sciences*, *21*(11), 835–851. https://doi.org/10.1016/j.tics.2017.09.003

Read, R. (2022). Critical Notice: Iain McGilchrist, *The Matter With Things: Our Brains, Our Delusions, and the Unmaking of the World* (Perspectiva, 2021). 2 volumes, 1500 pages, no price. *Philosophical Investigations*, *45*(4), 528–539. https://doi.org/10.1111/phin.12349

Schmidt, R., & Austin, S. (n.d.). *Adaptable Architecture*. Routledge.

Simon, H. (1972). Theories of Bounded Rationality. *Decision and Organization*.

Smith, C. P. (2021). *The Climate Change Hoax Argument*. Frederick Printing.

Spuybroek, Lars (Ed.). (2009). *The architecture of Variation*. Thames and Hudson.

Troiani, I., & Carless, T. (2021). The Death and Life of UK Universities and the Cultural Spaces They Consume. *Architecture_MPS*, *19*(1). https://doi.org/10.14324/111.444.amps.2021v19i1.002

Waite, R. (2022, November 8). Grenfell architect bears primary responsibility for fire, barrister tells inquiry. *Architects' Journal*.

Wall Street Journal. (2024, May 14). *Flood of Fake Science Forces Multiple Journal Closures Wiley to shutter 19 more journals, some tainted by fraud*. https://www.wsj.com/science/academic-studies-research-paper-mills-journals-publishing-f5a3d4bc?reflink=desktopwebshare_permalink

Wilson, C., Hargreaves, T., & Hauxwell-Baldwin, R. (2017). Benefits and risks of smart home technologies. *Energy Policy*, *103*, 72–83. https://doi.org/10.1016/j.enpol.2016.12.047

Wong, L. (2017). *Adaptive Reuse: Extending the Lives of Buildings*. Birkhauser.

https://doi.org/10.52202/080684-0020

LUND UNIVERSITY | LTH FACULTY OF ENGINEERING

CIRCULAR IN, CIRCULAR OUT: APPLYING CIRCULAR DESIGN IN A POP-UP PUBLIC BUILDING IN LUND

Marwa Dabaieh[1] and Anas Lila[2]

[1] Malmo University, Department of Urban studies

[2] University of West England, School of Architecture and Environment

[2]Tanta University, Department of Architecture Engineering

*Corresponding Author: marwa.dabaieh@mau.se

ABSTRACT

Background and Aim: Circular design, emphasizing sustainability and resource efficiency across a building's lifecycle, has become increasingly critical in addressing global environmental challenges. This paper examines the implementation of circular principles in a temporary kindergarten as a public building project in Lund, Sweden. The aim is to demonstrate how these principles can be integrated from conceptualization through to end-of-life considerations, highlighting their potential impact on environmental sustainability.

Methods and Data: The project's approach involved research into circular design methodologies and sustainable building practices. Methods included design-for-disassembly and design-for-circularity strategies to enhance material reuse and minimize waste generation. Data collection encompassed regulatory compliance, material sourcing, and stakeholder engagement processes.

Findings: Innovative solutions in designing light weight and affordable temporary modular construction techniques and using parametric modelling and lifecycle impact as a tool to ensure low carbon circular design. The findings contribute to a deeper understanding of practical applications of circular design in urban development contexts. The study highlighted the importance of holistic design approaches that prioritize environmental responsibility.

Theoretical/Practical/Societal Implications: Theoretical implications highlight the transformative potential of integrating circular principles into architectural practices, fostering sustainable urban development. Practically, this study demonstrates the feasibility of adopting circular design strategies within existing regulatory frameworks, promoting environmental stewardship and community well-being. Societally, the study advances discourse on sustainable architecture by showcasing Lund, Sweden, as a model for implementing comprehensive circular design strategies in public infrastructure projects.

KEYWORDS: Circular design, design-for-disassembly, public building, sustainable architecture

1 INTRODUCTION

As cities grow and evolve, so does the need for flexible and sustainable infrastructure to meet the demands of their inhabitants. In Lund, Sweden, the increasing demand for kindergartens poses a significant challenge. Constructing temporary facilities to address this need is often resource-intensive, time-consuming, and environmentally unsustainable due to the limited lifespan of such buildings. This research introduces an innovative architecture solution: a circular pop-up kindergarten designed to provide adaptable, sustainable, and efficient early childhood education spaces.

The proposed design focuses on modularity and circularity, allowing the kindergarten to be constructed within days and easily disassembled, relocated, or repurposed for other functions across neighbourhoods in Lund. By maximizing reuse and flexibility, the structure

https://doi.org/10.52202/080684-0021

can transition from a temporary to a permanent facility or even serve as a different type of building, such as a school, student housing, or a clinic.

Critical infrastructure components, including kitchens and bathrooms, are integrated as eco-cycle systems, enabling seamless assembly, disassembly, and potential off-grid operation. The design employs a parametric approach to explore alternative proposals and optimize flexibility, ensuring the system adapts to diverse spatial and functional requirements while maintaining environmental sustainability. This research demonstrates the potential for scalable, circular design solutions to address the pressing need for temporary educational facilities in growing urban contexts, with broad implications for sustainable urban development.

1.1 APPLYING CIRCULARITY FOR TEMPORARY USE

The concept of the circular economy (CE) has gained prominence as a paradigm aimed at mitigating environmental degradation by transitioning from linear "take-make-dispose" models to regenerative systems. (Pomponi & Moncaster, 2017) emphasize that CE principles, such as resource efficiency and lifecycle design, have started to influence the construction sector, though their integration remains in its infancy. These principles advocate for the reuse and adaptability of building materials, promoting designs that support disassembly and repurposing, which align with sustainable development goals. Circular design within architecture involves creating flexible and adaptable structures to extend their lifecycle and functionality (Dabaieh, 2023). Scholars such as (Velenturf & Purnell, 2021) argue that incorporating modular and reusable elements in temporary constructions, like in our case a pop up kindergartens, can address environmental concerns and reduce resource consumption. Generally according to (Dabaieh & Alwall, 2018) temporary buildings are particularly problematic in their traditional forms, as their limited lifespan often leads to significant waste and inefficiencies.

Innovative approaches, such as modular construction using straw and reed panels, exemplify CE's potential in architecture. These materials, being locally sourced and biodegradable, align with nature-inspired principles, including biomimicry. Biomimetic circular economy models advocate for designs inspired by ecological cycles, emphasizing adaptability, efficiency, and sustainability (Soliman & Bo, 2023). Such methods can transform temporary buildings into versatile structures capable of fulfilling various functions, from kindergartens to clinics, aligning with emerging circular design methodologies. Despite growing interest, the application of CE principles in architectural education and practice is limited. Research by (Pomponi & Moncaster, 2017) highlights that many studies focus on material reuse and waste management but often neglect comprehensive lifecycle approaches that integrate economic and environmental metrics.

The concept of circular architecture and its application in temporary buildings is gaining increasing attention in the context of urban development due to issues with environmental impacts. Circular design principles emphasize the reuse of materials, adaptability of spaces, and minimization of waste, aligning well with the challenges posed by temporary structures such as kindergartens. Lund's growing need for kindergartens, combined with the environmental and logistical limitations of traditional temporary construction methods, highlights the relevance of circularity and low impact design approaches.

2 LITERATURE REVIEW

2.1 CIRCULARITY IN ARCHITECTURE

Circular architecture is centred on designing structures that are resource-efficient, modular, and capable of being reused or reconfigured over time. According to (Pomponi & Moncaster, 2017), circular construction prioritizes closed-loop material cycles, where components are reused or recycled to reduce environmental impact. This aligns with the proposed modular pop-up kindergarten, where materials such as straw and reed panels offer renewable and adaptable solutions. Previous studies, such as (Ghisellini et al., 2016), emphasize the potential of bio-based materials in sustainable design, particularly for their low embodied energy and capacity for reuse.

Despite its potential, circular design in temporary urban infrastructure faces challenges, including cost, scalability, and public acceptance. Studies by Kirchherr et al. (2017) point to a lack of standardization and policy support as significant barriers to the widespread adoption of circular practices. However, projects like this proposed kindergarten serve as valuable case studies, demonstrating the feasibility of combining modularity, sustainability, and urban adaptability.

2.2 MODULAR DESIGN AND FLEXIBILITY

The adaptability of modular systems has been widely explored as a solution for temporary structures. Research by (Smith, 2010) demonstrates that modular buildings can be rapidly constructed, easily transported, and reconfigured to serve various functions, from housing to healthcare. The flexible nature of modular components in this project allows the kindergarten to transition into other uses, such as student housing or clinics, echoing similar principles in existing studies. This adaptability also addresses the challenge of temporary facilities becoming obsolete, a common criticism in urban planning.

The integration of eco-cycle systems, such as off-grid kitchens and bathrooms, reflects the broader trend of self-sustaining urban infrastructure. Scholars like (Timmeren, 2006) highlight the importance of decentralized and sustainable systems in reducing dependency on traditional utilities. These systems not only lower the environmental footprint but also enhance the resilience and autonomy of temporary structures.

https://doi.org/10.52202/080684-0021

2.3 LCA AND PARAMETRIC DESIGN IN CIRCULAR PROJECTS

Design methods like parametric design, which allow for rapid prototyping and scenario testing, offer promising avenues for addressing issues with circularity gaps. Such methods facilitate the design of adaptable and relocatable structures, as evidenced in recent parametric models for sustainable urban planning. Parametric design has emerged as a powerful tool for exploring alternative design scenarios in sustainable architecture. (Al-Azzawi & Al-Majidi, 2021) describes an approach that leverages emerging computer-aided technologies and advanced manufacturing methods to produce highly intricate forms. It operates by defining a set of variables or parameters, with any adjustment to these inputs automatically modifying the resulting design. Similar approaches have been used in experimental urban projects, demonstrating their utility in creating adaptable and sustainable designs (Bielik et al., 2012).

According to (Roberts et al., 2020) situating Life Cycle Assessment (LCA) within the Royal Institute of British Architects (RIBA) Plan of Work highlights opportunities to integrate environmental considerations throughout the design process. While many studies address the synergy of LCA with Building Information Modelling (BIM) or Life Cycle Cost (LCC), as well as environmentally oriented parametric design, challenges persist when attempting to conduct LCA prior to full BIM implementation (Röck et al., 2018). In contrast, parametric methodologies—particularly those enabled by visual scripting tools such as Grasshopper—offer significant advantages in the early design stages by allowing designers to rapidly generate and evaluate multiple alternatives. This iterative framework supports multi-objective optimization, facilitating prompt feedback on both material choices and overall building performance before core design decisions are locked in.

The proactive incorporation of LCA at this conceptual phase thus has greater potential to reduce environmental impacts, as opposed to reactive measures taken once a design is already finalized. Although parametric tools require further refinement, localization, and validation to achieve broader industry acceptance, their capacity to inform holistic, performance-driven decision-making at the project's earliest stages underscores their critical role in advancing sustainable design practices.

(Säwén et al., 2022) explores how Life Cycle Assessment (LCA) can be embedded within the early phases of building design, emphasizing the advantages of parametric workflows in delivering immediate, iterative feedback on environmental impacts. While underscoring the overarching value of parametric LCA, the authors propose a characterization method for different LCA tools based on their functionality, data integration, and requisite expertise. To illustrate this framework, the study analyses four specific applications—BHoM LCA Toolkit, Bombyx, Tortuga, and Ardinal LCA—examining the limitations, opportunities, and user agency each tool provides. This comparison reveals how factors such as ease of use, learning curve, and database comprehensiveness shape a tool's suitability for early conceptual design.

Although the investigation omits certain available tools and relies, in part, on trial versions of software, it nonetheless furnishes an instructive overview of how these systems accommodate diverse project requirements and user backgrounds. In doing so, the authors (Säwén et al., 2022) highlight not only the potential for parametric LCA to steer sustainable choices long before designs become entrenched, but also the practical constraints—such as data quality, interface complexity, and workflow integration—that determine whether such methods can be widely adopted. Ultimately, their conclusions reiterate that bringing LCA into the early stages of design can produce more proactive environmental strategies, provided that tools are appropriately matched to the design team's skill set, project phase, and performance objectives.

3 KINDERGARTEN DESIGN PROPOSAL

The temporary pop-up kindergarten is designed using sustainable, locally sourced materials to create a healthy and eco-friendly learning environment. The main modular load-bearing walls are constructed from prefabricated compressed straw and reeds panels, providing thermal insulation as high thermal mass and structural stability while maintaining a low carbon footprint. The ceilings and roof are made from wood, ensuring a lightweight yet durable framework that harmonizes with the natural aesthetic of the design. Yet no excessive use of wood as it is not an abundant local material in south Sweden, for both the interior and exterior cladding, clay plaster is applied, offering breathability, humidity regulation, and a toxin-free environment ideal for young children. A final layer of Terra blocks and linseed oil for the exterior clay plastering for water resistivity. The key components and principles for modular design for disassembly is shown in figure (1).

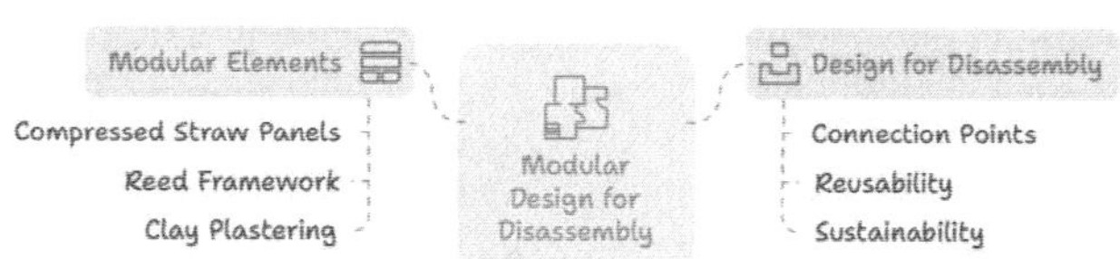

Figure (1) The key concept of circular modularity in the design proposal.

The design also integrates an eco-cycle system, ensuring resource efficiency and minimal environmental impact. A rainwater harvesting system is incorporated for irrigation and greywater reuse, while composting toilets contribute to waste reduction. Passive design strategies are used for heating, cooling, and ventilation, including thermal mass from straw and reed with clay plaster walls to regulate indoor temperatures, large overhangs for solar shading, and strategic window placement to maximize daylight and cross-ventilation. These features create a comfortable indoor climate year-round, reducing reliance on mechanical systems and promoting a self-sustaining, energy-efficient learning space. Figures 2 and 3 shows the architecture design proposal. Tables 1 and 2 present the material list and their corresponding U-values for the proposed kindergarten and a conventional one respectively.

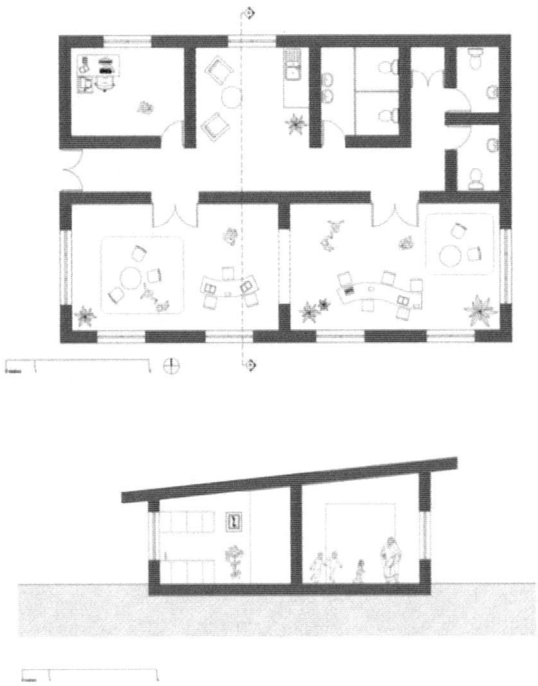

Figure (2) The design proposal (plan and section) for the school features a circular and modular pop-up building, emphasizing low-impact temporary architecture design proposal.

Table (1) List of materials used in pop up kindergarten design proposal

	Material	Thickness /cm	U -value
Roof	Solid wood	5 cm	0.1649
	Rock wool insulation membrane	20 cm	
	Cellulose membrane insulation	4 cm	
Wall	Strawbale	40 cm	0.1559
	Terra cladding bricks	12 cm	
	Flax fibre boards	5 cm	
	Clay plastering	2 cm	
Floor	Solid wood	5 cm	0.2676
	Flax seed insulation boards	15 cm	
	Compressed earth	10 cm	
Windows	Wood frame	2 cm	1.7592
	Double glazing	0.18	

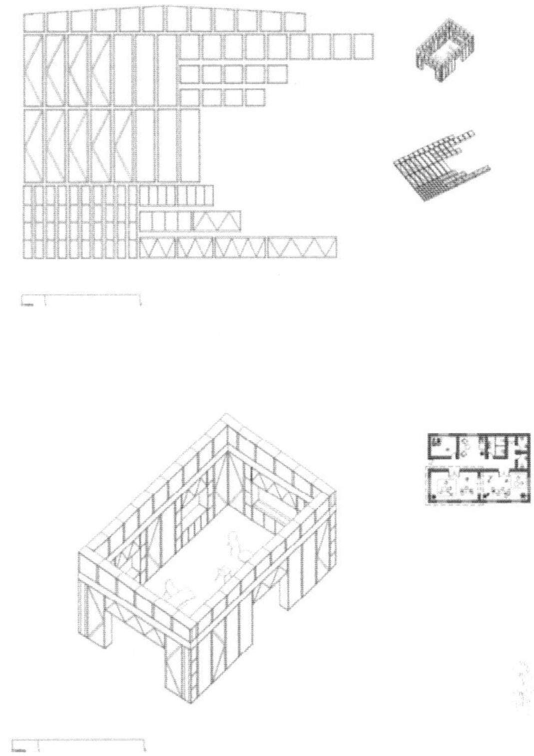

Figure (3) Modular school design featuring assembly and disassembly details for efficient construction, adaptability, and sustainable reuse.

Table (2) List of conventional materials used in temporary kindergarten

	Material	Thickness /cm	U -value
Roof	Precast concrete, high-strength concrete, ex works	20 cm	0.1801
	Rock wool (30 kg/m3)	20 cm	
	Bituminous waterproofing membrane, 50% recycled.	4 cm	
Wall	Precast concrete, high-strength concrete, ex works	20 cm	0.1765
	Rock wool, Flumroc (32 kg/m3)	20 cm	
	Gypsum lime plaster	4 cm	
Floor	Precast concrete, high-strength concrete, ex works	20 cm	0.331
	Rock wool (30 kg/m3)	20 cm	
	Natural stone slab cut	15 cm	
Windows	Wood frame	2 cm	1.7592
	Double glazing	0.18	

https://doi.org/10.52202/080684-0021

4 METHODOLOGY

This study adopts a structured methodology comprising four sequential steps to investigate and implement circular building design principles. These steps include a qualitative inquiry through expert interviews, design and development processes, building modelling and simulation, and life cycle assessment (LCA) calculations. Each step is described in detail below and show in figure 4.

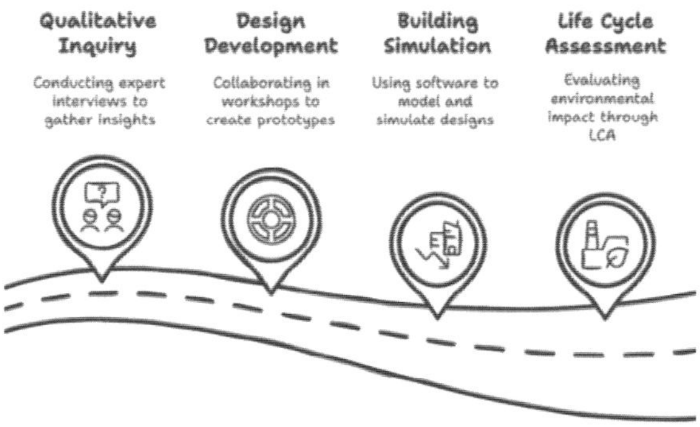

Fig. (4) The 4 steps methodological approach followed in this study.

Step 1: Qualitative Approach Using Expert Interviews

The initial phase involved a qualitative research approach aimed at gathering insights from professionals with expertise in circular design and prefabrication of modular building elements. 6 semi-structured interviews were conducted with six experts representing a diverse range of perspectives relevant to the field of circular building design. These included an architect, an engineer, a researcher, a municipality official, an investor and a contractor specialised in prefabricated construction. The interviews focused on assessing the feasibility of implementing circular design principles and modularity, as well as identifying potential challenges, constraints, and opportunities in both the design and construction processes using natural materials. The semi-structured nature of the interviews allowed for flexibility in exploring the unique insights of each expert while ensuring consistency in addressing the core study objectives. The data gathered during this phase provided critical context and informed subsequent stages for the kindergarten design.

Step 2: Design and Design Development

The second stage involved an iterative design process aimed at developing a circular kindergarten proof of concept. This phase was executed through two design workshops among the study team. These workshops facilitated collaborative brainstorming and the integration of circular principles into the design.
Following the workshops, the initial design concepts underwent systematic refinement through a process of parametric design rectification. This iterative approach allowed for the identification and resolution of design inefficiencies and inconsistencies while ensuring alignment with circular building principles. The design development phase was instrumental in translating theoretical concepts into actionable and practical design strategies.

Step 3: Building Modelling and Simulation

The third phase focused on the technical evaluation of the design through building modelling and simulation. Using advanced modelling software. In developing the modular design scenario, the authors implemented a parametric workflow in Grasshopper and coupled it with BombyX for real-time LCA calculations. The process began by establishing a baseline model—measuring 15 × 15 meters—with a 50% window-to-wall ratio (WWR) uniformly distributed across all four façades. The material specifications for this configuration were drawn from a predefined dataset, as detailed in Table 1. By systematically varying these materials through BombyX's parametric controls, the authors evaluated how modest changes in composition and assembly could yield measurable differences in overall environmental performance. To further probe design sensitivity, they introduced an additional variable—reducing the WWR to 40% in the final two iterations—thereby demonstrating how iterative refinements to fenestration ratios can influence LCA outcomes, even in a comparatively simple building massing.

Step 4: Life Cycle Assessment (LCA) Calculations

The fourth step is conducting a comprehensive life cycle assessment (LCA) to evaluate the environmental impact of the proposed building design, with a specific focus on carbon footprint. The LCA calculations adhered to established standards and methodologies, including ISO 14040 and ISO 14044, ensuring methodological rigor and comparability.
Bombyx is most frequently deployed for analysing upstream production impacts (A1–A3), which play a pivotal role in early-stage design decisions. This emphasis aligns with broader academic discourse highlighting how adjustments to geometry and material specifications at the conceptual and schematic phases can affect meaningful reductions in a building's overall environmental footprint. The software was chosen because of it's open-source foundation. It also permits bespoke extensions: as practitioners (as this study team) with Python coding skills can adapt the tool to include additional stages, thereby ensuring a more comprehensive life cycle assessment (Basic et al., 2019; Hollberg et al., 2022). However, the present study restricts its scope to the default cradle-to-gate functionalities (A1–A3) provided by Bombyx.
Bombyx's seamless integration with Grasshopper facilitates the modelling of fundamental building surfaces—walls, floors, roofs—through user-defined

geometries, which the tool links to regionalized databases like KBOB and Ökobaudat. By assigning impact factors based on each material's type and density, Bombyx automatically calculates key LCA indicators, such as global warming potential. Any change to the parametric model triggers an immediate recalculation, enabling designers to visualize how subtle modifications in building massing or component selections can reshape overall environmental performance in real time.

This near-instant feedback loop grants our study the flexibility to compare multiple design alternatives, optimize configurations, and explore an array of materials with minimal manual intervention. Existing research confirms Bombyx's capacity to embed sustainability considerations into iterative design processes, thereby reducing the likelihood of reactive changes later. Yet, it also underscores that reliance on standardized data sets and a predominant focus on the cradle-to-gate scope can limit the tool's utility for exhaustive life cycle tracking. Despite these constraints, Bombyx's potential to guide environmentally responsible choices remains substantial, particularly when applied at the earliest stages of building design which match this study scope. For those reasons, it was the most suitable tool to use for this study experimental work.

In this study, embodied energy is quantified by multiplying the mass or area of each construction material by its unit production-phase energy factor and incorporating additional replacement cycles determined by the 60-year reference study period and each component's service life. Operational energy is derived by summing annual useful energy demands for space heating, domestic hot water, lighting, and appliances, converting these to final energy via performance factors, and extrapolating the total over the same 60-year horizon. The total life-cycle energy footprint is then obtained by summing the embodied and operational energy contributions, with all calculations performed across multiple environmental indicators—including non-renewable and renewable primary energy use as well as global warming potential—to provide a comprehensive assessment of the building's environmental impacts.

5 RESULTS

By following this four-step methodology, the study systematically integrated qualitative insights from interviews , iterative design processes, technical modelling, and the quantitative environmental assessment to advance the understanding and implementation of circular building design. Each step contributed to the development of a holistic and sustainable building model, addressing both theoretical and practical aspects of circularity in the built environment.

5.1 THE OUTCOME OF THE SEMI STRUCTURED INTERVIEWS AND DESIGN WORKSHOP

The semi-structured interviews conducted with six professionals and practitioners —an architect, an engineer, a researcher, a municipality official, an investor and a contractor—provided valuable insights into the current state of circular design practices in architecture and construction. The findings highlighted the following key themes:

1. **Limited Awareness of Bio-based and Circular Materials**
   Across all stakeholder groups, there was a notable lack of familiarity with many sustainable, bio-based materials, such reeds, mycelium composites, and other agricultural waste-based products like straw. This knowledge gap was most pronounced among stakeholders directly involved in construction and municipal regulation.
2. **Perceptions of Feasibility and Implementation Challenges**
   While the concept of circular design was generally acknowledged as important, several interviewees expressed scepticism about its practical application. Challenges cited included concerns about material availability, regulatory ambiguity, and limited case studies demonstrating successful implementations.
3. **Institutional and Policy Barriers**
   The municipality official highlighted that regulatory frameworks or incentives for circular design are still in infant stages, making it difficult to advocate for the adoption of bio-based materials in public or large-scale projects.

Some of the key quotes that highlighted the challenges and the openness for the market to change towards alternative unconventional materials; from a factory CEO perspective working with precast concrete walls for prefabricated building *'Cement used in concrete, especially reinforced concrete, is a proven material. It's durable, low-maintenance, and long-lasting—all critical qualities in our line of work. While clay has its advantages, cement-based systems have been refined and optimized over decades'.* While from investor perspective *'I'm currently working on a multi-story rental building project in Lund, with funding from pension funds. The construction sector is very conservative, and shifting to unproven materials is a significant risk—especially for small investors like myself'.* As for an architect practitioner *'As an architect, I haven't been trained to design with clay and natural fibers. These materials aren't part of our architectural education, and most design tools and engineering support are tailored for concrete, steel, and other conventional systems'.*

https://doi.org/10.52202/080684-0021

For the design workshops, the interview outcomes were used as a supportive foundation for developing design sketches. The workshops served as a platform for brainstorming the kindergarten's simple and flexible architectural program, which includes two classrooms, a kitchen, bathrooms, and an administrative room. One of the primary requirements was to ensure the building could be adaptively reused for other functions if needed or continue serving as a kindergarten. Additionally, the design allows for future expansion by adding extra modules as necessary. The modular system was developed for disassembly and reassembly, with a targeted assembly time of eight hours for the entire kindergarten structure. The design was further refined following the parametric design and simulation phase to optimize performance and adaptability.

5.2 THE PARAMETRIC MODELLING AND LCA CALCULATIONS

The parametric modelling and Life Cycle Assessment (LCA) were conducted within a "cradle-to-gate" system boundary, which encompasses material extraction, processing, and manufacturing phases up to the point where materials leave the production site. The analysis excluded construction, use, and end-of-life stages to maintain a focused comparison of material choices in building design. The assessment revealed that the proposal, which incorporated a carefully selected list of bio-based natural materials, including wood, straw, reeds in addition to clay as a natural material, achieved a significant reduction in the building's carbon footprint. Compared to the conventional base case model relying on standard construction materials such as concrete and steel, the carbon footprint was reduced by more than 50%, indicating the immense potential of sustainable material strategies for mitigating environmental impact in architecture. The key findings from the LCA calculations are:

- **Carbon Emission Reduction:** The incorporation of natural materials reduced embodied carbon emissions to 6.148 (kg CO_2-eq/m^2 a) compared to 15.554 (kg CO_2-eq/m^2 a) for the base case. While embodied green house gasses is 3.32 (kg CO_2-eq/m^2 a) compared to 12.583 (kg CO_2-eq/m^2 a).
- **High Carbon Sequestration Potential:** Bio-based materials like reed and straw demonstrated the ability to sequester atmospheric carbon, contributing positively to the building's overall carbon balance. The Biogenic Carbon Storage is calculated to be 3.959 (kg CO_2-eq/m^2 a) compared to zero for the base case.
- **Material Efficiency:** Parametric modelling allowed for the optimization of material use, minimizing waste and enhancing structural efficiency while adhering to circular design principles.

5.3 MATERIAL CHOICES FOR MODULAR DESIGN DETAILING

The outcome of the simulation was informative to enhance the design detailing. The final choice for the materials suggested for the walls are primarily compressed straw bales panels and reeds, which serve as a load-bearing modular system, covered externally with pressed earth and treated Terra blocks and linseed oil for water resistance. The interior surfaces are plastered with clay. For the roof, solid wood is proposed, combined with rock wool thermal insulation boards and a cellulose membrane for waterproofing. The flooring structure consists of solid wood with flax fibers for thermal insulation and compressed earth for cladding. The modular elements are intentionally crafted from sustainable, biodegradable materials designed for adaptability, ensuring they can be assembled and reconfigured for varied purposes. For the base case comparison, the walls are assumed to consist of typical temporary precast concrete, with roofs using standard rock wool insulation and bitumen for floor waterproofing.

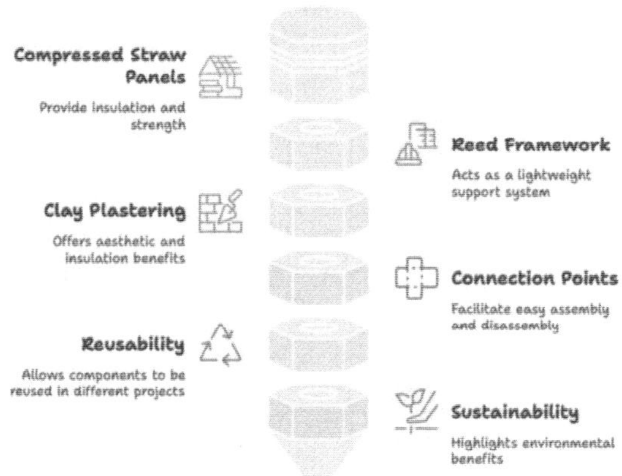

Fig. (5) The proposed circular and modular thinking for the pop-up kindergarten design proposal.

6 DISCUSSIONS

The findings from the interviews revealed still critical barriers to the adoption of circular and bio-based materials in contemporary architectural practice, as well as opportunities for advancing sustainable design practices. The consistent lack of familiarity with bio-based materials from practitioners' side underscores the need for industry-wide awareness campaigns. That is very aligned with what (Kanters, 2020) concluded in his study as well. Knowledge-sharing platforms and collaborative research between academic institutions and industry professionals can provide stakeholders with the necessary technical knowledge to confidently integrate these materials into projects. When it comes to addressing perceptions of feasibility, demonstration projects showcasing the successful application of bio-based materials in circular designs are crucial. They can help build confidence in these materials by providing evidence of their structural performance, durability, and environmental benefits as discussed and validated by (Pearlmutter et al., 2019). As for policy and market transformation, the regulatory support and incentive programs are essential for fostering innovation in circular building practices. Policies that encourage the use of bio-based materials, along with certification systems for their quality and safety, can create a more favourable environment for their adoption. Annually, agricultural systems around the world produce about 570 MT of waste, providing a vast amount of material with very high potential for processing into bio-based products (Puglia et al., 2021).

The results of the parametric modelling and Life Cycle Assessment (LCA) highlight the significant environmental benefits of incorporating bio-based materials such as wood, straw, reed and low impact materials like clay in building construction. The cradle-to-gate analysis shown a carbon footprint reduction of more than 50% compared to conventional construction materials, demonstrating the potential of bio-based alternatives in mitigating the environmental impact of the built environment. The carbon sequestration properties of these materials, combined with their renewable nature, contribute positively to sustainable design objectives. The use of parametric modelling further optimized material allocation and minimized waste, showcasing the efficiency of computational tools in sustainable architecture.

However, despite these promising outcomes, the adoption of bio-based materials faces practical challenges, including limited supply chains, concerns about material durability, and the lack of standardized construction practices. Overcoming these barriers will require collaboration between architects, engineers, policymakers, and material suppliers, as well as the establishment of supportive regulations and incentives. The findings of this study emphasize the feasibility of adopting circular and sustainable design principles, offering a pathway toward lower-carbon building practices.

A more detailed cradle-to-cradle study is needed as a follow-up to this pilot experimental work, including comprehensive modelling of building performance and energy consumption. This will provide a complete overview of the building's impact after its end of life. Additionally, a life cycle cost analysis would serve as a valuable complement, offering insights into the building's costs compared to conventional structures. Additionally, investigating the scalability of these principles in various building types and contexts could yield valuable insights for sustainable urban development.

7 CONCLUSIONS

This study emphasises the transformative potential of circular design principles in architecture, particularly through the integration of bio-based materials and innovative construction methods. The significant reduction in carbon footprint—over 50%—when utilizing sustainable materials compared to conventional options highlights the urgent need for the architectural community to embrace these strategies. The findings from the modular pop-up kindergarten study in Lund further illustrate the feasibility of implementing circular design in urban settings, despite challenges such as regulatory constraints and material sourcing. However, barriers to widespread adoption remain, including limited awareness among stakeholders and the need for supportive policies. To address these challenges, the study advocates for enhanced knowledge-sharing initiatives, demonstration projects, and collaborative efforts among professionals in the field.

ACKNOWLEDGEMENTS

The authors would like to thank the participants in the interviews for their time and valuable input.

REFERENCES

Al-Azzawi, T., & Al-Majidi, Z. (2021). Parametric architecture: the second international style. *IOP Conference Series: Materials Science and Engineering*. https://doi.org/10.1088/1757-899X/1067/1/012019

Basic, S., Hollberg, A., Galimshina, A., & Habert, G. (2019). A design integrated parametric tool for real-time Life Cycle Assessment – Bombyx project. *IOP Conference Series: Earth and Environmental Science*, *323*(1), 012112. https://doi.org/10.1088/1755-1315/323/1/012112

Bielik, M., Schneider, S., & König, R. (2012). Parametric Urban Patterns. *International Conference on Education and Research in Computer Aided Architectural Design in Europe*. https://scholar.google.com/citations?view_op=view_citation&hl=en&user=Etb2-HEAAAAJ&citation_for_view=Etb2-HEAAAAJ:hqOjcs7Dif8C

Dabaieh, M. (2023). *Circular design for zero emission architecture and building practice : it is the green way or the highway*. Woodhead Publishing. Cambridge, MA

Dabaieh, M., & Alwall, J. (2018). Building now and building back. Refugees at the centre of an occupant driven design and construction process. *Sustainable Cities and Society*, *37*, 619–627. https://doi.org/10.1016/j.scs.2017.11.002

Ghisellini, P., Cialani, C., & Ulgiati, S. (2016). A review on circular economy: the expected transition to a balanced interplay of environmental and economic systems. *Journal of Cleaner Production*, *114*, 11–32. https://doi.org/10.1016/J.JCLEPRO.2015.09.007

Hollberg, A., Tjäder, M., Ingelhag, G., & Wallbaum, H. (2022). A Framework for User Centric LCA Tool Development for Early Planning Stages of Buildings. *Frontiers in Built Environment*, *8*,

https://doi.org/10.52202/080684-0021

744946.https://doi.org/10.3389/FBUIL.2022.744946/BIBTEX
Kanters, J. (2020). Circular Building Design: An Analysis of Barriers and Drivers for a Circular Building Sector. In Buildings (Vol. 10, Issue 4, p. 77). Multidisciplinary Digital Publishing Institute. https://doi.org/10.3390/buildings10040077
Pearlmutter, D., Theochari, D., Nehls, T., Pinho, P., Piro, P., Koroļova, A., Papaefthimiou, S., Mateo, M. C. G., Calheiros, C. S. C., Zluwa, I., Pitha, U., Schosseler, P. M., Florentin, Y., Ouannou, S., Gal, E., Aicher, A., Arnold, K., Igondová, E., & Pucher, B. (2019). Enhancing the circular economy with nature-based solutions in the built urban environment: green building materials, systems and sites. In Blue-Green Systems (Vol. 2, Issue 1, p. 46). UWA Publishing. https://doi.org/10.2166/bgs.2019.928
Pomponi, F., & Moncaster, A. (2017). Circular economy for the built environment: A research framework. *Journal of Cleaner Production*, *143*, 710–718. https://doi.org/10.1016/J.JCLEPRO.2016.12.055
Puglia, D., Pezzolla, D., Gigliotti, G., Torre, L., Bartucca, M. L., & Buono, D. D. (2021). The Opportunity of Valorizing Agricultural Waste, Through Its Conversion into Biostimulants, Biofertilizers, and Biopolymers. In Sustainability (Vol. 13, Issue 5, p. 2710). Multidisciplinary Digital Publishing Institute. https://doi.org/10.3390/su13052710
Roberts, M., Allen, S., & Coley, D. (2020). Life cycle assessment in the building design process – A systematic literature review. *Building and Environment*, *185*, 107274. https://doi.org/10.1016/J.BUILDENV.2020.107274
Röck, M., Hollberg, A., Habert, G., & Passer, A. (2018). LCA and BIM: Visualization of environmental potentials in building construction at early design stages. *Building and Environment*, *140*, 153–161. https://doi.org/10.1016/J.BUILDENV.2018.05.006
Säwén, T., Magnusson, E., Sasic Kalagasidis, A., & Hollberg, A. (2022). Tool characterisation framework for parametric building LCA. *IOP Conference Series: Earth and Environmental Science*, *1078*(1), 012090. https://doi.org/10.1088/1755-1315/1078/1/012090
Smith, R. E. . (2010). *Prefab architecture : a guide to modular design and construction*. John Wiley & Sons. https://books.google.com/books/about/Prefab_Architecture.html?id=BuZCQpIHRbAC
Soliman, M. E., & Bo, S. (2023). An innovative multifunctional biomimetic adaptive building envelope based on a novel integrated methodology of merging biological mechanisms. *Journal of Building Engineering*, *76*, 106995. https://doi.org/10.1016/J.JOBE.2023.106995
Timmeren, A. van. (2006). *Autonomie en heteronomie integratie en verduurzaming van essenti*le stromen in de gebouwde omgeving*. Eburon.
Velenturf, A. P. M., & Purnell, P. (2021). Principles for a sustainable circular economy. *Sustainable Production and Consumption*, *27*, 1437–1457. https://doi.org/10.1016/J.SPC.2021.02.018

Proceedings of the Circular Building Sector Conference 2025, Lund, Sweden, 1-3 June 2025

CIRCULAR STUDIO - COMBINING HERITAGE, TRANSFORMATION, RE-USE AND EMISSIONS TO A HOLISTIC STUDENT-ACTIVE LEARNING EXPERIENCE IN ARCHITECTURE

Pasi Aalto[1*] and Nora Linnea Ådnanes[1]

[1]Department of Architecture and Technology, Norwegian University of Science and Technology

*Corresponding Author: pasi.aalto@ntnu.no

ABSTRACT

Background and aim. Norway has approximately 6000 professional architects in its workforce. Their historical and current bulk of new-built projects will become scarcer as climate change, nature loss and societal considerations force increased use of existing buildings. It is therefore necessary that new architects have a solid, updated, and coherent education in efficiently working with existing buildings as they enter the workforce. This study follows the introduction of Circular Studio, a piloting curriculum development architectural studio course that focuses on existing buildings, re-use of materials and design development, aiming to report the identified concepts and perspectives as manifestations of experiential learning.

Methods and Data. The study utilised a before-after survey (N=19 first iteration, N=19 second iteration, of which 10 were matched pairs). Measured dimensions were correlation to NEP-15 environmental attitudes, BIDR Impression management, 20 statements about buildings focusing on resource use and circularity, 16 statements on the role of architects and 2 open questions focusing in the positive and negative impacts of architects as agents, as well as opportunities and barriers.

Findings. The explorative approach identified 5 concepts and perspectives as well as a multitude of indications on individual aspects of experiential learning in Circular Studio.

Practical implications. The study provides an initial test for a framework for the practical design of circularity curriculum in architectural education and suggestions for co-developing curriculum and evaluation research for evidence-based development for this shift in the architectural profession.

KEYWORDS: architectural education, Circular Economy, experiential learning, perception change

1 BACKGROUND AND AIM

Architectural education of the last decades differs from many higher educations by the central role of the design studio, understood as both a physical space, as well as a pedagogical approach to learning by doing (Corazzo, 2019; Schon, 1987). As a material space, the studio houses making objects, bridging contexts, building meaning, enabling activities as well as a background for learning and a space the enables the expression of disciplinary identity (Corazzo, 2019).The latter, forming an identity as a practitioner and the development of joint norms within the profession, is enabled through the studio as a space of immersion and expected behaviour (Boling et al., 2016; Corazzo, 2019; Thoring et al., 2018) and becoming insiders by iterative processes of solving problems, expressing solutions and shaping their own identities (Corazzo, 2019). In architectural education, the studio (as both physical space and pedagogical concept) is an integral part of the interplay between active learning, informal and creative learning spaces, participating in a collective with both fellow students and staff and forming ones identity (Aalto et al., 2023; Corazzo, 2019; Leijon et al., 2022; Lundahl et al., 2017; Thoring et al., 2018, 2019). Supported by smaller courses on theory, methods or knowledge building, *the studio* has also become synonymous with a teaching unit or a course in order to achieve an immersive, active learning experience in many schools of architecture within universities. We would therefore argue that Kolb's experiential learning cycle (Kolb, 2014) in its revised form (Morris, 2020) provides a sound theoretical lens to describe the architectural studio:

"...consists of contextually rich concrete experience, critical reflective observation, contextual-specific

https://doi.org/10.52202/080684-0022

abstract conceptualization, and pragmatic active experimentation." (Morris, 2020, p. 3)

It is in this context new societal challenges must be introduced to architecture students. Overarching concerns of new critical areas such as equality and sustainability force changes to curricula as the necessary knowledge base of future architects expands (Sopeoglou, 2024). As one core consideration, the building industry is currently using too many resources to be in line with agreed global pathways like the Paris Agreement and the Kunming-Montreal Nature Agreement. As a consequence, re-using building materials in accordance with circular economy principles has received much attention (Kanters, 2020; Kirchherr et al., 2017; Sopeoglou, 2024; Wuyts et al., 2022). We assume that one key competency of future architects is prolonging and increasing the benefit-to-society of the existing building stock. To re-think architectural education towards a circular building industry, Norwegian University of Science and Technology (NTNU) has replaced 10 master level electable courses in architecture with a single learning pathway– Circular Studio – that is built from the ground up to combine current knowledge about building transformation, heritage, re-use of materials, energy and emission calculations, analyzing existing qualities like light and color – as well as experimental practices and research to promote an explorative, knowledge-developing practice among students.

As the students work with circularity in this new studio, they are assumed to socially co-shape their architectural preferences with cultural influences from the profession and school, as well as their nearest peers (Wilson, 1996) and this development can be understood both as a professional and personal self-construction (Kararmaz, 2024). These identity shaping processes complement and challenge a multitude of previous perspectives, where some might be more susceptible to amendment or strengthened, while others would be re-buffed. In this context, the students relationship to nature as a foundation for their understanding of sustainability, biodiversity and ultimately circularity in their own profession can be contrasted with a more traditional image of the architectural profession where other values of architectural quality were more prominent.

In this study, we explore the students' current attitudes and views to circularity, existing buildings and the role of architects as well as the change in their perceptions about circularity and their own ambitions as a future architect as a result of completing 22,5 ECTS credits in the new learning pathway. Our focus is specifically on identifying potential changes so that emerging education in circularity can include these aspects into the planning of courses, learning exercises and other student activities. Specifically, we asked the following research questions:

1. Is there a correlation between the students overall environmental attitudes (NEP-15), their impression management (BIDR-IM) and their professional perception when framed with different practices, such as architects role, heritage concerns and new buildings?
2. Do these perceptions change during the course of a semester as they work hands-on with circular projects?
3. What could be relevant concepts and questions to consider when implementing circularity into educational practices in architecture?

We hope that by answering these questions, we can shed light on the design and development of necessary circular learning activities in universities and encourage cultural change among architects and building industry.

2 CONTEXTUAL MEASURES

As understanding the mechanisms of learning circularity in architecture schools are still at infancy, finding societal relations through explorative means is necessary. In this study, we included measures for two central uncertainties that have emerged during the curriculum development discussion for Circular Studio. Firstly, whether the environmental attitude of the student plays a clear role in self-constructing their professional and personal perspectives on architects. Secondly, whether or not the students leaned towards more socially acceptable answers as the questions moved towards more identity critical questions about the role of architects.

2.1 NEW ENVIRONMENTAL PARADIGM SCALE (NEP-15)

Environmental attitudes have been of great concern in their own right, as well as a background variable to better interpret surveys in other topics. While multiple scales and measures exist, three are widely used and only one, the New Environmental Paradigm (NEP) (Dunlap & Van Liere, 1978) scale does this without referring to specific issues that have since become dated (Hawcroft & Milfont, 2010). The original NEP scale has since been revised (Dunlap et al., 2000) and now includes 15 items that measure 5 different facets of environmental attitude (Dunlap et al., 2000; Hawcroft & Milfont, 2010), the version which is used in this study.

2.2 BALANCED INVENTORY OF DESIRABLE RESPONDING - IMPRESSION MANAGEMENT (BIDR-IM)

As the questions in this study explore the students perception of themselves as future architects, it is vital to examine the extent of Socially Desirable Reporting (SDR), i.e. the over reporting of positive behavior and under-reporting negative behavior. The Balanced Inventory of Desirable Responding (BIDR) (Paulhus, 2012) includes two separate subscales, Self-deceptive Enhancement (SDE) and Impressions Management (IM). The latter signifies a bias toward pleasing others, the school and employees administering the survey in this case. We specifically utilised the 8 IM questions from the BIDR-16 scale proposed by (Hart et al., 2015).

3 METHODS

This study used a two-stage survey design to examine the attitudes of a select group of students. Of the 29 students, 19 (15 female, average age 25, SD = 2.44) answered the first survey in the beginning of the semester, and 19 (13 female, average age 23, SD = 2.74) answered the survey at the end of the semester after teaching was completed. Of these, 10 replies were overlapping, i.e. the same students answered both the first and the second survey to enable before-after analysis. In addition, 3 teachers answered the first iteration of the survey, and 2 teachers answered the second iteration of the survey. These replies were used as a control and validity checks during data processing and statistics exploration but were not included in the final analysis.

3.1 SURVEY DESIGN

The survey consisted of 5 parts. The first part was designed to collect demographics (age and sex at birth), the track they participated in as well as 4 filler questions that were used to generate a unique code for matching before-after replies while ensuring anonymity (letters in first name, left/right handedness, last digit of phone number and first letter of mothers first name). The second and third parts consisted of the NEP-15 scale as well as the BIDR-IM scale randomized within their own respective sections. The fourth part focused on the students' perceptions of buildings and heritage in society, while the fifth part focused on the role of the architects. Most responses utilised 5-point Likert style responses for agreeableness, apart from BIDR-IM, which uses a 7-point scale and two open questions on architecture actions and positive/negative impacts. The statements about buildings included:

- B1. Buildings should use less energy and resources
- B2. The built environment is one of the most important things our society should use resources on
- B3. Energy efficient buildings are not worth the cost in most cases
- B4. Demolition of buildings should be illegal
- B5. I think historical buildings are the most valuable buildings we have
- B6. I feel connected to my own history when visiting historical buildings
- B7. New, contrasting buildings should not be allowed in historical contexts
- B8. The government shouldn't spend money on privately owned heritage buildings
- B9. We are completely dependent on new roads and buildings
- B10. It's better to leave an old building to decay than for it to lose its character in a refurbishment
- B11. Each generation needs to design their own, new surroundings as a society
- B12. Even if a school building does not work perfectly, we should still use it instead of building a new one
- B13. Energy upgrading is more important than authenticity in heritage buildings
- B14. We should use existing buildings as long as we can
- B15. Material Aesthetics, rather than technical performance, should be prioritised more.
- B16. We should re-use as much building materials as possible
- B17. I would like to use only old materials in my home
- B18. I think using new materials is an important way to show that a building is new
- B19. I think using old materials sometimes makes a building look too shabby
- B20. I do not think it is safe to re-use materials for structural components

Of these, B2, B3, B8, B9, B11, B13, B18, B19, B20 are negatively worded, i.e. more agreeableness signifies more resource use. The statemens regarding architects as a professional group were:

- A1. I think architects are more concerned about heritage buildings than the average citizen
- A2. I think architects are fully capable of transforming existing buildings without any additional education
- A3. I think architects reduce the quality of existing buildings through their design interventions
- A4. I do not think architects are very good at taking care of our built heritage
- A5. I do not think architects should work with heritage buildings, but rather leave these to conservationists and experts
- A6. I think most buildings around me could be improved by architects
- A7. Having the possibility of working with existing buildings is an important criterion for me when looking for job opportunities.
- A8. I am quite concerned about ending up in an architecture office that only designs new buildings
- A9. I use a lot of effort to educate myself about how to design interventions in existing buildings
- A10. I think my future design projects will mainly be new buildings like the ones being designed today
- A11. I would rather design a new home for myself than buy an existing home
- A12. I think used and old materials are a better starting point for a good design than new products.
- A13. I find it difficult to understand if the use of old and used materials are actually better for the environment

https://doi.org/10.52202/080684-0022

A14. In my design work, the choice of new or old materials is first and foremost be a question about emission savings
A15. I like to do calculations about material quantities and emissions
A16. I feel at loss about the re-use of materials in design projects

In this section, A3, A4, A5, A10, A11, A13, A14 and A16 were negatively worded, i.e. view architects negatively in light of the collective circularity efforts of society. In addition, two open questions were asked regarding the role of architects:

Q1. When you are working as an architect, what actions do you think will have the most positive or negative environmental impact?
Q2. What do you think architects should do to make the biggest positive contribution towards a just, environmentally sound society?

These were included to capture more nuanced perspectives on how the students viewed their role as a future architect. The BIDR-IM scale can be found in Hart et al. (2015), while the NEP-15 scale is detailed in Hawcroft and Milfront (2010). Altogether, the participants answered 68 questions in the first iteration of the survey and 45 in the second iteration of the survey.

3.2 CONDUCTING THE SURVEY

The survey was conducted twice by using nettskjema.no, a Norwegian academic survey portal. The first iteration was opened 1 week after the students started the course (September 2024) at which point they were familiar with the main concepts of circularity, heritage and re-use to answer the questions in the survey. This iteration included all the survey parts. The survey received 3 reminders and was left open for 1,5 weeks to gather replies.
The second iteration was opened 3 days before final course submission (December 2024) and received 3 reminders to submit. In this iteration, the NEP-15 scale, as well as the BIDR-IM scale were omitted.

3.3 INTERVENTION

Circular Studio is a work-in-progress curriculum development course pilot by the department of Architecture and Technology. The course ran during the fall semester of 2024 and combined four previous master-level courses: Building Conservation, Making is Thinking, Light and Color, and Integrated Energy Design. The professors from these four different courses collaborated and established the concept of *tracks*, which can be understood as different areas of perspectives, where each of the professors brought their own area of expertise to the table. The goal was for students to approach the same project from various perspectives, drawing on different academic backgrounds.

Through shared weeks across the tracks, collaboration in a common drawing studio, and open lectures, the aim was to provide students with insights into various working methods and tools for the further development of circular architecture. In practice, collaboration between tracks proved challenging at times, despite working in the same studio spaces and students focusing on the same area. This was mainly due to the limited time available for the students to both explore the depth of their own track perspective as well as engage in the other tracks' activities.

The Circular Studio course started with two intensive introductory weeks, aiming to increase students' knowledge of materials, demonstrate the potential of existing materials as resources, and train students to see the value in existing buildings. Over four days, the students, in collaboration with Ørlandet Municipality and the Circular Studio teaching team, marked, dismantled, and transported a wooden, log-built storehouse from Hårberg in Ørlandet to Vipetangen in Brekstad center. Additionally, a material catalog of the storehouse was created. This documentation formed the basis for a 3-day task, where students were asked to transform the relocated storehouse into a seaside sauna/bathhouse. Using the knowledge they had gained from the fieldwork, the students developed different project proposals for the new seaside sauna/bathhouse.

The knowledge gained during these intensive weeks was intended to be carried forward into the various tracks. Two of the tracks continued to focus on Ørlandet, one of them working on empty, abandoned, and historically valuable farmhouses that were planned to be moved to a new neighborhood in Ørlandet. The other track worked on reconstructing the storehouse, where students learned a new type of traditional craftsmanship. The two remaining tracks worked on sites in Trondheim. This difference in site and tasks contributed to a span in approaches but was also perceived as an organizational challenge in collaboration between tracks.

The final course projects were presented in December 2024 as a collaborative session between all the tracks. The projects clearly showed different approaches to circularity and the students, together with teaching staff, discussed the implications of these perspectives, lessons learned between tracks and alterations to Circular Studio for future iterations.

3.4 DATA PROCESSING OF RESPONSES

Both iterations of the survey received the same data processing steps:

1. Generate a unique code for each participant
2. Map the Likert responses to numeric values
3. Change direction of negatively worded questions
4. Checks for statistical assumptions, data exploration and sanity checks

5. Omit teacher scores
6. Calculate the scores for NEP-15 and BIDR-IM (first iteration of survey only)
7. Combine iterations for third dataset in perception change (N=10)

This resulted in 4 datasets available for the study: First iteration of the survey, including NEP-15 and BIDR-IM (N=19), second iteration of the survey (N=19), a dataset for the change between iterations (N=10) and a text response dataset for the two open questions from both iterations (N=68).

3.5 ANALYSIS

The analysis must consider several limitations and assumptions, even for an explorative study that is mainly focused on identifying concepts and questions for future use. Since the surveys utilise Likert scales with 5 or 7 items, the resulting variables are ordinal, and this limits the selection of statistical analytical methods. Kolmogorov-Smirnow and Shapiro-Wilk tests of normality on the first iteration of the survey shows that topic-level questions are normally distributed, but individual questions vary. Given the small sample size of each iteration and before-after data, normality was also assessed by viewing the histograms of each item and score. We conclude that overall, the results seem normally distributed, but due caution, we select analysis methods that are robust towards small violations in assumptions about normality. In addition, the selected methods are based on their robustness when dealing with small sample sizes. The responses show weak internal consistency of Cronbach's Alpha, with the exception of NEP-15 (see table 1). This is to be expected as the building and architect question sets are not developed as scales, but rather exploratory questions.

Table 1: Cronbach's Alpha reliability score for sets

Question Set	Cronbach's Alpha
BIDR-IM	.598
NEP-15	.798
Survey 1, Buildings	.406
Survey 1, Architects	.517
Survey 2, Buildings	.452
Survey 2, Architects	.462

An exploratory principal component (PCA) analysis shows that the buildings and architects question sets have 5 and 8 underlying components that explain over 5% of the variation, respectively, in the first iteration of the survey, and 7 components each in the second iteration of the survey. Both the Cronbach's Alpha and the PCA confirm that any assumptions about underlying scales in the questions sets is premature and that identifying individual questions that indicate change should be a priority at this stage of understanding perspective changes in circularity education.

3.5.1 NEP-15 and BIDR-IM correlation analysis

We calculated the Spearman's rank correlation coefficient to examine the correlation from NEP-15 and BIDR-IM to each of the survey questions in the first iteration of the survey. Three significant ($p<.05$, 2-tailed) correlations for NEP-15 and two significant ($p<.05$, 2-tailed) correlations for BIDR-IM were found.

3.5.2 Before-after analysis of Likert scales

To examine the change in perceptions for the 10 students that answered both iterations of the survey, we utilised both a Sign Rank test and a Wilcoxon matched-pair signed rank test. The latter assumes an underlying, hidden, continuous scale for the Likert responses and therefore includes magnitudes of change, while the former produces a more conservative result by only examining the rank and direction of responses without any assumptions about magnitude. Both tests are non-parametric and make no assumptions about normality.

Only one of the tests between iterations revealed a significance of $p<.05$ (2-sided), an expected outcome, given the small sample size, N=10. A ranking of the questions based on significance scores was used to identify the questions most likely to capture changes. This resulted in 4 questions evident with both test methods ($p=.109$ to .250 in sign rank test and $p=.043$ to .221 in Wilcoxon test), of which 1 was negatively correlated to the assumed direction of the question. In addition, the Wilcoxon test resulted in 7 additional questions of $p<.26$ that were noted.

Also noteworthy, the Sign rank test revealed 15 questions and the Wilcoxon test 3 questions with a significance of 1, which would indicate a completely random change in the responses.

3.5.3 Before-after analysis of written responses

The two open field questions were included in both iteration of the survey: *When you are working as an architect, what actions do you think will have the most positive or negative environmental impact?* (Q1) and *What do you think architects should do to make the biggest positive contribution towards a just, environmentally sound society?* (Q2). The written responses to these questions were analysed in NVivo using a coding of positive and negative aspects, as well as categorisation for the first question, and a categorisation of responses for the second question. To determine changes in perspectives, the analysed lists were compared to identify new elements or change in weights. This part of the analysis did not use code linking of responses (N=68) but examined the entire student group (N=29) for each iteration of the survey and individual statements were coded multiple times, i.e. a statement can be coded as both a positive action and focusing on old buildings. In total, 191 elements were coded across 3 areas: positive and negative impacts (N=56), opportunities and barriers (N=44) and topics (N=91).

 https://doi.org/10.52202/080684-0022

4 FINDINGS AND DISCUSSION

The collected data supports the exploratory phase of identifying concepts and questions and provides insights into potential future development.

4.1 ENVIRONMENTAL ATTITUDES AND IMPRESSION MANAGEMENT

Hawthorne and Milfront (2010, Appendix 1) report a mean score of 3.79 across 51 NEP studies when only considering student participants. These studies range from 1992 – 2006 and it is safe to assume that environmental attitudes have changes since then as public consciousness on sustainability has increased. In comparison, the 19 students that answered the first survey including the NEP-15 score, scored on average a similar 3.89 score (N=19, min. 3.27, max. 4.80, SD=.377). This indicates that the students have a high, but representative pro-environmental attitude.

When comparing the NEP-15 scores to the questions about buildings and architects, three questions (Table 2) show a significant correlation.

Table 2: Descriptive statistics and significant correlations (Spearman's rank correlation coefficient) with NEP-15 for the survey questions of buildings and architects.

Question	Correlation	Sig.	SD	Score $\bar{x}$
B6	.621	.005	.74	4.11
A7	.596	.007	1.03	3.95
A8	.602	.006	1.07	3.63

In relation to environmental attitude, the statement *I feel connected to my own history when visiting historical buildings* (B6) could be interpreted to tie into a general awareness of the role of historical buildings as part of a sustainable environment, i.e. they are already built. This might further tie into the concerns of ones own role in aiding and abiding the continued high use of resoureces when considering statements A7 and A8, which both reflect different aspects of working as an architect and making decisions about resource use. Specifically, A7 and A8 show that the students with an environmentally concerned attitude might let this influence their work decisions if an architectural office is perceived to be working against their enviromental convictions.

Hart et al. (2015, table 1) report 4 values for BIDR-IM scores, with a range from 3.65 to 4.59 in mean scores, but with individual scores spanning the entire range. Our scores are comparable (N=19, mean = 4.36, min. 2.88, max. 6.00, SD=.946). When looking at the correlation between individual questions and the BIDR-IM scores (Table 3), two statements, A7 and A14, are significantly ($p<.05$) negatively correlated, while one statement, B10, is not significant ($p=.12$) but warrants discussion for its positive correlation.

Table 3: Descriptive statistics and significant correlations (Spearman's rank correlation coefficient) with BIDR-IM for the survey questions of buildings and architects.

Question	Correlation	Sig.	SD	Score $\bar{x}$
B10	.504	.028	1.06	2.00
A9	.463	.046	.765	3.84

The positively correlated statement *It's better to leave an old building to decay than for it to lose its character in a refurbishment* (B10) is a value laden statement. It is therefore fair to assume that while students might be at unease about the way in which they shift in this question, they would likely give more definitve answers should the students be presented with a concrete case to evaluate and to do evaluation as part of professional practice.

The other statement, *I use a lot of effort to educate myself about how to design interventions in existing buildings* (A9) is also a very relative question as it does not say anything about actual time used, just perceived effort. While some students might be exaggerating their efforts, others might simply have a feeling of "not learning enough" or simply have different notions of what a lot signifies.

The average signifigance between the individual questions and BIDR-IM correlations is .415, a very high number. This indicates that the students replies are not subject to impression management and are represent true and faithful responses. This is likely also influenced by the use of coupling codes to link the two iterations of the survey in such a way that full anonymity is guaranteed.

Exploring the relationship between the survey statements together with the established NEP-15 and BIDR-IM scales, few aspects seem to be influenced by overall environmental attitudes or the need for impression management. A reasonable assumption is that for some of the students, their environmental attitudes might influence theit choice of workplace, given the opportunity to choose freely.

4.2 PERSPECTIVES AND CHANGE

Ten students answered both iterations of the survey, consisting of 36 questions in each iteration. This resulted in 720 responses that can be analysed. Of the 36 statements in total, 11 had a positive change over .2 points from iteration 1 to 2, while only 3 had a negative change above .2 points (B15, B19 and A14). 22 had only small changes below .2 points. The overall means increased with 4.3 points or an average of .12 per question. There are therefore indications of slight positive tendency between the iterations overall. The Sign Rank and Wilcoxon Matched-pair Signed Rank tests identified 3 questions that had significant positive changes and 6 that had small changes that warrant more exploration.

The statement *Buildings should use less energy and resources* (Question B1, 1. Survey mean 3.7, SD=1.06; 2. Survey mean 4.4, SD=.7; Change=.7, SR sig=.219, T=1.225, WMPSR sig.=.221, T=1.225) indicates that the students increased their awareness of the role of buildings in both energy and resource use, although it seems many students were well aware of this during the first iteration and the change in scores is mainly due to the students who disagreed with the statement shifting their perspective. This is indicated by the min value shifting from *strongly disagree* to *neutral*. Five students had a positive shift on this statement, while only 1 had a negative shift.

The biggest change is found in the statement *Each generation needs to design their own, new surroundings as a society* (Question B11, 1. Survey mean 2.2, SD=1.23; 2. Survey mean 3.8, SD=1.03; Change=1.6, SR sig=.109, T=1.581, WMPSR sig.=.043, T=2.019). This question is negatively worded, signifying that the students alloted less importance to the newness of their surroudnings as they participated in the course. This could indicate a growing awareness of the potential of existing buildings and a growing appreciation of them for 8 of the students while the remaining 2 had a negative shift to the statement.

The final statement with a clear indication of change, *We should use existing buildings as long as we can* (Question B14, 1. Survey mean 4.4, SD=0.7; 2. Survey mean 4.9, SD=.32; Change=.5, SR sig=.125, T=1.500, WMPSR sig.=.059, T=1.890), has an increase from an already very high score that might indicate verifying already held strong beliefs. This shift is due to 4 students regarding the statement more positively. This statement also combines the sustainability and architectural quality narratives, being in favor of contributing to sustainability with a long lifespan while at the same time ensuring that the work of architects (of undescribed quality) is given a long-lasting place in society. It is therefore a win-win statement that seems easy to agree with, but also disregards the operational costs of a building in use.

The other 6 questions that showed small changes (B6, B9, B13, A1, A13 and A15) indicate nuanced shifts concerning energy upgrading and existing buildings in general, architects concerns as well as raised understanding of the technical necessities of working with existing buildings.

In total, there were 104 positive shifts and 78 negative shifts to the statements, indicating a change in perspective that allots a larger role on existing buildings in society and more awareness about the resource use of buildings, see Table 4 for an overview.

Table 4: Descriptives and changes in statements between the first and second iterations of the Survey.

Q	+	-	1. $\bar{x}$	1. SD	2. $\bar{x}$	2. SD	$\bar{x}$ change
B1	5	1	3.70	1.06	4.40	0.70	0.70
B2	1	2	2.40	0.84	2.30	0.82	-0.10
B3	2	3	3.20	0.79	3.20	0.92	0.00
B4	1	1	3.00	0.94	3.10	0.99	0.10
B5	3	1	3.30	0.68	3.50	0.71	0.20
B6	5	2	4.00	0.67	4.30	0.48	0.30
B7	3	3	3.40	0.84	3.40	1.27	0.00
B8	2	1	3.70	0.95	3.90	0.74	0.20
B9	4	1	3.50	0.71	3.80	0.92	0.30
B10	4	3	1.80	0.79	2.10	1.20	0.30
B11	8	2	2.20	1.23	3.80	1.03	1.60
B12	2	3	4.20	0.79	4.10	0.32	-0.10
B13	2	0	3.00	0.47	3.30	0.82	0.30
B14	4	0	4.40	0.70	4.90	0.32	0.50
B15	1	4	3.60	0.84	3.30	0.82	-0.30
B16	3	1	4.70	0.48	4.90	0.32	0.20
B17	2	5	3.00	0.67	2.80	1.03	-0.20
B18	2	4	4.50	0.71	4.40	0.52	-0.10
B19	0	3	4.20	0.63	3.90	0.57	-0.30
B20	3	1	4.20	0.63	4.40	0.52	0.20
A1	3	2	3.10	1.20	3.60	0.84	0.50
A2	2	1	2.50	0.85	2.40	0.97	-0.10
A3	2	2	3.70	0.82	3.60	0.52	-0.10
A4	2	2	3.30	0.68	3.10	0.88	-0.20
A5	1	2	4.50	0.53	4.40	0.52	-0.10
A6	2	4	4.00	0.67	3.80	0.42	-0.20
A7	2	2	3.70	1.16	3.80	1.03	0.10
A8	3	2	3.30	0.95	3.30	1.16	0.00
A9	2	4	3.80	0.63	3.60	0.84	-0.20
A10	5	2	3.60	0.84	3.80	0.92	0.20
A11	3	3	4.00	0.94	3.90	0.99	-0.10
A12	5	2	3.40	0.84	3.60	0.70	0.20
A13	3	1	3.80	0.79	4.10	0.88	0.30
A14	2	4	3.10	0.74	2.70	0.68	-0.40
A15	4	1	2.50	1.27	2.80	1.32	0.30
A16	6	3	2.90	1.10	3.20	1.32	0.30

4.3 PERCEIVED IMPACTS AND ACTIONS

The open questions yielded 68 responses across the study. Focusing on perspectives and concepts using explorative coding, 3 dimensions were identified.

Firstly, the survey explicitly asks for positive and negative impacts, naturally giving the initial dimension. The positive responses from iteration 1 show that the students already have a good grasp of sustainability and can describe these in relation to their own profession. They also naturally include many distinct actions, such as re-use of buildings and materials. The negative actions almost uniformly focus on resource waste, either as not designing for a long enough lifespan, not prioritising quality or building a new building altogether. An interesting aspect is that some responses distinctly raise building budgets as a culprit, making "*only thinking about*

https://doi.org/10.52202/080684-0022

the short-term cost of a project" (participant) a negative impact, although this is largely thought to be outside the sphere of influence of the architects (although it seems sometimes used as an excuse for less-than-optimal architectural work). The responses to the second iteration of the survey shows a clear change in perspective. Only one response in the first iteration brought up being an agent of change:

"Working in teams with people from multiple fields knowledge and constantly making sure to be working in a way that is open to explore new methods is very important towards making a positive environmental impact." (Participant)

Contrastingly, the second survey revealed 4 responses that focus on this aspect as an avenue to positive impact, although actions on individual design decisions are still prevalent. The responses focusing on negative impacts also revealed what might be a growing awareness of one's own role, highlighting business-as-usual for the architectural profession in somewhat stark terms:

"Less ego, less starchitects, less visibility. Less is more. More anonymity, discretion and care for context." (Participant)

"Any building is a negative environmental impact, less work is less environmental impact, simple." (Participant)

"Everything that leads to overconsumption." (Participant)

Secondly, responses overwhelmingly focus on opportunities (N=41) over barriers (N=3). Interestingly all barriers are attributed to others (clients, laws and people experiencing the architecture), a somewhat un-critical stance. As opportunities, learning and evolving is prevalent in the first iteration, as well as changing the norm:

"The role of an architect is not the same as it was just a couple of years ago." (Participant)

It seems clear that the students are well aware of the changes in the profession as the knowledge base expands but the responses clearly indicate that the necessary change is still in the future and that they are yet to reach it. In the second iteration, a new notion is identified. Listening to others, as opposed to many responses viewing the work of architects as more of a one-way communication for the betterment of society, is stated. It seems that the students are still overwhelmingly focused on the contributions of the architect and see themselves as being in the forefront of change but the responses in the second iteration show that this perception might be changing among few of the students during the course.

Thirdly, individual professional and process topics were identified. Building topics included well documented aspects within environmental sustainability, such as new vs old buildings, re-use or virgin materials and building lifespan. When describing building aesthetics, a more diverse set of responses emerged. It seems aesthetics are considered almost universally important and thought to also affect the lifespan of the building through attractiveness, but the relationship between technical aspects and aesthetics is seen as potentially opposites by some. It is still clear, however, that the students view aesthetics as a core contribution to sustainability from architects:

"When it is necessary to build something new, it should be built to last hundreds of years, and be so beautiful that people will protect it and take care of it" (Participant)

Some actions focus more on processes related to the design, rather than the design itself. Between iterations, two perspectives show a large change. The students increasingly mention teamwork as well as an increased need to supplement their own learning. This might indicate a growing understanding of the complexity of circularity and therefore the need to seek knowledge both through interdisciplinarity as well as self-study.

In sum, the two open questions imply that the students have a sound (and sometimes critical) understanding of the relation between their own profession and its inherent challenges with regards to sustainability and circularity. During the course, this understanding seems to develop from a solution-oriented towards more process- and collaboration-oriented.

4.4 SUMMARY OF FINDINGS

To construct a coherent narrative of findings, we would argue that the students are environmentally conscious and have a good grasp of the inherent sustainability challenges of being a business-as-usual architect, working on new buildings with virgin materials and little concern for energy use. They are critical about the architect's contribution into overuse of resources which they seem to perceive as an impending shift within the profession, but one that is still in the future. They see clear opportunities for win-win solutions, especially combining aesthetics with technical knowledge to extend building lifespan. These attitudes might influence career choices as they increasingly see themselves as agents of change as they participate in Circular Studio. This manifests itself in increased awareness about the role of existing buildings as an opportunity for sustainable designs, but also in viewing others initially as barriers to sustainability, but shifting into appreciating multiple, cross-disciplinary perspectives and an increased focus on their own learning needs. From the data, we have explicitly identified the following perspectives and concepts:

1. The student's critical perception of architects as a contributor to non-sustainable practices by designing new buildings with virgin materials
2. The perceived win-win opportunities of aesthetic quality and increased lifespan of buildings.
3. The shift from solution-oriented to process- and collaboration oriented as learning about circularity.
4. Increased self-perception as agent of change.
5. Perceived increase in need to learn more about sustainability while learning.

5 IMPLICATIONS AND LIMITATIONS

Implementing circularity into studio based courses in architectural education seems to support experiential learning (Kolb, 2014; Morris, 2020). We see clear indications of the students learning concepts, approaches and frameworks that are then used to re-frame, reject and amend personal and professional identity in self-construction (Kararmaz, 2024). The 5 identified concepts and perspectives can be seen as different manifestations of this learning process. It is however clear that the sample sizes, exploratory questions and limited case in this study is far from providing sufficient evidence for systematic conclusions. It is therefore vital to support a continued research effort and exploration alongside circular learning activities. There are clear indicators, and it is also the subjective opinion of the authors, that a continuous development of measures to examine the perception change in students as they are increasingly introduced to circularity in architectural education can be immensely valuable to support developing curriculum.

Should the research community be able to identify and hone precise and dependent measures for students' perception of circularity in their own design practice and beyond, a significant contribution could be made. Specifically, developing courses that not only include the systematic complexities and multitude of perspectives of circularity, but also ensure that students can actively use this learning to develop their architectural knowledge, values and approaches.

For these reasons, we recommend that architecture schools looking to implement circularity into their curriculum ensure their studio setting, course description and practical teaching support experiential learning principles and that they implement a before-after measure, interviews or other, replicable, evaluation methods that specifically target the 5 identified concepts and perspectives, but also try to modify them and uncover additional perspectives that should be published. Additionally, researchers should strive to develop coherent scales of questions for any identified perspectives and concepts and verify these through the means of statistical tools such as principal component analysis. In this way, the community can begin solidifying a more coherent and comprehensive evidence base to support the necessary introduction of circularity into architectural education.

ACKNOWLEDGEMENTS

The authors would like to acknowledge the much-appreciated help from Berit Therese Nilsen and Lucia Liste in the development of the survey questions.

REFERENCES

Aalto, P., Alsos, O. A., Haneberg, D. H., Steinert, M., Ege, D., & Sivertsen, I. B. (2023). THE EFFECTS OF LEARNING ENVIRONMENTS ON STUDENT ACTIVE LEARNING. *Proceedings of the International CDIO Conference ISSN 2002-1593*, 975–988. https://www.ntnu.edu/documents/1308247420/0/CDIO+2023+Proceedings.pdf/6989d202-864d-52e5-fee7-15d84a281f41?t=1697116952671

Boling, E., Schwier, R. A., Gray, C. M., Smith, K. M., & Campbell, K. (2016). Studio teaching in higher education. *Routledge. Https://Doi. Org/10, 4324*(97813), 15697.

Corazzo, J. (2019). Materialising the Studio. A systematic review of the role of the material space of the studio in Art, Design and Architecture Education. *The Design Journal*, *22*(sup1), 1249–1265. https://doi.org/10.1080/14606925.2019.1594953

Dunlap, R. E., & Van Liere, K. D. (1978). The "New Environmental Paradigm." *The Journal of Environmental Education*, *9*(4), 10–19. https://doi.org/10.1080/00958964.1978.10801875

Dunlap, R. E., Van Liere, K. D., Mertig, A. G., & Jones, R. E. (2000). New Trends in Measuring Environmental Attitudes: Measuring Endorsement of the New Ecological Paradigm: A Revised NEP Scale. *Journal of Social Issues*, *56*(3), 425–442. https://doi.org/10.1111/0022-4537.00176

Hart, C. M., Ritchie, T. D., Hepper, E. G., & Gebauer, J. E. (2015). The Balanced Inventory of Desirable Responding Short Form (BIDR-16). *SAGE Open*, *5*(4), 215824401562111. https://doi.org/10.1177/2158244015621113

Hawcroft, L. J., & Milfont, T. L. (2010). The use (and abuse) of the new environmental paradigm scale over the last 30 years: A meta-analysis. *Journal of Environmental Psychology*, *30*(2), 143–158. https://doi.org/10.1016/j.jenvp.2009.10.003

Kanters, J. (2020). Circular Building Design: An Analysis of Barriers and Drivers for a Circular Building Sector. *Buildings*, *10*(4), 77. https://doi.org/10.3390/buildings10040077

Kararmaz, Ö. (2024). An investigation of the self-construction experience in architect's education. *MEGARON / Yıldız Technical University, Faculty of Architecture E-Journal*, 526–538. https://doi.org/10.14744/MEGARON.2024.86658

Kirchherr, J., Reike, D., & Hekkert, M. (2017). Conceptualizing the circular economy: An analysis of 114 definitions. *Resources, Conservation and Recycling*, *127*, 221–232. https://doi.org/10.1016/j.resconrec.2017.09.005

Kolb, D. A. (2014). *Experiential learning: Experience as the source of learning and development*. FT press.

Leijon, M., Ivar Nordmo, Åse Tieva, & Rie Troelsen. (2022). Formal learning spaces in Higher Education – a systematic review. *Teaching in Higher Education*, 1–22. https://doi.org/10.1080/13562517.2022.2066469

Lundahl, L., Gruffman Cruise, E., Malmros, B., Sundbaum, A.-K., & Tieva, Å. (2017). Pedagogisk rum-tid och strategier för aktivt lärande i högre utbildning. *Utbildning och Lärande / Education and Learning*, *11*(1), 16–32. DiVA.

Morris, T. H. (2020). Experiential learning – a systematic review and revision of Kolb's model. *Interactive Learning Environments*, *28*(8), 1064–1077. https://doi.org/10.1080/10494820.2019.1570279

Paulhus, D. L. (2012). *Balanced Inventory of Desirable Responding* [Dataset]. https://doi.org/10.1037/t08059-000

Schon, D. A. (1987). *Educating the Reflective Practitioner. Toward a New Design for Teaching and Learning in the Professions. The Jossey-Bass Higher Education Series.* ERIC.

Sopeoglou, E. (2024). Tomorrow's architects: The emerging professional and education context in the UK and new regulatory framework. *E3S Web of Conferences*, *585*, 04004. https://doi.org/10.1051/e3sconf/202458504004

Thoring, K., Desmet, P., & Badke-Schaub, P. (2018). Creative environments for design education and practice: A typology of creative spaces. *Design Studies*, *56*, 54–83. https://doi.org/10.1016/j.destud.2018.02.001

Thoring, K., Desmet, P., & Badke-Schaub, P. (2019). Creative Space: A Systematic Review of the Literature. *Proceedings of the Design Society: International Conference on Engineering Design*, *1*(1), 299–308. https://doi.org/10.1017/dsi.2019.33

Wilson, M. A. (1996). THE SOCIALIZATION OF ARCHITECTURAL PREFERENCE. *Journal of Environmental Psychology*, *16*(1), 33–44. https://doi.org/10.1006/jevp.1996.0003

Wuyts, W., Miatto, A., Khumvongsa, K., Guo, J., Aalto, P., & Huang, L. (2022). How Can Material Stock Studies Assist the Implementation of the Circular Economy in Cities? *Environmental Science & Technology*, acs.est.2c05275. https://doi.org/10.1021/acs.est.2c05275

Proceedings of the Circular Building Sector Conference 2025, Lund, Sweden, 1-3 June 2025

LUND UNIVERSITY | LTH FACULTY OF ENGINEERING

CIRCULAR ARCHITECTURE MEETS CIRCULAR ECONOMY: A PILOT EXPERIENCE IN INTERDISCIPLINARY TEACHING FOR CLIMATE-NEUTRAL BUILDING PRACTICES

Marwa Dabaieh [1*] and Rebecka Lundgren[1]

[1]Urban Studies, Malmö University

*Corresponding Author: marwa.dabaieh@mau.se

ABSTRACT

Background and aim. This study evaluates the implementation and effectiveness of a pilot course aimed at integrating circular principles into architectural education to respond to the built environment's pivotal role in the climate crisis. The course introduces undergraduate students to the foundational concepts of circular design, promoting climate-neutral practices. Co-taught by specialists in circular architecture and circular economy, it blends a design project with practical training in life cycle calculations. By analysing the course structure and outcomes, the study highlights how circular design, and economic considerations can be effectively incorporated into academic learning.

Methods and Data. The study employs a mixed-method approach that includes qualitative project analysis and quantitative student surveys. Reflexivity and self-critical discussions were central to minimize bias, critically assess teaching practices, and ensure a balanced evaluation of learning outcomes and course impact.

Findings. The course revealed the necessity of a holistic approach to teaching sustainable architecture, demonstrating a strong correlation between understanding circular design and effectively applying LCA and LCC tools. The dual approach enhanced students' design skills while equipping them with practical abilities to assess both the environmental impact and economic viability of their designs.

Theoretical / Practical / Societal implications. This experience highlights the importance of interdisciplinary teaching in architectural education. The findings suggest that future courses should continue to integrate design with environmental and economic analysis, better preparing students for sustainable practice. The course offers a model that can be adapted in other contexts, contributing to the broader goal of climate-neutral buildings.

KEYWORDS: circularity, circular design, circular economy, interdisciplinary teaching.

1 INTRODUCTION

The urgency of climate action has never been more critical, as the world faces unprecedented environmental challenges, including rising temperatures, resource depletion, and biodiversity loss (IPCC, 2023). The built environment plays a pivotal role in this crisis, accounting for a significant portion of global carbon emissions and waste generation (Krausmann et al., 2017). The sector accounts for up to 40% of carbon dioxide emissions globally (World Economic Forum, 2016) and over 35% of the waste generation in the EU (European Commission, 2020). However, there is a large reduction potential in the built environment industries (IPCC, 2023). To address this potential, the integration of circular principles in both architecture and the economy is vital. Circular economy practices focus on reducing waste, reusing materials, and creating systems that regenerate natural resources, which is crucial for achieving climate-neutral goals in the construction sector (Lundgren, 2023).

As addressing climate change intensifies, the need for sustainable practices in architecture has never been more apparent. The concept of a circular economy presents a transformative approach to achieving sustainability. However, despite the growing importance of circularity, there exists a significant gap in how circular economy principles are integrated into architectural education. This

https://doi.org/10.52202/080684-0023

gap hinders the ability of future architects to fully grasp and implement these principles in their design processes (Dabaieh, 2023a; Kanters, 2020). The architectural profession, which is responsible for a substantial portion of global carbon emissions and material waste, has a significant opportunity to contribute to a more sustainable future by adopting circular strategies. This includes designing for adaptability, reusing materials, and employing sustainable construction methods that prioritize longevity and reduce environmental impact. Research indicates that the application of circular principles in architecture can drastically reduce resource consumption and improve the efficiency of building materials, contributing to the reduction of the carbon footprint in the construction sector (Kibert, 2022). Addressing this gap is crucial for fostering climate-neutral practices and developing a built environment that can contribute to the broader goals of sustainability and environmental responsibility.

On one hand, despite the clear potential of circular economy and circular design practices in mitigating climate change, architectural education often falls short in fully integrating these principles into its curricula. Traditional architectural education tends to focus heavily on aesthetic, functional, and technical aspects of design, with sustainability often relegated to a peripheral topic or addressed primarily through energy efficiency measures (Barton, 2021). Although many institutions have begun to introduce sustainability into their courses (Jürgens et al., 2023; Silva et al., 2023; Viere et al., 2024), the specific intersection of circular economy and architectural design remains insufficiently explored. Architectural students are typically taught about sustainability in terms of energy efficiency, material selection, and environmental impact assessments, but circularity—the idea of designing buildings and systems that can perpetually renew and adapt—often remains an abstract or niche subject (Dabaieh, 2023b).

On the other hand, one more of such core challenges contributing to this gap in architectural education is the lack of interdisciplinary integration between architecture and circular economy principles. Circular economy is inherently interdisciplinary, requiring knowledge not only in design but also in economics, material science, and systems thinking (Geissdoerfer et al., 2017a). However, architectural programs traditionally operate in siloed disciplines, focusing as mentioned above on design aesthetics, structural engineering, and environmental impact with limited interaction with disciplines that focus on circular economy and resource management. This separation limits students' ability to appreciate how economic factors—such as the life cycle costs of materials and buildings—affect the sustainability of their designs. The absence of this interdisciplinary approach results in an incomplete understanding of the potential for circularity in the built environment.

In this paper we aim to address these gaps, through showing the outcome of an experimental course in circular economy and circular architecture for bachelor students in sustainable architecture and urban design. We started this course with the belief that architectural education must evolve to incorporate a more holistic framework that integrates circular economy principles throughout the design process. One approach is to introduce project-based learning that encourages students to explore the practical applications of circular economy within architectural projects. As (Fahlstedt et al., 2024)discussed in their work, this would involve not only designing buildings but also analysing the economic and environmental impact of these designs over their entire life cycle, including reuse, renovation, and end-of-life management. Moreover, (Gomes et al., 2022) stressed in their study that integrating courses that focus on life cycle assessment (LCA) and life cycle cost (LCC) calculations would provide students with the tools to make informed decisions regarding the material and energy choices they make in their designs, while understanding the long-term economic implications of those choices.

The course aims to equip students with foundational knowledge and practical skills in circular economy and sustainable architecture, emphasizing the integration of environmental and economic considerations into design processes. Its interdisciplinary approach is reflected in both the diverse teaching team, which includes experts from architecture, real estate science, engineering, economics, and sustainability fields, and the course content, which combines theoretical lectures, practical workshops, and real-world case studies. This structure ensures students gain a holistic understanding of circular design principles, fostering collaboration across disciplines to address complex design challenges. Our hypotheses are that when students learn applying circular design principles, that can help in architectural practice transition from the traditional linear model of "take, make, dispose" to one that promotes sustainability through the circular thinking and reuse of materials, energy efficiency, and long-term durability. This shift not only can contribute to reducing carbon footprints but also fosters more sustainable economic models that align environmental health with economic growth.

2 BACKGROUND AND THEORETICAL FRAMEWORK

Circular design is the broader process of creating products, systems, and services that adhere to the principles of a circular economy. In architecture, it refers to the intentional design of products and systems to enable their continual reuse, repair, or recycling. The goal is to eliminate waste and ensure that every material and resource is either biodegradable or capable of being reincorporated into the system in a way that does not harm the environment (Minunno et al., 2020a). Circular design includes product life extension, modularity, and cradle-to-cradle design, which ensures that the end-of-life of a building or product is considered at the design stage. Designers focus on creating systems that can be easily

https://doi.org/10.52202/080684-0023

maintained, repaired, or upgraded rather than replaced. For example, the use of demountable building systems or reusable components means that when the building reaches the end of its life cycle, its parts can be recycled or reused in new projects (Geissdoerfer et al., 2017b).

Circular economy strategies are frequently categorized into R-imperatives, as outlined by Reike et al. (2018). These strategies are organized based on efficiency, wherein the closer a product remains to its original user and intended function, the higher its level of efficiency is deemed to be. Reike et al. (2018) propose ten R-imperatives, which are categorized into short, medium, and long loops, with shorter loops being considered more efficient. Refuse, reduce, resell/reuse, and repair are classified as short loops, while refurbish, remanufacture, and repurpose fall under medium loops. In contrast, recycling materials, recovering energy, and re-mining are categorized as long loops.

In the built environment context, Akhimien et al. (2021) elicited seven main CE strategies prevalent in extant literature, namely, design for disassembly, design for recycling, building materiality, building construction, building operation, building end of life, and building optimisation. In the first six, the main focus is to enable reuse and recycling of building materials. Building optimisation on the other hand concerns extending the life of a building or a component, e.g., maintenance, repair, replacement and refurbishment. The strategies presented by Akhimien et al. (2021) are not ordered in a hierarchy of efficiency.

To further enhance the understanding of the efficiency of circular strategies in the built environment, life cycle assessments can be effectively utilised. There are currently three main life cycle assessments corresponding to the three sustainability dimensions, namely LCA for the environment, LCC for economic, and S-LCA for the social dimension. Of the three the LCA and LCC are the most commonly employed. All three frameworks are structured in a similar manner, allowing them to be used complementarily to one another (Lundgren, 2023).

The LCA framework is an international standard (ISO14044:2006) and is commonly applied in the built environment context (e.g., Berglund et al., 2018; Deschamps et al., 2018; Eberhardt et al., 2019; Minunno et al., 2020). However, Bragança et al. (2010) emphasize the inherent complexity of conducting LCAs at the building level. The LCA is employed to assess environmental impact, with the most common impact employed in the built environment being global warming (Andersen et al., 2022). In addition to the LCA, the LCC framework is also recognised as an international standard (ISO 15686-5:2008). LCC is an assessment of cost and income generation of a product or system over its life. It is commonly combined with LCA and provides the economic perspective. In the context of the built environment, the LCC is conducted as a cash flow analysis, resulting in the determination of the net present value (Bejrum, 1991). The S-LCA framework has not yet been established as an international standard. However, the United Nations Environment Programme has developed guidelines that have been applied to the built environment, albeit on a limited scale (UNEP, 2020). Furthermore, a European standard for assessing the social impact of buildings exists; however, this standard currently focuses exclusively on the use phase of the building life cycle (EN 16309:2014+A1:2014).

3 COURSE CONTEXT

This interdisciplinary course introduces first-year bachelor students in Sustainable Architecture and Urban Design program to the principles of circular economy and circular architecture, with a strong focus on design. Students engage in a blend of theoretical knowledge, hands-on experience, and practical applications through lectures, workshops, guest lectures, and site visits. Through individual assignments and group projects, students learn how to apply LCA and LCC calculations to evaluate the environmental and economic impacts of their design choices. A core project challenges students to design a 20-square-meter temporary shelter based on circular design principles.

Students select a site in a specific climate and local materials, develop modular construction methods that allow for efficient assembly and disassembly, and create a detailed design manual and physical model. Emphasis is placed on integrating sustainability, material efficiency, and adaptability into their designs. The course culminates in the presentation of the design process, technical aspects, and a comprehensive report detailing LCA and LCC analyses. While the students' designs centre on circular new buildings, the theoretical aspects of the course, along with other courses in the program, cover additional topics in circular construction, such as adaptive reuse.

The course achieves a thoughtful balance between theory and practice, ensuring students develop both a solid foundation of knowledge and practical application skills. The lectures in fundamentals of circular economy and circular design, human centric circular architecture, social sustainability, circular metabolism, circular architecture for adaptive reuse, lifecycle assessments, and circular building certification were among the theoretical topics taught in the course. They provide students with the theoretical foundation needed to understand the principles of circular economy and sustainable design, equipping them to approach design challenges critically and analytically. Site visits for pilot projects offer direct exposure to real-world applications of circular practices, inspiring innovative approaches and contextual understanding. Hands-on workshops focus on developing practical skills, enabling students to perform accurate LCA and LCC calculations while exploring circular design techniques, such as modular construction and material efficiency, to address real-world constraints effectively.

Both the LCA and the LCC were taught in two consecutive workshops. The LCA was performed using

https://doi.org/10.52202/080684-0023

global warming potential as the selected impact category, with carbon dioxide equivalents serving as the primary indicator. The LCC analysis was conducted as a cash flow evaluation, with the outcomes expressed in terms of net present value. The students applied the assessments to their circular design. They were then asked to reflect on the application of the life cycle assessments as tools to evaluate designs in an individual written assignment. Further, the students needed to reflect on the outcome of the assessments in relation to their design, both in the individual assignment and in the final presentation of their designs.

The LCA was performed utilizing the online tool OneClick. The students had access to the teaching version of the tool which has limited capabilities, such as the inability to add new materials to the list of available materials. The LCC was performed in Excel using the principles of cashflow analysis, typical for use in the built environment context as described by Bejrum (1991). The students carried out one LCA and one LCC for their building design, covering the entire life cycle of the building.

Data collection for this study employed a combination of methods, including qualitative feedback from students, analysis of their design projects, and observations of their interaction with the course material. This multi-method approach enabled a comprehensive evaluation of the course's impact on student learning outcomes and the development of their competencies in circular design. Furthermore, it provided us, as course designers, with insights into potential areas for improvement and opportunities for the course's future development, including its possible expansion into a more comprehensive course. Figure 1 shows the structure for the course methodological structure.

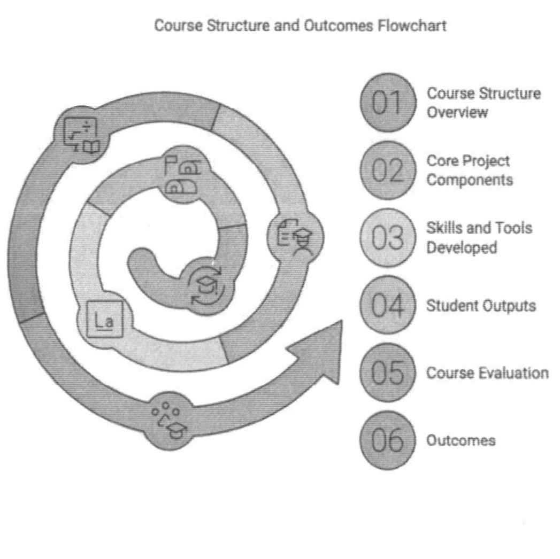

Figure 1: The course methodological steps.

4 METHODS

This study employs a mixed-method approach as shown in figure (2), integrating both qualitative and quantitative research methods to analyse the effectiveness of teaching circular architecture and circular economy within an academic setting. Given that the authors of this study were also the primary instructors of the course, particular emphasis was placed on reflexivity to minimize bias and ensure a critical, objective analysis.

The data collection process was structured into two primary phases. First, students' projects were analysed to assess their understanding and application of circular architectural principles. This provided tangible evidence of learning outcomes and design integration. Second, a structured student survey was conducted to gather direct feedback on their learning experiences, challenges, and perceptions regarding circularity in architecture. The survey included both close-ended and open-ended questions, allowing for a combination of statistical analysis and qualitative insights.

To strengthen the reliability of the findings, an analytical and self-critical approach was applied. The authors engaged in structured discussions and iterative reflections to critically assess both the course design and their own roles as educators. This process included reviewing students' feedback in relation to the intended learning outcomes and identifying potential areas where instructional methods may have influenced responses. Reflexivity was embedded in this phase by systematically questioning assumptions, considering alternative interpretations, and acknowledging the instructors' positionality in shaping students' learning experiences.

By combining quantitative data from surveys with qualitative insights from project analysis and reflexive discussions, this methodology aims to provide a balanced and critically informed evaluation of the course's impact. This approach ensures that the study does not merely validate pre-existing assumptions but instead offers a rigorous and transparent assessment of how circular principles can be effectively integrated into architectural education.

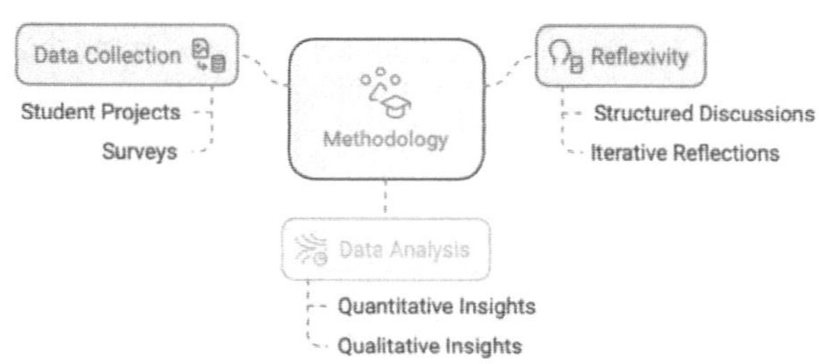

Figure 2: The study methodological steps.

5 RESULTS

This section presents the results of the study and is divided into two sub-sections, namely, teaching staff and student experience.

5.1 TEACHING STAFF EXPERIENCE

Overall, the students displayed a deep understanding of circular design principles and their economic and environmental impact. Students crafted designs which adhered to circular principles, including many of the R-imperatives. This became evident in the LCAs where the pavilion had a very small impact on the environment, in some cases almost non-existent. However, the limitations of the online LCA tool occasionally presented challenges in effectively visualising these results as the locally sourced materials were not available in the database. The students thereby gained first-hand experience of the challenges associated with conducting LCAs. Furthermore, the students demonstrated creativity in addressing the economic aspects, as evidenced by the results of the LCCs. Several students utilized volunteer labour and locally sourced natural materials, which significantly minimized the cost.

Students successfully bridged design creativity with environmental and economic assessment by integrating innovative approaches with rigorous analytical methods. They explored creative concepts in designing 20-square-meter shelters, focusing on adaptability, modularity, and the use of locally sourced materials. Through the application of LCA and LCC, they critically evaluated the environmental and financial impacts of their design choices, ensuring practicality and sustainability without compromising aesthetic or functional goals. This process led to diverse and innovative design solutions that addressed site-specific needs, material efficiency, and construction feasibility, demonstrating the harmony between imaginative design and responsible decision-making.

The interdisciplinary approach employed in the course fostered a comprehensive understanding among students by highlighting connections between various disciplines. This was evidenced by their ability to adapt design strategies based on LCA and LCC results, effectively balancing design considerations with environmental and economic impacts. Furthermore, this approach catalysed innovative designs, with students integrating circular measures into their projects as informed by the life cycle assessments. The course facilitated the development of skills spanning a broad spectrum of interconnected aspects of building design, enabling students to comprehend the relationship between design choices and their broader implications in terms of environment and economy. Notably, many students extended their designs to include social impacts, despite this aspect being covered only in theoretical sessions rather than through a dedicated S-LCA workshop.

The interdisciplinary nature of the course also equips students with essential skills for careers in increasingly collaborative and interdisciplinary professional environments. This is particularly significant in the context of the built environment and sustainability, both of which are inherently complex and require multifaceted approaches. Further, it enhanced adaptability by exposing students to different ways of thinking and problem-solving. This was evident in the iterative refinement of the students' designs, informed by the feedback loop generated from their life cycle assessment results.

Additionally, the interdisciplinary approach has yielded professional growth for the educators involved. Through collaboration with colleagues across different disciplines, sharing resources, methodologies, and perspectives, each teacher has gained a deeper understanding of the complexities inherent in this field. However, designing interdisciplinary courses that effectively balance depth and breadth remains a challenge. Student feedback indicated that while the learning experience was perceived as broad, it sometimes lacked the desired depth, particularly concerning life cycle assessments.

However, a clear correlation was observed between the students' understanding of circular design principles and their effective application of life cycle assessments.

The integration of LCA and LCC offered the students immediate, quantifiable feedback on both the environmental impact and economic feasibility of their circular design proposals. This process established a feedback loop, which the students themselves were actively responsible for initiating and managing, thereby enhancing their critical evaluation and iterative design capabilities.

The students demonstrated significant improvement in their ability to balance aesthetics with environmental and economic considerations following the completion of their LCAs. At the beginning of the course, they applied various circular strategies to their designs based on the theoretical knowledge they had acquired. However, after conducting the LCAs, there was a notable shift in their approach, as they began to critically assess the actual impact of these strategies. They gained an understanding of how certain strategies are more efficient than others and explored opportunities to combine them for enhanced overall efficiency.

Through the integration of LCA and LCC, the students gained awareness of the dual impact of circular strategies on both environmental outcomes and project costs. Notably, they observed, to their satisfaction, that these strategies often exhibited a positive correlation, demonstrating that environmentally conscious design could simultaneously contribute to cost efficiency. Furthermore, the students encountered the inherent challenges of performing life cycle assessments, gaining insight into the current limitations of existing frameworks and tools, enabling the students to identify areas for improvement within the field. Examples of students' work is show in figure (3)

https://doi.org/10.52202/080684-0023

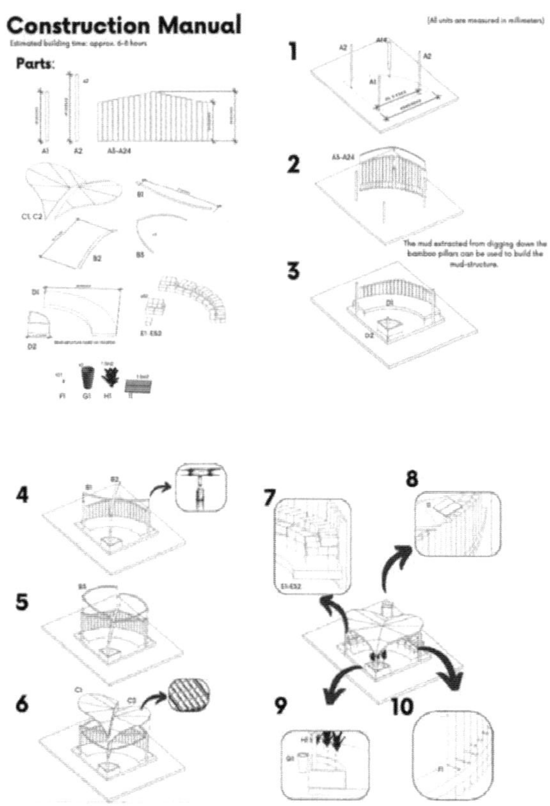

Figure 3: Sample from students' projects.

5.2 STUDENT EXPERIENCE

This section provides results from student course evaluations and student communication throughout the course. Number of respondents to the course evaluation online survey was 23 out of 45 students (51,11%). Results from the course evaluation survey are presented in Figure 1a-d.

Student feedback and perception of theoretical component

Many students felt that the LCA and LCC component of the course should have been larger, with requests mainly being for additional lectures on the topic, with students noting in the section to mention something that can be done better with the course that they wish for: "more LCA lectures", "more focus on LCA", "additional run-through of LCA", "more description and learning about LCC", and "LCA and LCC are very large, you cannot learn that much in the little time we had, but I wish we could have had more on the topic, more lectures and exercises".

Similarly, students expressed a desire for a deeper exploration of the theoretical foundations of circular architecture design. While they appreciated the introduction to concepts such as material reuse, design for disassembly, and regenerative building strategies, many felt that these topics could have been expanded further. Common feedback included requests for "more detailed hands-on explanations of circular design principles" and "additional case studies showcasing real-world applications." Several students also suggested incorporating more hands-on workshops to better bridge the gap between theory and practice, noting that "practical examples help in understanding how to apply circular design in real projects." Overall, students found the theoretical components valuable but felt that more time and focus on these aspects would enhance their ability to implement circular design thinking in architectural practice.

Student feedback and perception of practical component

Group project: Group projects usually tend to receive not so good feedback by some students. The practical component being a group project received similar critique. In regards to the examination type giving the opportunity to achieve course objectives one student notes in the course survey: "not the group assignment due to a bad constellation of members". Another student comments in the section of what can be improved in the course: "there are too many group projects in the programme in general. It is hard to collaborate when so many are unmotivated". Several more comments are similar to these, however some students also provided positive feedback regarding the group project; "the group assignment was inspiring and exciting to carry out".

Individual assignment: The students found the LCA framework complex and struggled with various aspects, such as defining the scope, establishing system boundaries, and selecting impact categories. They expressed a desire for a definitive guide on how to proceed, however, due to the nature of the rather novel method and speed of it evolving this is not possible. This challenge was particularly evident in the individual assignment, where students were required to conduct an LCA of their building design. Frustration arose from the unavailability of certain materials within the LCA tool. Nevertheless, most students successfully developed the skills to critically assess their LCA results, which was the primary objective of the assignment.

To enhance this learning experience, students proposed incorporating a benchmarking exercise into the assignment; "another aspect is that we didn't have anything to compare our results with, and it would have been nice to make a comparison to how a normal building would be built". This would involve conducting an LCA of a building similar to their design but constructed using conventional methods and materials. Such a benchmarking exercise would enable students to compare the two LCA results, facilitating a clearer understanding of the differences and similarities. Additionally, this approach would help students acknowledge and critically examine the results of the circular design LCA, particularly when circular materials are not available in the LCA tool, resulting in outcomes that may more closely resemble those of traditional building designs than expected.

Student feedback and perception of interdisciplinary mixed approach

Overall, the students perceived the subject of circular economy and circular architecture as extensive and challenging to grasp within a limited timeframe. The concepts of circular economy, circular building design, and lifecycle assessments are each substantial and relatively novel topics, contributing to their complexity and making them less straightforward and more difficult to understand. Students felt that each component was not given enough focus. For example, one student states in the course survey: "LCA and LCC were not given enough focus". Despite this, most students felt they had achieved the course objectives, with the majority feeling they had achieved them to a high or very high extent. Similar results can be seen in the students' perceptions of the learning activities and approaches, as well as examinations, in achieving the course objectives. One student notes that "despite the [small] size of the course we have gained an introductory understanding of circular economy tools. It was obvious that there was much more to learn, but I perceive that I have achieved a good basic understanding [of the topic]". However, another student notes that there was "too much individual work and more lectures would have been appreciated". In the comment section of the survey where students can mention something they think was good in the course the lectures were brought up to a larger extent than any other component.

Numerous students noted that the mixed approach to learning was beneficial. One student comments that "the [practical] group assignment was inspiring and exciting to carry out, it gave large artistic freedom whilst giving important context for the preliminary work with the LCA, LCC, and circular assessment". Further, most students felt that the course provided them the opportunity to take responsibility for their own learning to a high or very high extent. The majority of students felt that their expectations of the course were met, with 26,1% and 21,7% feeling expectations were met to a high or very high extent, respectively. Some students did however feel that their expectations of the course were not met.

The students generally appreciated the mix of practical activities in the course, such as study visits, hands-on workshops for Life Cycle Assessment (LCA) and Life Cycle Costing (LCC), and the pin-up critique and feedback sessions for their projects. They found that the practical and theoretical components of the course complemented each other well. Specifically, the lectures, which were primarily focused on theoretical knowledge and fundamental principles of circular architecture and the circular economy, were perceived as valuable. Given that this field of education—sustainable architecture and urban design—is largely practical, we, as instructors, strive to maintain a balance between theoretical and practical elements in our courses.

While the course appears to be well-structured, blending practical activities with theoretical instruction, the differing student perspectives suggest that there is room for improvement in balancing these components. The students' appreciation of hands-on activities highlights the effectiveness of experiential learning in fields like sustainable architecture, where practical applications of theory are crucial for developing skills. By offering study visits and workshops on LCA and LCC, the course provides a realistic context that likely enhances student understanding of sustainability challenges in architecture. The feedback regarding the theoretical component is especially noteworthy. Some students are calling for an increase in theory, indicating that they may feel underprepared in terms of foundational knowledge or conceptual understanding. This suggests that the course could benefit from more integration of theory into the practical activities. For instance, linking theory more directly to design tasks might provide a clearer rationale for decision-making in practical projects, bridging the gap between abstract principles and their application.

Additionally, concerns over group work versus individual contributions in design highlight a potential tension between collaborative learning and individual expression. While group work encourages teamwork, which is critical in real-world architectural practice, it may also obscure individual creativity and skill, especially in a field where personal design vision is highly valued. In response, instructors might consider incorporating more flexible project structures, where students can demonstrate both individual capabilities and their capacity to collaborate effectively. This could involve a mix of group assignments with specific individual roles or opportunities for solo projects alongside group work.

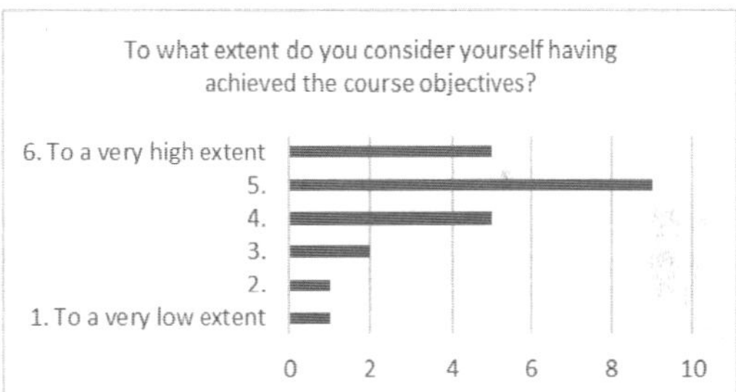

Figure 4a: Course evaluation, achieving objectives

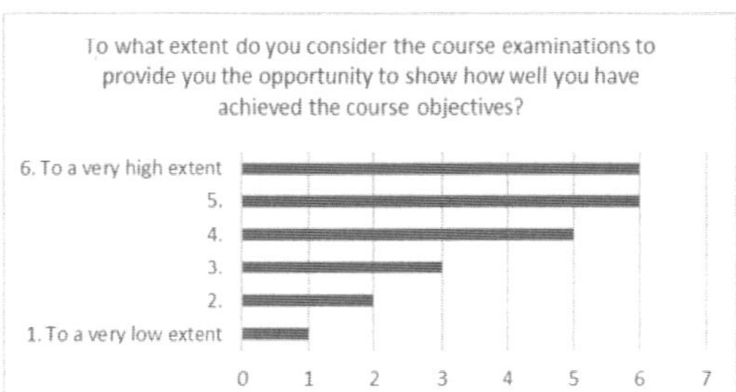

Figure 4b: Course evaluation, examinations

https://doi.org/10.52202/080684-0023

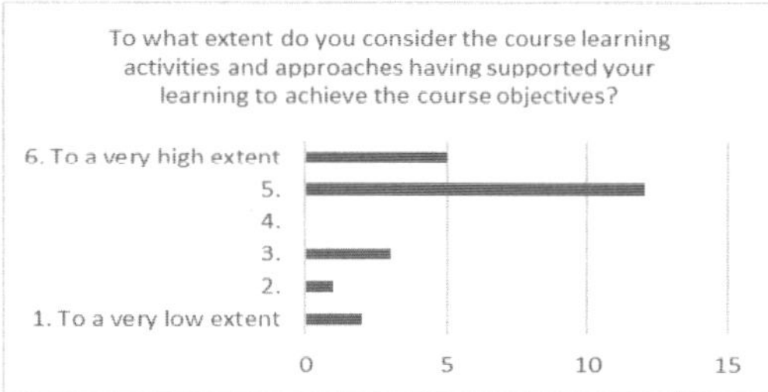

Figure 4c: Course evaluation, learning activities and approaches

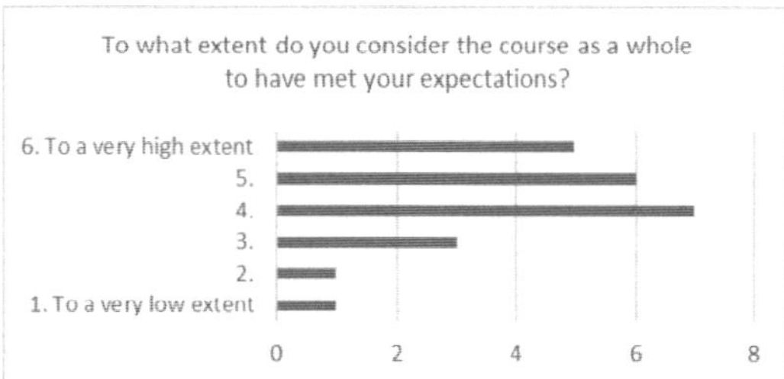

Figure 4d: Course evaluation, meeting expectations

6 DISCUSSION

This pilot study highlights the gap in architectural education regarding circular economy principles and the critical challenge of integrating sustainability into the profession. As other studies that have shown similar critical challenges in including circularity in higher education (Kopnina, 2019). The interdisciplinary course aimed to equip students with skills to make sustainable and economically viable decisions by blending design innovation with LCA and LCC tools. A key feature was the practical application of these assessments, allowing students to analyze environmental and economic impacts while refining their designs through iterative feedback loops. This approach enhanced critical thinking, adaptability, and the ability to navigate trade-offs between sustainability, functionality, and aesthetics.

By mirroring real-world challenges, the course prepared students for interdisciplinary collaboration, emphasizing the balance between sustainability goals and economic constraints. That is tangent to other studies showing that kind of real-life wicked problems (Kanters, 2020). Students not only gained hands-on experience but also identified limitations in existing LCA frameworks, underscoring the need for continuous advancement in sustainable design methodologies. However, challenges arose in balancing creative freedom with technical assessments and navigating diverse disciplinary perspectives. Strong teamwork and guidance were essential in addressing these complexities.

This paper underscores the value of integrating circular economy principles into architectural education, bridging the gap between theory and practice. By fostering collaboration across disciplines such as architecture, engineering, and economics, the course equips students with the necessary skills to address sustainability in the built environment. This model serves as a blueprint for evolving architectural education to meet the demands of climate-neutral and sustainable practices.

Furthermore, the course bridges the gap between circular economy theory and architectural practice, providing empirical insights into how students interpret and apply circular principles. It contributes to discussions on material lifecycles, adaptive reuse, and regenerative design, positioning circularity as a systemic rather than purely technical solution. Additionally, the paper highlights the role of pedagogical reflexivity, emphasizing how educators' biases and assumptions influence sustainability education. By challenging linear design models, the course promotes design for disassembly and materials-as-a-service frameworks, advocating for the evolution of architectural education to incorporate lifecycle thinking and systemic sustainability.

7 RECOMMENDATIONS AND IMPLICATIONS

Enhancing sustainable design education requires a comprehensive, interdisciplinary approach that integrates diverse expertise, practical application, and a strong emphasis on sustainability. This can be achieved by forming interdisciplinary teaching teams from architecture, engineering, economics, and environmental science, providing students with a broad foundation in sustainable design complexities.

Project-based learning should engage students in real-world challenges, combining creative design with environmental and economic analysis while reflecting local contexts. Access to LCA and LCC tools, along with targeted workshops, builds proficiency in sustainability assessments. A balanced curriculum should blend theoretical and practical learning through lectures, guest presentations, site visits, and hands-on workshops. Group projects simulating professional environments foster collaboration, peer learning, and problem-solving.

To balance creativity with analytical rigor, students should produce varied deliverables such as design manuals, models, and technical reports, guided by structured timelines, mentoring, and feedback. Embedding circular design principles across the curriculum ensures sustainability is a core design consideration.

When it comes to direct self-critical thoughts on how the course could be even developed further, the course could be further scaled to incorporate additional disciplines, such as building technology, allowing for a more comprehensive examination of student designs. This expansion would require students to engage more deeply with aspects of building physics and construction methodologies. Such an inclusion would enrich the interdisciplinary nature of the course by prompting students to address not only the environmental and economic and the aesthetic considerations, but also feasibility of design and potential construction challenges. This approach would create a synergistic feedback loop where issues from a construction technology perspective provide more detailed data for the life cycle assessments which in turn inform building designs.

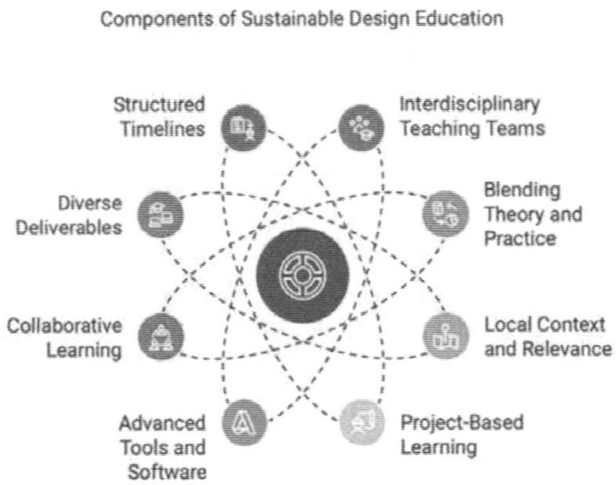

Figure 5: The vision for how sustainable design education can be.

8 CONCLUSIONS

By adopting the recommended strategies, design education can better prepare students to tackle the complexities of sustainable design, equipping them with the skills and knowledge needed to create innovative and climate-resilient solutions. Training future architects and urban designers capable of addressing climate neutrality has profound societal long-term implications. These practitioners play a pivotal role in designing buildings and urban spaces that minimize carbon emissions, promote energy efficiency, and integrate renewable resources. By embedding sustainability and circular economy principles into their practice, they contribute to the global transition toward resilient, low-carbon societies. Their ability to balance aesthetic, functional, and environmental considerations fosters communities that are both livable and sustainable. Moreover, these practitioners drive innovation in material use, construction techniques, and policy advocacy, influencing industries and inspiring systemic change. In the long term, their expertise supports climate action goals, enhances ecological preservation, and creates equitable, adaptive environments for future generations.

This study has illustrated how circular economy principles in architecture can be effectively integrated into educational practices. Embracing circular economy and circular architecture principles are essential for future-proofing architecture and ensuring that urban development is resilient, resource-efficient, and capable of mitigating climate change impacts. It must start in early stages in architecture education and not perceived as a commodity or things to learn during architecture practice. It needs to be impeded in the DNA of future architects and practitioners in general. Finally, we can conclude that this kind of educational transformation is not just about improving curricula but is about preparing future architects to embrace a new paradigm—one where circularity is as integral to the design process as aesthetics, functionality, and structure.

ACKNOWLEDGEMENTS

We would like to extend our heartfelt thanks to the students who participated in this course, which is an integral part of our research study. Your engagement, feedback, and contributions have provided invaluable insights that will advance our understanding and help improve future educational practices. Your efforts are deeply appreciated, and we are grateful for your role in shaping this important work.

REFERENCES

Akhimien, N.G., Latif, E., Hou, S.S., 2021. Application of circular economy principles in buildings: A systematic review. Journal of Building Engineering. https://doi.org/10.1016/j.jobe.2020.102041

Andersen, S.C., Birgisdottir, H., Birkved, M., 2022. Life Cycle Assessments of Circular Economy in the Built Environment—A Scoping Review. Sustainability (Switzerland). https://doi.org/10.3390/su14116887

Bejrum, H., 1991. Livscykelekonomiska kalkyler för byggnader och fastigheter.

Berglund, D., Kharazmi, P., Miliutenko, S., Björk, F., Malmqvist, T., 2018. Comparative life-cycle assessment for renovation methods of waste water sewerage systems for apartment buildings. Journal of Building Engineering 19, 98–108. https://doi.org/10.1016/j.jobe.2018.04.019

Bragança, L., Mateus, R., Koukkari, H., 2010. Building sustainability assessment. Sustainability 2, 2010–2023. https://doi.org/10.3390/su2072010

Dabaieh, M., 2023a. Circular design for zero emission architecture and building practice : it is the green way or the highway. Woodhead Publishing, Cambridge, MA.

Dabaieh, M., 2023b. Chapter 3 - Circular architecture and building design, in: Circular Design for Zero Emission Architecture and Building Practice. Woodhead Publishing, pp. 87–132. https://doi.org/https://doi.org/10.1016/B978-0-12-820079-7.00002-6

Deschamps, J., Simon, B., Tagnit-Hamou, A., Amor, B., 2018. Is open-loop recycling the lowest preference in a circular economy? Answering through LCA of glass powder in concrete. J Clean Prod 185, 14–22. https://doi.org/10.1016/j.jclepro.2018.03.021

Eberhardt, L., Birgisdottir, H., Birkved, M., 2019. Comparing life cycle assessment modelling of linear vs. circular building components, in: IOP Conference Series: Earth and Environmental

 https://doi.org/10.52202/080684-0023

Science. Institute of Physics Publishing. https://doi.org/10.1088/1755-1315/225/1/012039
EN 16309:2014+A1:2014, n.d. Sustainability of construction works - Assessment of social performance of buildings - Calculation methodology.
European Commission, 2020. A new Circular Economy Action Plan.
Fahlstedt, O., Rasmussen, F.N., Temeljotov-Salaj, A., Huang, L., Bohne, R.A., 2024. Building renovations and life cycle assessment - A scoping literature review. Renewable and Sustainable Energy Reviews 203, 114774. https://doi.org/https://doi.org/10.1016/j.rser.2024.114774
Geissdoerfer, M., Savaget, P., Bocken, N.M.P., Hultink, E.J., 2017a. The Circular Economy – A new sustainability paradigm? J Clean Prod 143, 757–768. https://doi.org/10.1016/j.jclepro.2016.12.048
Geissdoerfer, M., Savaget, P., Bocken, N.M.P., Hultink, E.J., 2017b. The Circular Economy – A new sustainability paradigm? J Clean Prod 143, 757–768. https://doi.org/https://doi.org/10.1016/j.jclepro.2016.12.048
Gomes, V., da Silva, M.G., Kowaltowski, D.C.C.K., 2022. Long-Term Experience of Teaching Life Cycle Assessment and Circular Design to Future Architects: A Learning by Doing Approach in a Design Studio Setting. Sustainability (Switzerland) 14. https://doi.org/10.3390/su14127355
IPCC, 2023. Climate Change 2023: Synthesis Report. Contribution of Working Groups I, II and III to the Sixth Assessment Report of the Intergovernmental Panel on Climate Change. https://doi.org/10.59327/IPCC/AR6-9789291691647
ISO 15686-5, 2008. ISO 15686-5:2008(en) Buildings and constructed assets — Service-life planning — Part 5: Life-cycle costing.
ISO14044, 2006. Environmental management - Life cycle assessment - Requirements and guidelines.
Jürgens, M., Hartmann, K., Endres, H.-J., Spierling, S., 2023. Life cycle assessment of higher education institutions – method and case study. J Clean Prod 430. https://doi.org/10.1016/j.jclepro.2023.139649
Kanters, J., 2020. Circular building design: An analysis of barriers and drivers for a circular building sector. Buildings 10. https://doi.org/10.3390/BUILDINGS10040077
Kibert, C.J., 2022. Sustainable construction : green building design and delivery. Wiley, Hoboken, NY.
Krausmann, F., Wiedenhofer, D., Lauk, C., Haas, W., Tanikawa, H., Fishman, T., Miatto, A., Schandl, H., Haberl, H., 2017. Global socioeconomic material stocks rise 23-fold over the 20th century and require half of annual resource use. Proc Natl Acad Sci U S A 114, 1880–1885. https://doi.org/10.1073/pnas.1613773114
Kopnina, H., 2019. Green-washing or best case practices? Using circular economy and Cradle to Cradle case studies in business education. *Journal of Cleaner Production*, *219*, 613–621. https://doi.org/https://doi.org/10.1016/j.jclepro.2019.02.005
Lundgren, R., 2023. Adaptive reuse and shared spaces as circular business models. Lund University, Lund.
Minunno, R., O'Grady, T., Morrison, G.M., Gruner, R.L., 2020a. Exploring environmental benefits of reuse and recycle practices: A circular economy case study of a modular building. Resources, Conservation & Recycling 160, N.PAG-N.PAG. https://doi.org/10.1016/j.resconrec.2020.104855
Minunno, R., O'Grady, T., Morrison, G.M., Gruner, R.L., 2020b. Exploring environmental benefits of reuse and recycle practices: A circular economy case study of a modular building. Resour Conserv Recycl 160. https://doi.org/10.1016/j.resconrec.2020.104855
Reike, D., Vermeulen, W.J.V., Witjes, S., 2018. The circular economy: New or Refurbished as CE 3.0? — Exploring Controversies in the Conceptualization of the Circular Economy through a Focus on History and Resource Value Retention Options. Resour Conserv Recycl 135, 246–264. https://doi.org/10.1016/j.resconrec.2017.08.027
Silva, D.A.L., Giusti, G., Rampasso, I.S., Farrapo Junior, A.C., Anholon, R., 2023. Life Cycle Assessment Teaching Innovation: Experiences from a Brazilian Higher Education Institution. International Journal of Sustainability in Higher Education 24, 449–461.
UNEP, 2020. Guidelines for SOCIAL LIFE CYCLE ASSESSMENT OF PRODUCTS AND ORGANIZATIONS.
Viere, T., Lehmann, J., Miao, Z.C., Weyand, S., Harding, K., Strothmann, P., Wright, L., Chitaka, T.Y., Sonnemann, G., 2024. Global state of the art of teaching life cycle assessment in higher education. International Journal of Life Cycle Assessment 29, 1290–1302. https://doi.org/10.1007/s11367-024-02319-5
World Economic Forum, 2016. Industry Agenda Shaping the Future of Construction A Breakthrough in Mindset and Technology Prepared in collaboration with The Boston Consulting Group.

AUTHOR INDEX